T0318366

Computational Analysis of Structured Media

Mathematical Analysis and its Applications Series

Computational Analysis of Structured Media

Simon Gluzman
Bathurst St. 3000, Apt 606, ON M6B3B4 Toronto, Canada

Vladimir Mityushev
Pedagogical University,
Faculty of Mathematics, Physics and Technical Science
Institute of Computer Sciences,
ul. Podchorazych 2, 30-084 Krakow, Poland

Wojciech Nawalaniec
Pedagogical University,
Faculty of Mathematics, Physics and Technical Science
Institute of Computer Sciences,
ul. Podchorazych 2, 30-084 Krakow, Poland

Series Editor
Themistocles M. Rassias

ELSEVIER

ACADEMIC PRESS
An imprint of Elsevier

Academic Press is an imprint of Elsevier
32 Jamestown Road, London NW1 7BY, UK
525 B Street, Suite 1800, San Diego, CA 92101-4495, USA
225 Wyman Street, Waltham, MA 02451, USA
The Boulevard, Langford Lane, Kidlington, Oxford OX5 1GB, UK

British Library Cataloguing in Publication Data
A catalogue record for this book is available from the British Library

Library of Congress Cataloging-in-Publication Data
A catalog record for this book is available from the Library of Congress

ISBN: 978-0-12-811046-1

For information on all Academic Press publications
visit our website at https://www.elsevier.com/

Working together
to grow libraries in
developing countries

www.elsevier.com • www.bookaid.org

Publisher: Janco Candice
Acquisition Editor: Graham Nisbet
Editorial Project Manager: Susan Ikeda
Production Project Manager: Poulouse Joseph
Designer: Victoria Pearson

Typeset by SPi Books and Journals

For Our Families

CONTENTS

ACKNOWLEDGMENT

We thank our friends and colleagues, Pierre Adler, Igor Andrianov, Leonid Berlyand, Piotr Drygaś, Barbara Gambin, Leonid Filshtinsky, Alexander G. Kolpakov, Pawel Kurtyka, Natalia Rylko, Ryszard Wojnar, Vyacheslav Yukalov, for all their support and useful discussions. We are also indebted to the whole Elsevier team for their help and guidance throughout the publishing process. We are grateful to the editor for carefully reviewing the book during this process. The authors (V.M. and W.N.) were partially supported by Grant of the National Centre for Research and Development 2016/21/B/ST8/01181.

Simon Gluzman, Vladimir Mityushev, and Wojciech Nawalaniec
April 2017

PREFACE

The present book may be considered as an answer to the question associated to the picture on the last front matter page. Why does James Bond prefer shaken, not stirred martini with ice?[1] Highly accurate computational analysis of structural media allows us to explain the difference between various types of random composite structures.

We are primarily concerned here with the effective properties of deterministic and random composites. The analysis is based on accurate analytical solutions to the problems considered by respected specialists as impossible to find their exact solutions. Consensual opinions can be summarized as follows: " It is important to realize that solution to partial differential equations, of even linear material models, at infinitesimal strains, describing the response of small bodies containing a few heterogeneities are still open problem. *In short, complete solutions are virtually impossible*" (see [5, p. 1]).

Of course, it is impossible to resolve all the problems of micromechanics and their analogs, but certain classes such as boundary value problems for Laplace's equation and bi-harmonic two-dimensional (2D) elasticity equations can be solved in analytical form.

At least for an arbitrary 2D multiply connected domain with circular inclusions our methods yield analytical formulae for most of the important effective properties, such as conductivity, permeability, effective shear modulus and effective viscosity. Randomness in such problems is introduced through random locations of non-overlapping disks. It is worth noting that any domain can be approximated by special configurations of packed circular disks.

Many respectful authors apply various self-consistent methods (SCM) such as effective medium approximation, differential scheme, Mori–Tanaka approach, etc. [2]. They claim that such methods give general analytical formula for the effective properties. Careful analysis though shows their restriction to the first- or second- order approximations in concentration.

Actually SCM perform elaborated variations on the theme of the celebrated Maxwell formula, Clausius–Mossotti approximation, and so forth [3]. All of them are justified

---

[1]The complete answer on the question is yet to be found, and most likely after many experiments. But the mathematical answer is given on 268.

rigorously only for a dilute composites when interactions among inclusions are neglected. In the same time, exact and high-order formulae for special regular composites which go beyond SCM were derived.

Despite a considerable progress made in the theory of disordered media, the main tools for studying such systems remain numerical simulations and questionable designs to extend SCM to high concentrations. These approaches are sustained by unlimited belief in numerics and equal underestimation of constructive analytical and asymptotic methods. They have to be drastically reconsidered and refined. In our opinion there are three major developments which warrant such radical change of view.

1. Recent mathematical results devoted to explicit solution to the Riemann–Hilbert and $\mathbb{R}$–linear problems for multiply connected domains, see Chapter 1, [4].[2]

2. Significant progress in symbolic computations (see MATHEMATICA®, MAPLE®, MATLAB® and others) greatly extends our computational capacities. Symbolic computations operate on the meta-level of numerical computing. They transform pure analytical constructive formulae into computable objects. Such an approach results in symbolic algorithms which often require optimization and detailed analysis from the computational point of view. Moreover, symbolic and numeric computations do integrate harmoniously [1].

But we can not declare a victory just yet, because even long power series in concentration and contrast parameters are not sufficient because they won't allow to cover the high-concentration regime. Sometimes the series are short, in other cases they do not converge fast enough, or even diverge in the most interesting regime. Your typical answer to the challenges is to apply an additional methods powerful enough to extract information from the series. But in addition to a traditional Padé approximants applied in such cases, we would need a

3. New post-Padé approximants for analysis of the divergent or poorly convergent series, including different asymptotic regimes discussed in Chapter 5.

In the present book, we demonstrate that the theoretical results [4] can be effectively implemented in symbolic form that yields long power series. Accurate analytical formulae for deterministic and random composites and porous media can be derived employing approximants, when the low-concentration series are supplemented with information on the high-concentration regime where the problems we encounter are characterized by power laws.

As to the engineering needs we recognize the need for an additional fourth step. The engineer would like to have a convenient formula but also to incorporate in it all available information on the system, with a particular attention to the results of numerical simulations or known experimental values.

---

[2]Poisson's type formula for an arbitrary circular multiply connected domain is one of its particular cases.

Method of "regression on approximants" consists in applying different multivariate regression techniques to the results of interpolation and (or) extrapolation with various approximants within the framework of supervised learning paradigm. In this case approximants by themselves are treated as new variables and experimental, exact or numerically "exact" data, or else training data set, are used to construct regression on approximants. Some statistical learning procedure should be included in order to select the best regression and to make sure it is better indeed than choosing the best performer among the approximants.

Consider a non-dimensional effective property $\vartheta(f)$. Here f stands for the concentration of particles. Let the asymptotic expansion of $\vartheta(f)$ in the weak-coupling limit be

$$\vartheta(f) \simeq a_0 + a_1 f + a_2 f^2 + a_3 f^3 + a_4 f^4 + \cdots, \quad \text{as } f \to 0. \qquad (0.0.1)$$

In addition to the expansion, let us consider as available to us, "exact" numerical values $\vartheta(f_i)$ (or "labels"), for the effective property for some typical values of $f = f_i$, $i = 1, 2, \ldots, K$.

This information to be incorporated into the algorithm:

(a) construct all possible approximants for such expansion, such as $Ap_j^*(f)$, $j = 1, 2, \ldots, M$. The approximants are also assumed to incorporate the information from the high-concentration regime whenever such information is available.

(b) make "predictions", i.e., calculate with all constructed approximants for all f_i the values of $Ap_j^*(f_i)$.

Let us consider all predictions as a j-dimensional vector $\mathbf{Ap}^*(f_i)$ and organize $\mathbf{Ap}^*(f_i)$ and $\vartheta(f_i)$ into pairs $\{\mathbf{Ap}^*(f_i), \vartheta(f_i)\}$, $i = 1, 2, \ldots, K$, thus creating a training data set.

(c) based on information contained in the training data set, we attempt to learn the (multivariate) mapping $\vartheta^* = F(\mathbf{Ap}^*)$ (regression model), allowing to make predictions for arbitrary f (since $\mathbf{Ap}^*$ depends on f);

(d) most simple regression models, such as multivariate linear regression and k-Nearest Neighbors could be used.

(e) for learning, i.e., selection of the best regression based on prediction error incurred by particular regression within a given training data set, one can use, e.g., a jackknife (or leave-one-out cross-validation) error estimates. Starting from the whole training data set, the jackknife begins with throwing away the first label, leaving a re-sampled data set, which is used to construct regression and " prediction" is made for the " missing " label. In turn, such obtained prediction is compared with the true label. Resampling is continued till predictions and comparison are performed for each and every label from the training data set.

(f) as a cumulative measure for the prediction error one can take mean absolute percentage error calculated over all labels.

Such approach is amenable to automation and true predictions of the sought physical quantity for any f can be generated.

REFERENCE

1. Czapla R, Nawalaniec W, Mityushev V, Effective conductivity of random two-dimensional composites with circular non-overlapping inclusions, Comput. Mat. Sci. 2012; 63: 118–126.
2. Kanaun SK, Levin VM, Self-consistent methods for composites. Dordrecht: Springer-Verlag; 2008.
3. Landauer R, Electrical conductivity in inhomogeneous media. ed. Garland JC, Tanner DB, Electrical Transport and Optical Properties of Inhomogeneous Media, 1978.
4. Mityushev VV, Rogosin SV, Constructive methods to linear and non-linear boundary value problems of the analytic function. Theory and applications. Boca Raton etc: Chapman & Hall / CRC; 1999/2000. [Chapter 4].
5. Zohdi TI, Wriggers P, An introduction to Computational mechanics. Berlin etc: Springer-Verlag; 2008. [Chapter 1].

Simon Gluzman, Vladimir Mityushev, and Wojciech Nawalaniec

Kraków, Poland

April 2017

Nomenclature

Geometry

a_k stands for center (complex coordinate) of the disk $\mathbb{D}_k$

$\mathbb{D}$ stands for circular multiply connected domain

$\mathbb{D}_k$ stands for disk (inclusion) in $\mathbb{R}^2$ or ball in $\mathbb{R}^3$

$\mathbb{D}^+$ stands for $\cup_{k=1}^{n}\mathbb{D}_k$

D stands for a domain in $\mathbb{R}^2$ or in $\mathbb{R}^3$

D_k stands for simply connected domain (inclusion) in $\mathbb{R}^2$ or in $\mathbb{R}^3$

∂D stands for boundary of D

$E_k(z)$ stands for Eisenstein function of order k

$e_{m_1,m_2,m_3,\ldots,m_n}$ stands for basic sum of the multi-order $\mathbf{m} = (m_1,\ldots,m_n)$

f stands for volume fraction (concentration) of the inclusions

$\mathbf{n} = (n_1, n_2)$ normal vector to a smooth oriented curve L

$n = n_1 + in_2$ complex form of the normal vector $\mathbf{n} = (n_1, n_2)$

$\mathbf{s} = (-n_2, n_1)$ tangent vector

$\mathbb{R}^d$ stands for the Euclidean space

S_m stands for Eisenstein–Rayleigh lattice sums

$\mathbb{U}$ stands for unit disk on plane

BCC means body-centered cubic lattice

FCC means face-centered cubic lattice

HCP means hexagonal close-packed lattice

SC means simple cubic lattice

ω_1 and ω_2 stand for the fundamental translation vectors on the complex plane

Conductivity

$\sigma(\mathbf{x})$ stands for scalar local conductivity

σ_k is used for conductivity of the kth inclusion

σ is used for conductivity of inclusions of two-phase composites when the conductivity of matrix is normalized to unity

σ_e stands for effective conductivity tensor

σ_{ij} stands for components of the effective conductivity tensor

σ_e stands for scalar effective conductivity of the macroscopically isotropic composites

$\sigma^*, \sigma_{e,n}(f), \sigma_{add}, \sigma_1^{ad}$ etc stands for various approximations of σ_e

$\rho = \frac{\sigma^+ - \sigma^-}{\sigma^+ + \sigma^-}$ stands for contrast parameter between media with the conductivities σ^+ and σ^-; for two-phase composites $\rho = \frac{\sigma-1}{\sigma+1}$

s   stands for the critical index for superconductivity

t   stands for the critical index for conductivity

Elasticity

μ stands for shear modulus and viscosity

k stands for bulk modulus

ν stands for Poisson's ration

κ stands for Muskhelishvili's constant

$\sigma_{xx}, \sigma_{xy} = \sigma_{yx}, \sigma_{yy}$ the components of the stress tensor

$\epsilon_{xx}, \epsilon_{xy} = \epsilon_{yx}, \sigma_{yy}$ the components of the strain tensor

Other symbols and abbreviations

$\sim$ indicates asymptotic equivalence between functions, i.e., indicates that functions are similar, of the same order

$\simeq$ indicates that the two functions are similar or equal asymptotically

A and B stand for the critical amplitudes

$\mathbb{C}$ stands for the complex numbers

$\mathbf{C}$ stands for the operator of complex conjugation

error   measures the error as a ratio of the exact value

$\mathcal{H}^\alpha(L)$ stands for functions Hölder continuous on a simple smooth curve L

$\mathcal{H}^\alpha(D) \equiv \mathcal{H}(D)$ stands for functions analytic in D and Hölder continuous in its closure

PadeApproximant$[F[z], n, m]$ stands for the Padé (n, m)-approximant of the function $F(z)$

$\mathbb{R}$ stands for the real numbers

$\wp(z)$ stands for the Weierstrass $\wp$-function

$\zeta(z)$ stands for the Weierstrass ζ-function

1D, 2D, and 3D mean one-, two-, and three- dimensional

LHS and RHS mean the expressions "left-hand side" and "right-hand side" (of an equation), respectively

EMA   means effective medium approximations

RS   means random shaking

RSA   means Random Sequential Addition

RVE   stands for representative volume element

RW   means random walks

SCM   means self-consistent methods

CHAPTER 1

Introduction

*Man's quest for knowledge is an expanding
series whose limit is infinity, but philosophy
seeks to attain that limit at one blow, by a short
circuit providing the certainty of complete and
in-alterable truth. Science meanwhile
advances at its gradual pace, often slowing to
a crawl, and for periods it even walks in place,
but eventually it reaches the various ultimate
trenches dug by philosophical thought, and,
quite heedless of the fact that it is not supposed
to be able to cross those final barriers to the
intellect, goes right on.*

—Stanislav Lem, His Master's Voice

The terms *closed form solution, analytical solution*, etc., are frequently used in literature in various contexts. We introduce below some definitions for various solutions with different levels of "exactness".

Consider an equation $Ax = b$. One can write its solution as $x = A^{-1}b$ where A^{-1} is an inverse operator to A. Let a space where the operator A acts be defined and existence of the inverse operator in this space be established. Then, a mathematician working on the level of functional analysis can say that the deal is done. This is a standard question of the existence and uniqueness in theoretical mathematics. In the present book, we consider problems when the existence and uniqueness take place and we are interested in a constructive form of the expression $x = A^{-1}b$.

1. We say that $x = A^{-1}b$ is a *closed form solution* if the expression $A^{-1}b$ consists of a finite number of elementary and special functions, arithmetic operations, compositions, integrals and derivatives.

2. A solution is an *analytical form solution* if the expression $A^{-1}b$ consists of a finite number of elementary and special functions, arithmetic operations, compositions, integrals, derivatives and **series**.

The difference between items 1 and 2 lies with the usage of the infinite series for a analytical solution. We use the term *constructive solution* for items 1 and 2 following the book [20]. In the second case, we have to cut the considered infinite series to get an expression convenient for *symbolic* analytical and numerical analysis. This book steadfastly adheres to the analysis of obtained truncated series. In particular, to

Computational Analysis of Structured Media
http://dx.doi.org/10.1016/B978-0-12-811046-1.50001-0

constructive investigation of their behaviour near divergence points when the physical percolation effects occur. The main feature of such a constructive study is in retention of the fundamental physical parameters, concentration for instance, in symbolic form. The term analytical solution means not only closed form solution, but also a solution constructed from asymptotic approximations.

Asymptotic methods [1, 2, 3, 5, 15] are assigned to analytical methods when solutions are investigated near the critical values of the geometrical and physical parameters. Hence, asymptotic formulae can be considered as analytical approximations.

In order to distinguish our results from others, we proceed to classify different types of solutions.

3. *Numerical solution* means here the expression $x = A^{-1}b$ which can be treated only numerically. *Integral equation methods* based on the potentials of single and double layers usually give such a solution. An integral in such a method has to be approximated by a cubature formula. This makes it pure numerical, since cubature formulae require numerical data in kernels and fixed domains of integrations. Fredholm's and singular integral equations frequently arise in applications [6, 20, 22, 23]. Effective numerical methods were developed and systematically described in the books [6, 10, 11, 12, 23] and many others. Nevertheless, some integral and integro-differential equations can be solved in a closed form, e.g., [25].

Central for our study, *series method* arises when an unknown element x is expanded into a series $x = \sum_{k=1}^{\infty} c_k x_k$ on the basis $\{x_k\}_{k=1}^{\infty}$ with undetermined constants c_k. Substitution of the series into equation can lead to an infinite system of equations on c_k. In order to get a numerical solution, this system is cut short and a finite system of equations arise, say of order n. Let the solution of the finite system tend to a solution of the infinite system, as $n \to \infty$. Then, the infinite system is called regular and can be solved by the described *truncation method*. This method was justified for some classes of equations in the fundamental book [14]. The series method can be applied to general equation $Ax = b$ in a discrete space in the form of infinite system with infinite number of unknowns. In particular, Fredholm's alternative and the Hilbert–Schmidt theory of compact operators can be applied [14]. So in general, the series method belongs to numerical methods. In the field of composite materials, the series truncation method was systematically used by Guz et al. [13], Kushch [17], and others.

The special structure of the composite systems or application of a low-order truncation can lead to an approximate solution in symbolic form. Par excellence examples of such solutions are due to Rayleigh [26] and to McPhedran's et al, [18, 24] where classical but incomplete analytical approximate formulae for the effective properties of composites were deduced.

It is worth noting that methods of integral equations and methods of series are often well developed computationally. So that many mathematicians would stop on a formula which includes numerically computable objects without actually computing.

For instance, the Riemann–Hilbert problem for a simply connected domain D was solved in a closed form and up to a conformal mapping [8]. Its solution for a disk is given in a closed form by Poisson's type integral. According to the famous Riemann theorem any simply connected domain whose boundary contains more than one point can be conformally mapped onto the unit disk $\mathbb{U}$. The solution of the Riemann–Hilbert problem can be written as a composition of Poisson's type integral and the mapping $\varphi : D \to \mathbb{U}$. Construction of the map φ is considered as a separate computational problem.

Analogous remark concerns the Riemann–Hilbert problem for a multiply connected domain. In accordance with [27] this problem is solved in terms of the principal functional on the Riemann surface which could be found from integral equations. So, realization of the theory [27] could be purely numerical, but again remains undeveloped into concrete algorithm.

In the same time, the Riemann–Hilbert problem for an arbitrary circular multiply connected domain $\mathbb{D}$ was solved analytically in [20], more precisely in terms of the uniformly convergent Poincaré series [19]. Therefore, it is solved for any multiply connected domain D up to a conformal mapping $\varphi : D \to \mathbb{D}$ (see demonstration of its existence in [9]).

Similar remark concerns the Schottky–Klein prime function [4]. This function was constructively expressed in terms of the Poincaré α-series for an arbitrary circular multiply connected domain $\mathbb{D}$ [21]. Formulae of [21] include centers and radii of holes of $\mathbb{D}$ in symbolic form. Therefore, the formulae obtained in [21] are referred to constructive solutions from item 2. Analogous particular formulae [7] are valid under an additional separation condition on $\mathbb{D}$, which means that the holes are far away from each other. In the same time, a numerical algorithm based on the series method can be applied to an arbitrary domain [7]. However, this algorithm does not retain geometric parameters of $\mathbb{D}$ in a symbolic form.

4. *Discrete numerical solution* refers to applications of the finite elements and difference methods. These methods are powerful and their application is reasonable when the geometries and the physical parameters are fixed. In this case the researcher can be fully satisfied with numerical solution to various boundary value problems. Many specialists perceive a pristine computational block (package) as an exact formula: just substitute data and get the result! They have a good case if such a block allows to investigate symbolic dependencies and, perhaps by fitting methods to understand a process or an object. However, a sackful of numbers is not as useful as an analytical formulae. Pure numerical procedures can fail as a rule for the critical parameters and analytical matching with asymptotic solutions can be useful even for the numerical computations.

Moreover, numerical packages sometimes are presented as a remedy from all deceases. It is worth noting again that numerical solutions are useful if we are interested in a fixed geometry and fixed set of parameters for engineering purposes. Different geometry or parameters would require to relaunch the numerical procedure every time properly generating random data and conducting Monte-Carlo experiment [16].

This book deals with constructive analytical solutions. In other words with exact and approximate analytical solutions, when the resulting formulae contain the main physical and geometrical parameters in symbolic form. The obtained truncated series, actually are considered as polynomials. They are supposed to "remember their infinite expansions", so that with a help of some additional re-summation procedure one can extrapolate to the whole series. A significant part of the book is devoted to restoration of these infinite series by means of special constructive forms called approximants, expressions that are asymptotically equivalent to the truncated series. But the approximants are richer in a sense that they also suggest an infinite additional number of the coefficients $\{c_k\}_{k=n+1}^{\infty}$. The coefficients are supposed to be in a good enough agreement with unknown exact c_k. Their quality should be tested against the real data or with respect to the error incurred to original equations. The quality of approximants significantly improves when more than one asymptotic regime can be studied and incorporated into the approximant.

REFERENCE

1. Andrianov IV, Manevitch LI, Asymptotology: Ideas, Methods, and Applications. Dordrecht, Boston, London: Kluwer Academic Publishers; 2002.
2. Andrianov IV, Awrejcewicz J, Danishevs'kyy VV, Ivankov AO, Asymptotic Methods in the Theory of Plates with Mixed Boundary Conditions. Dordrecht, Boston, London: John Wiley & Sons; 2014.
3. Awrejcewicz J, Krysko VA, Introduction to Asymptotic Methods. New York etc: Chapman & Hall/CRC; 2006.
4. Baker HF, Abel's theorem and the allied theory, including the theory of the theta functions. Cambridge: Cambridge University Press; 1897.
5. Berlyand L, Kolpakov AG, Novikov A, Introduction to the Network Approximation Method for Materials Modeling. Cambridge: Cambridge University Press; 2012.
6. Colton D, Kress R, Integral Equation Methods in Scattering Theory. New York etc: John Wiley & Sons; 1983.
7. Crowdy D, The Schwarz problem in multiply connected domains and the Schottky–Klein prime function. Complex Variables and Elliptic Equations 2008; 53: 221–236.
8. Gakhov FD, Boundary Value Problems. 3rd Moscow: Nauka; 1977. Engl. transl. of 1st ed.: Pergamon Press, Oxford, 1966.
9. Golusin GM, Geometric Theory of Functions of Complex Variable. 2nd Moscow: Nauka; 1966. Engl. transl.: AMS, Providence, RI 1969.
10. Grigolyuk EI, Filshtinsky LA, Perforated Plates and Shells. Moscow: Nauka; 1970 [in Russian].
11. Grigolyuk EI, Filshtinsky LA, Periodical Piece–Homogeneous Elastic Structures. Moscow: Nauka; 1991 [in Russian].
12. Grigolyuk EI, Filshtinsky LA, Regular Piece-Homogeneous Structures with defects. Moscow: Fiziko-Matematicheskaja Literatura; 1994 [in Russian].
13. Guz AN, Kubenko VD, Cherevko MA, Diffraction of elastic waves. Kiev: Naukova Dumka; 1978 [in Russian].
14. Kantorovich LV, Krylov VI, Approximate methods of higher analysis. Groningen: Noordhoff; 1958.
15. Kolpakov AA, Kolpakov AG, Capacity and Transport in Contrast Composite Structures: Asymptotic Analysis and Applications. Boca Raton etc: CRC Press Inc.; 2009.

16. Krauth W, Statistical Mechanics: Algorithms and Computations. Oxford: Oxford University Press; 2006.
17. Kushch I, Micromechanics of Composites. Butterworth-Heinemann: Elsevier; 2013.
18. McPhedran RC, McKenzie DR, The Conductivity of Lattices of Spheres. I. The Simple Cubic Lattice. Proc. Roy. Soc. London 1978; A359: 45–63.
19. Mityushev VV, Convergence of the Poincaré series for classical Schottky groups. Proceedings Amer. Math. Soc. 1998; 126(8): 2399–2406.
20. Mityushev VV, Rogosin SV, Constructive methods to linear and non-linear boundary value problems of the analytic function. Theory and applications. Boca Raton etc: Chapman & Hall / CRC; 1999/2000. [Chapter 4].
21. Mityushev V, Poincaré α-series for classical Schottky groups and its applications. ed. Milovanović GV and Rassias MTh, Analytic Number Theory, Approximation Theory, and Special Functions, 827–852, 2014.
22. Muskhelishvili NI, Some Basic Problems of Mathematical Elasticity Theory. 5th Moscow: Nauka; 1966. English translation of the 1st ed.: Noordhoff, Groningen, 1953.
23. Mikhlin SG, Integral Equations and Their Applications to Certain Problems in Mechanics, Mathematical Physics and Technology. 2nd New York: Macmillan; 1964.
24. Nicorovici NA, Movchan AB, McPhedran RC, Green's tensors and lattice sums for elastostatics and elastodynamics. Proc. Roy. Soc. London 1978; A453: 643–662, doi:10.1098/rspa.1997.0036.
25. Samko S, Kilbas AA, Marichev OI, Fractional Integrals and Derivatives: Theory and Applications. Switzerland; Philadelphia: Gordon & Breach Science Publishers; 1993.
26. Rayleigh Lord, On the influence of obstacles arranged in rectangular order upon the properties of a medium. Phil. Mag. 1892; 34: 481–502.
27. Zverovich EI, Boundary value problems of analytic functions in Hölder classes on Riemann surfaces. Uspekhi Mat. nauk 1971; 26: 113–179 [in Russian].

CHAPTER 2

Complex Potentials and $\mathbb{R}$-linear problem

...the danger of proving too much would not have been great if there had not been in Penguinia, as there are, indeed, everywhere, minds framed for free inquiry, capable of studying a difficult question, and inclined to philosophic doubt. They were few; they were not all inclined to speak, and the public was by no means inclined to listen to them. Still, they did not always meet with deaf ears.

— Penguin Island, Anatole France

1. Complex potentials

In this chapter we establish an equivalence between the $\mathbb{R}$-linear problem with constant coefficients and the *perfect contact problem* (*transmission problem*) from the theory of composites. Rigorous proof requires some extensive groundwork with forays into the different parts of Complex Analysis.

In many interesting cases, the general problems of continuum mechanics can be reduced to boundary value problems for 2D Laplace's equation

$$\frac{\partial^2 u}{\partial x_1^2} + \frac{\partial^2 u}{\partial x_2^2} = 0 \qquad (2.1.1)$$

in a domain $D \subset \mathbb{R}^2$. The function $u(x_1, x_2)$ is called harmonic in D. Let $z = x_1 + ix_2$ denote a complex variable. It is well known that any function $u(x_1, x_2)$ harmonic in a simply connected domain D can be presented as the real part of the function $\varphi(z)$ analytic in D

$$u(x_1, x_2) = \text{Re } \varphi(z), \quad z \in D. \qquad (2.1.2)$$

The function $\varphi(z)$ is called *complex potential*. This equation relates the important in application class of partial differential equations and analytic functions. This is the main reason why we pay attention to boundary value problems stated and solved in terms of Complex Analysis.

Computational Analysis of Structured Media
http://dx.doi.org/10.1016/B978-0-12-811046-1.50002-2

BOX A.2.1 Cauchy–Riemann equations and analytic functions

Let two real functions $u(x_1, x_2)$ and $v(x_1, x_2)$ be continuously differentiable at a neighbourhood V_0 of a point $z_0 = x_{10} + ix_{20}$ and satisfy there the Cauchy–Riemann equations

$$\frac{\partial u}{\partial x_1} = \frac{\partial v}{\partial x_2}, \quad \frac{\partial u}{\partial x_2} = -\frac{\partial v}{\partial x_1}, \quad (x_1, x_2) \in V_0. \tag{2.1.3}$$

Then, the function $\varphi(z) = u(x_1, x_2) + iv(x_1, x_2)$ is called analytic at z_0. It follows from (2.1.3) that the complex derivative $\varphi'(z)$ does not depend on the direction of differentiation and

$$\varphi'(z) = \frac{\partial u}{\partial x_1} + i\frac{\partial v}{\partial x_1} = \frac{\partial u}{\partial x_1} - i\frac{\partial u}{\partial x_2}. \tag{2.1.4}$$

An ordinary complex differentiation in the neighborhood of z_0 implies existence of all derivatives of φ which yields infinite differentiability of the real functions u and v.

Application of the operator $\frac{\partial}{\partial x_1}$ to the first relation (2.1.3), $\frac{\partial}{\partial x_2}$ to the second one and their summation yields Laplace's equation (2.1.1).

—Cauchy–Riemann equations [13, 43]

1.1. Domains and curves on complex plane

The *complex plane* is isomorphic to the real plane $\mathbb{R}^2$ and denoted as $\mathbb{C}$. The *extended complex plane* (Riemann sphere) is obtained from $\mathbb{C}$ by addition of the infinite point: $\widehat{\mathbb{C}} = \mathbb{C} \cup \{\infty\}$. Any complex number z can be presented in the form $z = |z|e^{i \arg z}$ where $|z|$ denotes the modulus of z and the argument $\arg z$ is chosen as $0 \le \arg z < 2\pi$. Sometimes, it is supposed that $-\pi < \arg z \le \pi$.

The following geometrical sets will be considered in the complex plane. Let a continuous curve $L \subset \widehat{\mathbb{C}}$ be parametrized by equation $z = g(t)$ ($0 \le t \le 1$) and $C > 0$, $0 \le \alpha \le 1$ be given constants. The curve belongs to the class $C^{m,\alpha}$ if $|g^{(m)}(t_1) - g^{(m)}(t_2)| < C \, |t_1 - t_2|^\alpha$, for all $t_1, t_2 \in [0, 1]$. If $m \ge 1$, then L is called *smooth*. A curve L of class $C^{1,\alpha}$ is called *Lyapunov curve* . Let a simple smooth curve L bound a domain D and be oriented in such a way that $L = \partial D$ leaves D to the left. The opposite orientation of L will be denoted by $-L$ or $-\partial D$.

The *complex conjugation* is denoted by bar as $\bar{z} = x_1 - ix_2$. It defines the operator of complex conjugation $\mathbf{C}z = \bar{z}$. The *inversion (reflection) with respect to* a circle $|z - a| = r$ is defined by the formula

$$z^* = \frac{r^2}{\overline{z - a}} + a. \tag{2.1.5}$$

Let $\varphi(z)$ be analytic in a disk $|z - a| < r$. Then, the function $\overline{\varphi(z^*)}$ is analytic in $|z - a| > r$. Vice versa, if $\varphi(z)$ is analytic in $|z - a| > r$, then the function $\overline{\varphi(z^*)}$ is analytic in $|z - a| < r$. The inversion preserves boundary properties of functions on the circle. An open half-plane can be treated as a disk of the infinite radius.

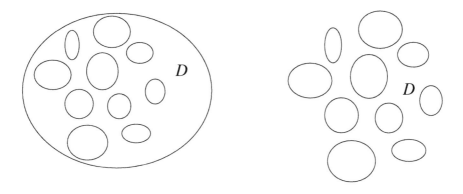

Figure 2.1 A bounded and an unbounded multiply connected domains.

Domain D is called *simply connected* if any curve Γ in D is homotopic to a point, i.e., it can be continuously reduced to a point within D. In the opposite case, domain D is *multiply connected*. If the boundary of D consists of n simple continuous mutually disjoint curves, the connection of D is equal to n. Infinitely connected domains with countable connectivity, for instance, doubly periodic domains will be also considered. Two types of domains are displayed in Fig. 2.1. For definiteness, it is supposed that every closed curve is oriented in counter-clockwise sense.

BOX A.2.2 Conformal Mapping

A rigorous definition of *Conformal Mapping* can be found in textbooks on Complex Analysis. For our purposes it is sufficient to characterize Conformal Mapping as a map $\varphi : D \to D'$ which is analytic in a domain D and $\varphi'(z) \neq 0$ in D. The conformal mapping is one-to-one correspondence of D and a domain D'. Conformal Mapping preserves the angles between smooth curves and the connectivity of the domain.

A Conformal Mapping φ transforms (locally near a point z_0) a small circle onto another small circle, as the linear function $w = \varphi(z_0) + \varphi'(z_0)(z - z_0)$ approximates map locally up to the terms $O(|z - z_0|^2)$. It is worth noting that an $\mathbb{R}$-differentiable function transforms a small ellipse onto another small ellipse.

Analytic functions and Laplace's equation (2.1.1) are invariant under conformal mappings. More precisely, let $u(x_1, x_2) \equiv u(z)$ satisfy Laplace's equation in D, φ conformally maps D onto D' and φ^{-1} be the inverse map of D onto D'. Then, φ^{-1} maps conformally D' onto D

and the function $w(\xi_1, \xi_2) := u[\varphi^{-1}(\xi_1 + i\xi_2)]$ satisfies Laplace's equation in D' in variables $\xi_1 = \text{Re } \varphi(x_1 + ix_2)$ and $\xi_2 = \text{Im } \varphi(x_1 + ix_2)$. This follows from the relation

$$\frac{\partial^2 u}{\partial x_1^2} + \frac{\partial^2 u}{\partial x_2^2} = |\varphi'(z)|^2 \left(\frac{\partial^2 w}{\partial \xi_1^2} + \frac{\partial^2 w}{\partial \xi_2^2} \right),$$

where the Jacobian of the $\mathbb{R}^2$-transformation of D onto D', i.e., $(x_1, x_2) \mapsto (\xi_1, \xi_2)$, is given by formula

$$|\varphi'(z)|^2 = \left(\frac{\partial U}{\partial x_1} \right)^2 + \left(\frac{\partial V}{\partial x_2} \right)^2.$$

Here, $U(x_1, x_2) = \text{Re } \varphi(z)$ and $V(x_1, x_2) = \text{Im } \varphi(z)$.

–Conformal Mapping [13, 43, 44]

1.2. Equations in functional spaces

Let L be a finite union of simple closed curves on $\widehat{\mathbb{C}}$. Functions continuous in L form the Banach space of *continuous functions* $C(L)$ endowed with the norm

$$\| \varphi \|_C := \max_{z \in L} | \varphi(z) | .$$

Analogous space is introduced for functions continuous in $D \cup \partial D$ with the norm

$$\| \varphi \|_C := \max_{z \in D \cup \partial D} | \varphi(z) | . \tag{2.1.6}$$

A function φ is called the Hölder continuous function on L (we write $\varphi \in \mathcal{H}^\alpha(L)$) with the power $0 < \alpha \le 1$ if there exists such a constant $C > 0$ that $|\varphi(x) - \varphi(y)| < C|x - y|^\alpha$ for all $x, y \in L$. $\mathcal{H}^\alpha(L)$ is a subspace of the continuous functions. It is a Banach space with the norm

$$\| \varphi \|_\alpha := \| \varphi \|_C + \sup_{x,y \in L, x \ne y} \frac{| \varphi(x) - \varphi(y) |}{| x - y |^\alpha}. \tag{2.1.7}$$

Let L be a smooth curve consisting of not more than finite collection of connected component. Let L divide the complex plane onto two domains D^+ and D^-. The set of all continuous functions on L contains two subsets of functions analytically extended into D^+ and D^-. These subsets will be denoted $C(D^\pm)$.

Analogous designations $\mathcal{H}^\alpha(D^\pm) \equiv \mathcal{H}(D^\pm)$ are used for Hölder continuous functions. It follows from Maximum Modulus Principle for analytic functions that convergence in the space $\mathcal{H}(D^\pm)$ implies the uniform convergence in the closure of $D^\pm$.

The *Lebesgue space* $\mathcal{L}_p(L)$ consists of functions f having the finite integral $\|\varphi\|_p := \left(\int_L | \varphi(t) |^p \, ds \right)^{\frac{1}{p}}$. The space $\mathcal{L}_p(L)$ for $p \ge 1$ is Banach. The *Hardy space* $\mathcal{H}_p(\mathbb{U})$ for $p > 0$ is defined for any $p > 0$ as the space of all analytic functions in the unit disk

$\mathbb{U} = \{z \in \mathbb{C} : |z| < 1\}$ satisfying the following condition

$$\sup_{0<r<1} \int_0^{2\pi} |f(re^{i\theta})|^p \, d\theta < \infty.$$

$\mathcal{H}_p(\mathbb{U})$ are Banach spaces for any $p, 1 \leq p < \infty$. Such spaces can be defined also for harmonic setting. For instance *harmonic Hardy spaces* $h_p(\mathbb{U}), 0 < p < \infty$, are the spaces of all harmonic functions u, satisfying the following condition

$$\sup_{0<r<1} \int_0^{2\pi} |u(re^{i\theta})|^p \, d\theta < \infty.$$

$h_\infty(\mathbb{U})$ is the space of all bounded harmonic functions on $\mathbb{U}$.

Consider an operator equation

$$x = \mathbf{A}x + b, \tag{2.1.8}$$

where $\mathbf{A}$ is a linear operator on a Banach space $\mathcal{B}$. Widely used method of the solution of the equation (2.1.8) is the *method of successive approximation*. Taking an arbitrary element $x_0 \in \mathcal{B}$ (called an *initial point* of approximation) one can determine the sequence of approximations

$$x_{n+1} = \mathbf{A}x_n + b, \; n = 0, 1, \ldots \tag{2.1.9}$$

The convergence of this sequence is connected with the convergence of the operator series

$$\mathbf{I} + \mathbf{A} + \mathbf{A}^2 + \cdots + \mathbf{A}^n + \cdots \tag{2.1.10}$$

Usually, the absolute convergence of the series (2.1.10) is considered when the following number series converges

$$1 + \|\mathbf{A}\| + \|\mathbf{A}\|^2 + \cdots + \|\mathbf{A}\|^n + \cdots \tag{2.1.11}$$

It is not always the case of our study. We shall use another type of convergence associated to the space $\mathcal{B}$. Let $\mathcal{B}$ be the space of continuous functions C on $D \cup \partial D$ endowed with the norm (2.1.6). The convergence in C means the uniform convergence on $D \cup \partial D$ which differs from the absolute convergence. It will be seen below that the method of successive approximations for the $\mathbb{R}$-linear problem converges uniformly for any location of inclusions and may become divergent when separation condition (2.3.106) is not satisfied. Therefore, the uniform convergence is preferable for our investigations. In order to study such a convergence we consider an equation dependent on a spectral parameter.

Theorem 1 ([21, p. 75]). *Let $\mathbf{A}$ be a linear bounded operator in a Banach space $\mathcal{B}$. If for any element $b \in \mathcal{B}$ and for any complex number v satisfying the inequality $|v| \leq 1$*

equation

$$x = \nu \mathbf{A} x + b \qquad (2.1.12)$$

has a unique solution, then the unique solution of the equation

$$x = \mathbf{A} x + b \qquad (2.1.13)$$

can be found by method of successive approximations. The approximations converge in $\mathcal{B}$ to the solution

$$x = \sum_{k=0}^{\infty} \mathbf{A}^k b. \qquad (2.1.14)$$

Corollary 1. *Let $\mathbf{A}$ be a compact operator in $\mathcal{B}$. If the equation*

$$x = \nu \mathbf{A} x \qquad (2.1.15)$$

has only trivial solution for all $\nu \in \mathbb{C}, |\nu| \leq 1$, then equation (2.1.12) has a unique solution for all $b \in \mathcal{B}$, and for all $|\nu| \leq 1$. This solution can be found by the method of successive approximations convergent in $\mathcal{B}$.

1.3. Relation between analytic and harmonic functions

Let D be a simply connected domain. The real part of a function analytic in D is a function harmonic in D. And conversely, each function $u(z)$ harmonic in D is the real part of a function $\varphi(z)$ analytic in D

$$u(z) = \operatorname{Re} \varphi(z), \; z \in D.$$

The function $\varphi(z)$ is determined by $u(z)$ up to an arbitrary purely imaginary additive constant due to the formula

$$\varphi(z) = u(z) + iv(z), \; \text{where } v(z) = \int_{w}^{z} -\frac{\partial u}{\partial y} dx + \frac{\partial u}{\partial x} dy + C, \qquad (2.1.16)$$

where w is an arbitrary point in D and C is a real constant.

Let now D be a multiply connected domain. The real part of a function analytic in D is a harmonic in D function. The converse assertion is no longer true. This situation is described in the following two statements

Theorem 2 (Logarithmic Conjugation Theorem [3, p. 179; 32, p. 23]). *Let D be a finitely connected unbounded domain $(\infty \in D)$. Let $D_1, D_2, \ldots, D_n$ be simply connected domains, the components of $\widehat{\mathbb{C}} \backslash (D \cup \partial D)$, and let $z_j \in D_j$ for $j = 1, 2, \ldots, n$. If u is real valued and harmonic on D, then there exist φ analytic in D and real con-*

stants $A_1, A_2, \ldots, A_n$ such that

$$u(z) = Re \; \varphi(z) + \sum_{j=1}^{n} A_j \log \left| z - z_j \right| \qquad (2.1.17)$$

for all $z \in D$. Moreover, $\sum_{j=1}^{n} A_j = 0$.

The principal value of logarithm is defined as follows $\log z := \log |z| + i \arg z$. For details, see also page 20.

Theorem 3 (Decomposition Theorem [32, p. 23]). *Let D and D_j be domains defined in the previous theorem. Any function u harmonic in D and continuous in its closure has the unique decomposition*

$$u(z) = Re \; \sum_{j=1}^{n} \left[\varphi_j(z) + A_j \log (z - z_j) \right] + A, \qquad (2.1.18)$$

where φ_j is analytic in $D \cup \partial D_j \cup D_j$ and $\varphi_j(\infty) = 0$ for all $j = 1, 2, \ldots, n$. The constants A_j and A are real and $\sum_{j=1}^{n} A_j = 0$.

For L being a closed curve the integral $\int_L d\varphi(z)$ is called the *period* of $\varphi(z)$ along L. Calculating periods of the function $\omega(z) := \sum_{j=1}^{n} \left[\varphi_j(z) + A_j \log \left(z - z_j \right) \right]$ along each component L_m of $(-\partial D)$, one can see by virtue of logarithmic function properties that

$$\int_{L_m} d\omega(z) = 2\pi i A_m.$$

Therefore the constants $2\pi A_m$ which appear in (2.1.17), (2.1.18) are imaginary parts of the periods of the multi-valued analytic function $\omega(z)$ along the components of $(-\partial D)$.

1.4. Cauchy type integral and singular integrals

The classical Cauchy integral formula [14] can be presented in the following way. Let L be a simple, closed, piece-wise smooth curve on the complex plane $\mathbb{C}$ dividing $\widehat{\mathbb{C}}$ onto two simply connected domains D^+ and $D^- \ni \infty$. If function $\Phi(z)$ is analytic in D^+ and continuous up to the boundary, it can be represented in the form of *Cauchy integral*

$$\frac{1}{2\pi i} \int_L \frac{\Phi(t)}{t - z} dt = \begin{cases} \Phi(z), & z \in D^+, \\ 0, & z \in D^-. \end{cases} \qquad (2.1.19)$$

If $\Phi(z)$ is analytic in D^- and continuous up to the boundary, we have

$$\frac{1}{2\pi i} \int_L \frac{\Phi(t)}{t - z} dt = \begin{cases} \Phi(\infty), & z \in D^+, \\ -\Phi(z) + \Phi(\infty), & z \in D^-. \end{cases} \quad (2.1.20)$$

Let L be a simple, rectifiable Jordan closed curve (or open arc) on $\mathbb{C}$ and $\phi \in \mathcal{H}^\alpha(L)$, then

$$\frac{1}{2\pi i} \int_L \frac{\phi(t)}{t - z} dt, \quad z \in \widehat{\mathbb{C}} \setminus L$$

is called *Cauchy type integral*. It gives two analytic functions $\Phi^+(z)$ and $\Phi^-(z)$ in the domains D^+ and D^- respectively if L is closed curve (or unique function $\Phi(z)$, analytic in $\widehat{\mathbb{C}} \setminus L$, if L is open arc).

Let again L be a simple, rectifiable Jordan closed curve on $\mathbb{C}$ and $\phi \in \mathcal{H}^\alpha(L)$. Let $B(t, \varepsilon)$ denote the disk with the center at t of radius ε. The following limit

$$\lim_{\varepsilon \to 0} \frac{1}{\pi i} \int_{L \setminus B(t, \varepsilon)} \frac{\phi(\tau)}{\tau - t} d\tau = v.p. \frac{1}{\pi i} \int_L \frac{\phi(\tau)}{\tau - t} d\tau \quad (2.1.21)$$

is called *Cauchy principal value of a singular integral* $\frac{1}{\pi i} \int_L \frac{\phi(\tau)}{\tau - t} d\tau$ or simply *singular integral* (with Cauchy kernel).

Let simple, closed, smooth curve L divide $\widehat{\mathbb{C}}$ onto two domains $D^+, D^- \ni \infty$. Then the boundary functions $\Phi^+(t)$, $\Phi^-(t)$ of the Cauchy type integral

$$\Phi^\pm(z) = \frac{1}{2\pi i} \int_L \frac{\phi(\tau) d\tau}{\tau - z}, \quad z \in D^\pm,$$

do satisfy the *Sochocki–Plemelj formulae*[1]

$$\Phi^\pm(t) = \pm \frac{1}{2} \phi(t) + \frac{1}{2\pi i} \int_L \frac{\phi(\tau) d\tau}{\tau - t}, \quad t \in L. \quad (2.1.22)$$

---

[1]Sokhotsky–Plemelj formulae in literature. The surname of the Polish-Russian mathematician Julian Sochocki (1842-1927), was originally spelled in Russian in scientific literature. Later it was written also in Latin in several different ways. One of the authors of this book (V.M.) goes by several different names as well (LOL). For example, Polish colleagues know him as Władimir Mitiuszew.

1.5. Poisson integral.

Any function from $h_p\,(\mathbb{U})$ $(1 < p \le \infty)$ can be represented in the form of *Poisson integral*, i.e., if $u \in h_p\,(\mathbb{U})$ $(1 < p \le \infty)$, there exists a function $\phi \in L_p\,(0, 2\pi)$ such that

$$u(re^{i\theta}) = \frac{1}{2\pi} \int_0^{2\pi} \frac{1 - r^2}{1 + r^2 - 2r\cos(\theta - t)}\phi(t)dt = \frac{1}{2\pi} \int_0^{2\pi} P_r(\theta - t)\phi(t)dt, \qquad (2.1.23)$$

$0 \le r < 1, \theta \in (-\pi, \pi]$, where $P_r(\tau) = \frac{1-r^2}{1+r^2-2r\cos\tau}$ is called *Poisson kernel*. It should be noted that Poisson kernel P_r is in fact the real part of the so called *Schwarz kernel* :

$$P_r(\theta - t) = \frac{1 - r^2}{1 + r^2 - 2r\cos(\theta - t)} = \mathrm{Re}\,\frac{e^{it} + re^{i\theta}}{e^{it} - re^{i\theta}} = \mathrm{Re}\,\frac{e^{it} + z}{e^{it} - z}. \qquad (2.1.24)$$

It is not always hold for an arbitrary harmonic function in $\mathbb{U}$, but it is true for bounded harmonic functions in $\mathbb{U}$, hence for any one continuous up to the boundary.

Let a harmonic function u be continuous in the closure of the unit disk. Then,

$$u(e^{it}) = \lim_{re^{i\theta} \to e^{it}} u(re^{i\theta}).$$

Moreover, if $\phi(t)$ is an arbitrary continuous function on $[-\pi, \pi]$ and $\phi(-\pi) = \phi(\pi)$, the Poisson integral solves the Dirichlet problem (see Section 2.5 of the book [32])

$$u(e^{it}) = \phi(t), \quad |t| = 1.$$

1.6. Schwarz operator

The problem of determination of an analytic function in a domain via boundary values of its real part is often called *Schwarz problem* (see, e.g., [14, 32]). In the case of simply connected domain it is simply reduced to the Dirichlet problem for harmonic functions. An operator that stays in correspondence to a given on ∂D real-valued function u and analytic in D function φ, such that $\mathrm{Re}\,\varphi_{|\partial D} = u$, is called *Schwarz operator* $\mathbb{T}: u \mapsto \varphi$. In the case of unit disc $\mathbb{U}$, the Schwarz operator has an explicit form

$$(\mathbb{T}u)(z) := \frac{1}{2\pi i} \int_{\partial\mathbb{U}} u(\zeta) \frac{\zeta + z}{\zeta - z} \frac{d\zeta}{\zeta}. \qquad (2.1.25)$$

For the upper half-plane, we have

$$(\mathbb{T}u)(z) := \frac{1}{\pi} \int_{-\infty}^{\infty} u(\xi) \frac{x_2}{(\xi - x_1)^2 + x_2^2} d\xi, \quad x_2 > 0.$$

One can also use the following (equivalent) form for the Schwarz operator in the unit disk

$$(\mathbb{T}u)(z) := \frac{1}{\pi i} \int_{\partial \mathbb{U}} u(\zeta) \frac{d\zeta}{\zeta - z} - \frac{1}{2\pi i} \int_{\partial \mathbb{U}} u(\zeta) \frac{d\zeta}{\zeta}, \quad z \in \mathbb{U}. \qquad (2.1.26)$$

For general simply connected domains, Schwarz operator can be introduced using the so called *complex Green function* (see, e.g., [27]). Recall that the Green function of the Dirichlet problem for a domain $D \subset \widehat{\mathbb{C}}$ is the function $G(z, \zeta)$ of two complex variables, satisfying the following properties: (i) $G(z, \zeta) = \log \frac{1}{|z-\zeta|} + g(z, \zeta)$, where g is continuous in $D \times D$, harmonic in $z \in D$ for any fixed $\zeta \in D$, and in $\zeta \in D$ for any fixed $z \in D$; (ii) $g(z, \zeta)$ is continuous in $\zeta \in D \cup \partial D$ for any fixed $z \in D$, and besides $g(z, t) = -\log \frac{1}{|z-t|}, z \in D, t \in \partial D$.

Using the Green function, one can solve Dirichlet problem, i.e., find a harmonic in D function $u(z)$, whose boundary values coincide with a given Hölder-continuous function $h(\tau)$ on ∂D

$$u(z) = \frac{1}{2\pi} \int_{\partial D} h(\tau) \frac{\partial G}{\partial \mathbf{n}}(z, \tau) \, ds_\tau, \quad z \in D,$$

where $\frac{\partial}{\partial \mathbf{n}}$ denotes the normal derivative to ∂D. It is known also that if $\omega(z)$ is an analytic function realizing the conformal mapping of the domain D onto unit disc $\mathbb{U}$, then the Green function can be presented in the form

$$G(z, \zeta) = \log \left| \frac{\omega(z) - \omega(\zeta)}{1 - \omega(z)\overline{\omega(\zeta)}} \right|,$$

Introduce a harmonic conjugate to $G(z, \zeta)$ in variable z (cf. (2.1.16))

$$H(z, \zeta) := \int_w^z -\frac{\partial G}{\partial x_2} dx_1 + \frac{\partial G}{\partial x_1} dx_2,$$

where w is an arbitrary point in D. Then the following formula gives a complex Green function for the domain D

$$M(z, \zeta) := G(z, \zeta) + iH(z, \zeta) = \log \frac{\omega(z) - \omega(\zeta)}{1 - \omega(z)\overline{\omega(\zeta)}}.$$

Thus, the Schwarz operator can be represented in the form

$$(\mathbb{T}u)(z) := \frac{1}{2\pi} \int_{\partial G} u(\tau) \frac{\partial}{\partial \mathbf{n}} M(z, \tau) ds_\tau, \quad z \in D.$$

The Schwarz operator is related to a conjugation problem stated in the theorem below where the simply connected domains D_k $(k = 1, 2, \dots, n)$ complement D to the extended complex plane: $\varphi = \varphi^- \in \mathcal{H}(D)$ and $\varphi^+ \in \mathcal{H}(\cup_{k=1}^n D_k)$.

Theorem 4 ([32, 34]). *The problem*

$$Re\ \varphi(t) = g(t), \quad t \in \partial D \tag{2.1.27}$$

is equivalent to the problem

$$\varphi^-(t) = \varphi^+(t) - \overline{\varphi^+(t)} + g(t), \quad t \in \partial D, \tag{2.1.28}$$

i.e., the problem (2.1.27) is solvable if and only if (2.1.28) is solvable. If (2.1.27) has a solution $\varphi(z)$, it is a solution of (2.1.28) in D and solution of (2.1.28) in D_k can be found from the following simple problem for the simply connected domain D_k with respect to function $2Im\ \varphi^+(z)$ harmonic in D_k

$$2Im\ \varphi^+(t) = Im\ \varphi^-(t) - g(t), \quad t \in \partial D_k, \quad (k = 1, 2, \ldots, n). \tag{2.1.29}$$

The problem (2.1.29) has a unique solution up to an arbitrary additive real constant.

The problem (2.1.27) is called the *Schwarz problem* for the domain D. Along the similar lines (2.1.29) is the Schwarz problem for the domain D_k. The operator solving the Schwarz problem is called the *Schwarz operator* (in appropriate functional space). The function $v(z) = 2Im\ \varphi(z)$ is harmonic in D_k. Therefore, the Schwarz problem (2.1.29) is equivalent to the Dirichlet problem

$$v(t) = Im\ \varphi^-(t) - g(t), \quad t \in \partial D.$$

For multiply connected domains D, the Schwarz problem (2.1.27) is not equivalent to a Dirichlet problem for harmonic functions, since any function harmonic in D is represented as the real part of a single-valued analytic function plus logarithmic terms (see Theorem 2 on page 12).

Denote $L_k = \partial D_k$, then $\partial D = -\cup_{k=1}^n L_k$. The problem

$$Re\ \varphi(t) = g(t) + c_k, \quad t \in L_k, \ k = 1, 2, \ldots, n, \tag{2.1.30}$$

with undetermined constants c_k is called the *modified Schwarz problem*. The problem (2.1.30) always has a unique solution up to an arbitrary additive complex constant [27].

2. $\mathbb{R}$-linear problem

2.1. Statement of the problem and modeling of the perfect contact

Let D be a multiply connected domain described above. Let D_k ($k = 1, 2, \ldots, n$) be a simply connected domains complementing D to the extended complex plane. In the theory of composites, the domains D_k are called inclusions. The $\mathbb{R}$-*linear conjugation problem* or simply $\mathbb{R}$-*linear problem* is stated as follows. Given Hölder continuous functions $a(t) \neq 0$, $b(t)$ and $c(t)$ on ∂D, how to find a function $\varphi(z)$ analytic (meromorphic) in $\cup_{k=1}^{n} D_k \cup D$, continuous in $D_k \cup \partial D_k$ and in $D \cup \partial D$ with the conjugation condition

$$\varphi^-(t) = a(t)\varphi^+(t) + b(t)\overline{\varphi^+(t)} + c(t), \quad t \in L := \cup_{k=1}^{n} \partial D_k = -\partial D. \qquad (2.2.31)$$

Here $\varphi^-(t)$ is the limit value of $\varphi(z)$ when $z \in D$ tends to $t \in L$, $\varphi^+(t)$ is the limit value of $\varphi(z)$ when $z \in D_k$ tends to $t \in \partial D_k$.

In the case of smooth boundary ∂D, the homogeneous $\mathbb{R}$-linear problem with constant coefficients

$$\varphi^-(t) = a\varphi^+(t) + b\overline{\varphi^+(t)}, \quad t \in L \qquad (2.2.32)$$

is equivalent to the *perfect contact problem* (*transmission problem*) from the theory of composites

$$u^+(t) = u^-(t), \quad \sigma^+ \frac{\partial u^+}{\partial \mathbf{n}}(t) = \sigma^- \frac{\partial u^-}{\partial \mathbf{n}}(t), \quad t \in L. \qquad (2.2.33)$$

Here the real function $u(z)$ is harmonic in $\widehat{\mathbb{C}} \backslash \partial D$ (except prescribed singularities) and continuously differentiable in $D_k \cup \partial D_k$ and in $D \cup \partial D$, $\frac{\partial}{\partial \mathbf{n}}$ is the normal derivative to ∂D. The conjugation conditions express the perfect contact between materials with different conductivities σ^+ and σ^-. For instance in the language of the heat conduction, the first equation (2.2.33) means the equality of temperature and the second equation (2.2.33) means the equality of the normal fluxes from the both sides of the interface curve. The "$\pm$" form of (2.2.32) and (2.2.33) corresponds to the local "$\pm$" orientation of the interface between different media. For dispersed composites, it is convenient to numerate each inclusion and to write the perfect contact condition in a slightly different form presented below.

BOX A.2.3 Local conductivity laws

Consider locally the 2D steady heat conduction in a domain U occupied by a conducting material with a constant conductivity $\sigma > 0$. The temperature distribution $u(x_1, x_2)$ and the heat flux $\mathbf{q}(x_1, x_2) = (q_1(x_1, x_2), q_2(x_1, x_2))$ satisfy in U the Fourier law

$$\mathbf{q} = -\sigma \nabla u \qquad (2.2.34)$$

and the conservation energy law

$$\nabla \cdot \mathbf{q} = 0 \iff \frac{\partial q_1}{\partial x_1} + \frac{\partial q_2}{\partial x_2} = 0. \qquad (2.2.35)$$

Here, $\nabla := (\frac{\partial}{\partial x_1}, \frac{\partial}{\partial x_2})$. Substitution of (2.2.35) into (2.2.34) yields Laplace's equation in the domain U

$$\frac{\partial^2 u}{\partial x_1^2} + \frac{\partial^2 u}{\partial x_2^2} = 0. \tag{2.2.36}$$

Let a smooth oriented curve $L \subset U$ be fixed. The normal heat flux across a point of L is given by formula

$$q_{\mathbf{n}} = \mathbf{q} \cdot \mathbf{n} = -\sigma \frac{\partial u}{\partial \mathbf{n}},$$

where $\mathbf{n}$ denotes the unit normal vector to L.

—Local conductivity laws [44]

Let the inclusions D_k has the conductivity $\sigma^+ = \sigma_k$ ($k = 1, 2, \ldots, n$) and the conductivity of D be normalized to unity as $\sigma^- = 1$. Such a normalization does not limit the generality of the problem. The coefficients σ_k then become dimensionless and are considered as the ratios of the conductivities of the kth inclusion to the conductivity of matrix. In these designations, (2.2.33) becomes

$$u(t) = u_k(t), \quad \frac{\partial u}{\partial \mathbf{n}}(t) = \sigma_k \frac{\partial u_k}{\partial \mathbf{n}}(t), \quad t \in L_k \quad (k = 1, 2, \ldots, n). \tag{2.2.37}$$

The subscript k pertains to the inclusions.

We now reduce (2.2.37) to an $\mathbb{R}$-linear problem. To this end, introduce the complex potentials $\varphi(z)$ and $\varphi_k(z)$ analytic (meromorphic) in D and D_k, respectively. The harmonic and analytic functions are related by the equalities

$$u(z) + iv(z) = \varphi(z), \quad z \in D, \tag{2.2.38}$$

$$u_k(z) + iv_k(z) = \frac{2}{\sigma_k + 1} \varphi_k(z), \quad z \in D_k \ (k = 1, 2, \ldots, n), \tag{2.2.39}$$

where $v(z)$ and $v_k(z)$ denote the imaginary parts of the analytic (meromorphic) functions $\varphi(z)$ and $\varphi_k(z)$, respectively. The multiplier in the RHS of (2.2.39) is selected for convenience of computations and does not influence the final result. If the real part of an analytic function is given, its imaginary part is determined by (2.1.16) uniquely up to an arbitrary additive constant. It is assumed that the functions $\varphi(z)$ and $\varphi_k(z)$ are continuously differentiable in the closures of the considered domains except prescribed singularities. Usually, it is assumed that

$$\varphi(z) = \varphi_0(z) + g(z), \quad z \in D, \tag{2.2.40}$$

where $\varphi_0(z)$ is analytic in D and $g(z)$ can have logarithmic singularities and poles in D that corresponds to sources, sinks and multipoles (see Section A.2.4 for details). For simplicity, it is assumed that $g(z)$ posses a finite number of poles in D and its limit values are continuously differentiable on ∂D. When $\varphi(z)$ has a simple pole at infinity, we will have $g(z) = z$. Then, the principal part of $\varphi(z)$ at infinity, simply the function z, models the external field described by the flux $\mathbf{q}_\infty = -(1, 0)$.

It will be shown below that $\varphi(z)$ from (2.2.38) is single-valued in the multiply connected domain D for finite σ_k, i.e., the logarithmic terms in the representations (2.1.17) and (2.1.18) vanish.

BOX A.2.4 Singular points

Singular points of the complex potentials are supposed to be poles and logarithms. The complex logarithm $\log z = \log |z| + i \arg z$ is introduced as a branch of the multi-valued function $\text{Log} z = \log |z| + i \arg z + 2\pi i k$ $(k \in \mathbb{Z})$ inverse to the exponential function. The argument $\arg z$ of the complex number z can be fixed, for instance, as $0 \leq \arg z < 2\pi$. Then, the function $\log z$ is uniquely defined in $\mathbb{C} \setminus [0, +\infty]$. It is assumed that the cut $[0, +\infty]$ has two sides considered as different lines. The real function $\log x$ is defined for positive x as the limit $\log x = \lim_{z \to x, Im\, z > 0} \log z$. From other side, we have $\lim_{z \to x, Im\, z < 0} \log z = \log x + 2\pi i$; hence, the complex function $\log z$ has the jump $2\pi i$ across the half-axis $(0, +\infty)$. The cut $(0, +\infty)$ can be replaced by another smooth simple curve Γ connecting the points $z = 0$ and $z = \infty$. Then, the logarithm is defined in $\mathbb{C} \setminus \Gamma$ by equation $\log z = \log |z| + i \arg z$ where the argument has the jump $2\pi i$ when z passes across Γ. Frequently, Γ is taken as $(-\infty, 0)$, then $-\pi < \arg z \leq \pi$. It is worth noting that the function $\text{Re} \log z = \log |z|$ is always single-valued contrary to $\text{Im} \log z = \arg z$. This is the reason why a single-valued harmonic function does not necessarily yield by (2.1.16) a single-valued analytic function in a multiply connected domain. The function $\frac{Q}{2\pi} \log z$ for real Q determines a source or sink at zero of intensity Q. This follows from the total normal derivative calculated through the integral

$$\frac{Q}{2\pi} \int_{|z|=r_0} \frac{\partial}{\partial \mathbf{n}} \log |z| \, ds = Q. \tag{2.2.41}$$

The derivative $(\log z)' = \frac{1}{z}$ is a single-valued function. This function models the dipole. The functions $\frac{1}{z^n}$ for $n = 2, 3, \ldots$ model multipoles.

It follows from the Laurent expansion near the isolated singularity $z = 0$ that any function analytic in $0 < |z| < \varepsilon$ can be represented as a linear combination of multipoles up to an arbitrary function $F^+(z)$ analytic at $z = 0$

$$\varphi(z) = \sum_{n=1}^{\infty} c_n z^{-n} + F^+(z) = \sum_{n=1}^{\infty} c_n z^{-n} + O(|z|^0). \tag{2.2.42}$$

The above described singularity is located at $z = 0$ and can be put at an arbitrary point by a translation $z - z_0$.

Multipoles at infinity are described by polynomials in z. For instance, the polynomial $P(z) = Bz$ yields a first-order pole at infinity, hence a dipole. This dipole describes a plane-parallel heat flow since the complex number $P'(z) = B = B_1 + iB_2$ is isomorphic to the vector (B_1, B_2). In the case $B = 1$ we obtain the vector $(1, 0)$, i.e., the physical flux $\mathbf{q} = (-1, 0)$ parallel to the real axis. Here, the conductivity near infinity is normalized to unity.

–Singular points [24]

BOX A.2.5 Cauchy–Riemann equations on a curve

Let functions $u(x_1, x_2)$ and $v(x_1, x_2)$ be continuously differentiable in the closure of the open simply connected domain V with the Hölder continuous boundary ∂V and satisfy the Cauchy–Riemann equations (2.1.3) in V. Consider the unit tangent $\mathbf{s} = (-n_2, n_1)$ and normal $\mathbf{n} = (n_1, n_2)$ vectors to ∂V. Continuous differentiability implies that the Cauchy–Riemann equations (2.1.3) are fulfilled in $V \cup \partial V$. Consider the directional derivatives

$$\frac{\partial u}{\partial \mathbf{n}} = n_1 \frac{\partial u}{\partial x_1} + n_2 \frac{\partial u}{\partial x_2}, \quad \frac{\partial v}{\partial \mathbf{s}} = -n_2 \frac{\partial v}{\partial x_1} + n_1 \frac{\partial v}{\partial x_2} \tag{2.2.43}$$

and $\frac{\partial u}{\partial \mathbf{s}}, \frac{\partial v}{\partial \mathbf{n}}$. Using (2.1.3) we obtain the Cauchy–Riemann equations on ∂V

$$\frac{\partial u}{\partial \mathbf{n}} = \frac{\partial v}{\partial \mathbf{s}}, \quad \frac{\partial u}{\partial \mathbf{s}} = -\frac{\partial v}{\partial \mathbf{n}}. \tag{2.2.44}$$

–Cauchy-Riemann conditions on a curve [14, 39]

Using the first relation (2.2.44) we can write the second equation (2.2.37) as

$$\frac{\partial v}{\partial \mathbf{s}}(t) = \sigma_k \frac{\partial v_k}{\partial \mathbf{s}}(t), \quad t \in L_k. \tag{2.2.45}$$

Integration on the natural parameter s yields

$$v(t) = \sigma_k v_k(t) + d_k, \quad t \in L_k, \tag{2.2.46}$$

where d_k is a constant of integration. Multiplying (2.2.46) by the imaginary unit and adding to (2.2.37) we arrive at the $\mathbb{R}$-linear condition (2.2.32) with additive constants C_k

$$\varphi(t) = \varphi_k(t) - \rho_k \overline{\varphi_k(t)} + C_k, \quad t \in L_k \ (k = 1, 2, \ldots, n), \tag{2.2.47}$$

where the parameter

$$\rho_k = \frac{\sigma_k - 1}{\sigma_k + 1}, \quad \text{on } L_k, \tag{2.2.48}$$

is called *the contrast parameter* between materials occupying D_k and D. Arbitrary additive complex constants C_k arise in (2.2.47) because of the integration constants in (2.2.46) and pure imaginary constants in (2.1.16).

Let us check now that $\varphi(t)$ is univalent in D when $|\rho_k| < 1$, i.e., its periods vanish. Let the curve $L_k = \partial D_k$ be fixed. The period of $\varphi_k(t)$ over L_k is equal to zero since $\varphi_k(z)$ is analytic in the simply connected domain. Hence, the period of $\overline{\varphi_k(t)}$ is also equal to zero. Then, the period of $\varphi(t)$ over L_k vanishes by (2.2.47).

Using (2.2.40) we can write the $\mathbb{R}$-linear condition (2.2.47) for analytic functions

$$\varphi_0(t) = \varphi_k(t) - \rho_k\overline{\varphi_k(t)} - g(t) + C_k, \quad t \in L_k \ (k = 1, 2, \ldots, n). \tag{2.2.49}$$

Without loss of generality we can assume that $\varphi_0(z)$ vanishes at infinity since in the opposite case we can include $\varphi_0(\infty)$ into C_k. In the case $|\rho_k| < 1$, the constant C_k can be included into $\varphi_k(z)$ through other constants c_k by means of the representation $C_k = c_k - \rho_k\overline{c_k}$ and introduction $\tilde{\varphi}_k(z) = \varphi_k(z) + c_k$. Therefore, C_k can be set to zero without loss of generality when $|\rho_k| < 1$. The limit case $|\rho_k| = 1$ is considered separately at the end of this section.

The $\mathbb{R}$-linear condition (2.2.49) relates the limit value of the complex potentials. We now deduce analogous condition for the complex fluxes

$$\psi(z) \equiv \varphi'(z) = \frac{\partial u}{\partial x_1}(z) - i\,\frac{\partial u}{\partial x_2}(z), \quad z \in D, \tag{2.2.50}$$

$$\psi_k(z) \equiv \varphi_k'(z) = \frac{\sigma_k + 1}{2}\left(\frac{\partial u_k}{\partial x_1}(z) - i\,\frac{\partial u_k}{\partial x_2}(z)\right), \quad z \in D_k \quad (k = 1, 2, \ldots, n). \tag{2.2.51}$$

It follows from equation (2.2.34) that the flux $\mathbf{q}$ in D is isomorphic to the complex function $-\overline{\psi(z)}$ and $\mathbf{q}$ in D_k to $-\frac{\sigma_k+1}{2}\overline{\psi_k(z)}$.

The conditions (2.2.37) imply that

$$\frac{\partial u}{\partial \mathbf{s}} = \frac{\partial u_k}{\partial \mathbf{s}}, \quad \frac{\partial u}{\partial \mathbf{n}} = \sigma_k\frac{\partial u_k}{\partial \mathbf{n}}, \quad \text{on } L_k \ (k = 1, 2, \ldots, n). \tag{2.2.52}$$

Using formulae (2.2.43) for the differential operators we obtain on every L_k

$$-n_2\frac{\partial u}{\partial x_1} + n_1\frac{\partial u}{\partial x_2} = -n_2\frac{\partial u_k}{\partial x_1} + n_1\frac{\partial u_k}{\partial x_2}, \quad n_1\frac{\partial u}{\partial x_1} + n_2\frac{\partial u}{\partial x_2} = \sigma_k n_1\frac{\partial u_k}{\partial x_1} + \sigma_k n_2\frac{\partial u_k}{\partial x_2}. \tag{2.2.53}$$

Equations (2.2.50) and (2.2.51) yield

$$\frac{\partial u}{\partial x_1} = \frac{1}{2}\left(\psi + \overline{\psi}\right), \quad \frac{\partial u}{\partial x_2} = \frac{i}{2}\left(\psi - \overline{\psi}\right) \tag{2.2.54}$$

and

$$\frac{\partial u_k}{\partial x_1} = \frac{1}{\sigma_k + 1}\left(\psi_k + \overline{\psi_k}\right), \quad \frac{\partial u}{\partial x_2} = \frac{i}{\sigma_k + 1}\left(\psi - \overline{\psi}\right). \tag{2.2.55}$$

After substitution of (2.2.54), (2.2.55) into (2.2.53) and simple transformations, we arrive at the $\mathbb{R}$-linear condition for the complex fluxes

$$\psi(t) = \psi_k(t) + \rho_k\overline{n(t)^2\psi_k(t)}, \quad t \in L_k \ (k = 1, 2, \ldots, n)\,, \tag{2.2.56}$$

where the complex unit normal vector $n(t) = n_1(t) + in_2(t)$ is identified with the geometric unit normal vector $\mathbf{n}(t) = (n_1(t), n_2(t))$. Using (2.2.40) we introduce the analytic in D function

$$\psi_0(z) = \psi(z) - g'(z) \tag{2.2.57}$$

and write the $\mathbb{R}$-linear condition (2.2.56) for analytic functions

$$\boxed{\psi_0(t) = \psi_k(t) + \rho_k \overline{n(t)^2} \overline{\psi_k(t)} - g'(t), \quad t \in L_k \ (k = 1, 2, \ldots, n)}. \tag{2.2.58}$$

Here, the function $\psi_0(z)$ vanishes at infinity.

Equation (2.2.58) can be also obtained from (2.2.49) by differentiation along L_k. We have

$$\frac{\partial \varphi_k(t)}{\partial \mathbf{s}} = \varphi_k'(t) \, t_s', \qquad \frac{\partial \overline{\varphi_k(t)}}{\partial \mathbf{s}} = \overline{\varphi_k'(t)} \frac{\overline{dt}}{dt} \, t_s'. \tag{2.2.59}$$

Using the relation $\frac{\overline{dt}}{dt} = \frac{\bar{s}}{s}$, where $s = -n_2 + in_1$ is the complex form of the unit tangent vector $\mathbf{s}$, and the relation $s = i\,\bar{n}$ where $n = n_1 + in_2$, we arrive at the formula

$$\frac{\partial \overline{\varphi_k(t)}}{\partial \mathbf{s}} = -\overline{\varphi_k'(t)} \, n^2(t) \, t_s'. \tag{2.2.60}$$

It is worth noting that in the above manipulations t and $\bar{t}$ are dependent since t lies on the curve L_k.

Consider now the limit cases of (2.2.37) when σ_k is equal to $+\infty$. This case $\sigma_k = +\infty \Leftrightarrow \rho_k = 1$ corresponds to the perfectly conducting inclusions. Then, the potential $u_k(z)$ becomes a constant c_k in D_k. This gives the modified Dirichlet problem, e.g., (2.1.30),

$$u(t) = c_k, \quad \text{on } L_k \quad (k = 1, 2, \ldots, n). \tag{2.2.61}$$

The modified Dirichlet problem (2.2.61) can be treated as the limit case of the $\mathbb{R}$-linear problem (2.2.47). The real part of (2.2.47) after substitution $\rho_k = 1$ yields (2.2.61) with $u(z) = \text{Re } \varphi(z)$ and $c_k = \text{Re } C_k$. It is worth noting that the function $\varphi(z)$ is single-valued in D, i.e., all $A_j = 0$ in the representation (2.1.17).

Another limit cases of (2.2.37) takes place when σ_k is equal to zero that is equivalent to non-conducting inclusions and to $\rho_k = -1$. In this case, we take the imaginary part of (2.2.47) and arrive at the problem

$$v(t) = \text{Im } C_k \quad \text{on } L_k. \tag{2.2.62}$$

Differentiate (2.2.62) along the curve L_k, i.e., apply the operator $\frac{\partial}{\partial s}$. Further, application of the first equation (2.2.44) yields the Neumann problem

$$\frac{\partial u}{\partial \mathbf{n}}(t) = 0 \quad \text{on } L_k. \tag{2.2.63}$$

In this case, the function $\varphi(z)$ is not necessary single-valued in D, i.e., A_j in the representation (2.1.17) can be non-zero.

Using the representation (2.2.40) we can state modified Dirichlet problem (2.2.61) for functions $u_0(z) = u(z) - \operatorname{Re} g(z)$ harmonic in D as

$$u_0(t) = -\operatorname{Re} g(t) + c_k, \quad \text{on } L_k \ (k = 1, 2, \ldots, n). \tag{2.2.64}$$

Along the similar lines the Neumann problem (2.2.63) can be written as the non-homogeneous Neumann problem

$$\frac{\partial u_0}{\partial \mathbf{n}}(t) = h(t), \quad \text{on } L_k \ (k = 1, 2, \ldots, n), \tag{2.2.65}$$

where

$$h(t) = -\frac{\partial \operatorname{Re} g}{\partial \mathbf{n}}(t) = \frac{\partial \operatorname{Im} g}{\partial \mathbf{s}}(t) \quad \text{on } L_k. \tag{2.2.66}$$

The constants A_j in the representation (2.1.17) can be calculated as in Section 31 of the book [27]. First, for each fixed j the circulation of $\varphi_0(z) = u_0(z) + iv_0(z)$ from (2.1.17) is equal to $2\pi i A_j$. This yields

$$A_j = \frac{1}{2\pi} \int_{L_j} \frac{\partial v_0}{\partial \mathbf{s}} \, ds = \frac{1}{2\pi} \int_{L_j} \frac{\partial u_0}{\partial \mathbf{n}} \, ds \tag{2.2.67}$$

by the Cauchy–Riemann equations. This implies that the function $u_0(z)$ satisfying the Neumann problem (2.2.65) can be represented in the form (2.1.17) with

$$A_j = \frac{1}{2\pi} \int_{L_j} h \, ds. \tag{2.2.68}$$

B.2.1 Historical notes on the general $\mathbb{R}$-linear problem

In 1932 using the theory of potentials Muskhelishvili [38] (see also [39, p. 522]) reduced the problem (2.2.33) to a Fredholm integral equation and proved that it has a unique solution in the case $\sigma^\pm > 0$, the most interesting in applications. In 1933 I. N. Vekua and Ruhadze [45, 46] constructed a solution of (2.2.33) in closed form for annulus and ellipse. Hence, the paper [38] published in 1932 contains the first result on solvability of the $\mathbb{R}$-linear problem, while [45] and [46] published in 1933 are the first papers devoted to exact solution of the $\mathbb{R}$-linear problem for annulus and ellipse. A little bit later Golusin [15, 17] considered the $\mathbb{R}$-linear problem in the form (2.2.33) using of the functional equations for analytic functions. Therefore, Golusin's paper [15] published in 1934 is the first paper concerning constructive solution to the $\mathbb{R}$-linear problem for special circular multiply connected domains.

In the subsequent works the above first results were not associated to the $\mathbb{R}$-linear problem even by their authors. Thus, in 1946 Markushevich [23] had stated the $\mathbb{R}$-linear problem in the form (2.2.31) and studied it in the case $a(t) = 0$, $b(t) = 1$, $c(t) = 0$

when (2.2.31) is not a Nöther problem. Later Muskhelishvili [39] (page 455 in Russian edition) did not recognize in (2.2.31) his earlier problem (2.2.33) and congratulated Markushevich for the statement of a "new boundary value problem".

In 1960 Bojarski [6] shown that in the case $|b(t)| < |a(t)|$ with $a(t)$, $b(t)$ belonging to the Hölder class $\mathcal{H}^{1-\varepsilon}$ with sufficiently small $\varepsilon > 0$, the $\mathbb{R}$-linear problem (2.2.31) qualitatively is similar to the $\mathbb{C}$-linear problem (see Theorem 5 below)

$$\varphi^+(t) = a(t)\varphi^-(t) + c(t), \ t \in \partial D. \tag{2.2.69}$$

Let $\kappa = wind_L a(t)$ denote the winding number (index) of $a(t)$ along L. Mikhailov [26] reduced the problem (2.2.31) to an integral equation and justified the absolute convergence of the method of successive approximation for the later equation in the space $\mathcal{L}^p(L)$ for $wind_L a(t) = 0$ and under the restriction

$$(1 + \|S_p\|)|b(t)| < 2|a(t)|, \tag{2.2.70}$$

where $\|S_p\|$ is the norm of the singular integral in $\mathcal{L}^p(L)$. Later Mikhailov [26] (first published in [25]) developed this result to continuous coefficients $a(t)$ and $b(t)$; $c(t) \in \mathcal{L}^p(\partial D)$. The case $|b(t)| < |a(t)|$ was called the elliptic case. It corresponds to the particular case of the real constant coefficients a and b considered by Muskhelishvili [38].

Theorem 5 (Bojarski [6] simply connected domains; [7] multiply connected domains). *Let the coefficients of the problem (2.2.31) satisfy the inequality*

$$|b(t)| < |a(t)|. \tag{2.2.71}$$

If $\kappa \geq 0$, the problem (2.2.31) is solvable and the homogeneous problem (2.2.31) $(c(t) = 0)$ has 2κ $\mathbb{R}$-linearly independent solutions vanishing at infinity. If $\kappa < 0$, the problem (2.2.31) has a unique solution if and only if $|2\kappa|$ $\mathbb{R}$-linearly independent conditions on $c(t)$ are fulfilled.

The condition (2.2.70) is stronger than (2.2.71) since always $\|S_p\| \geq 1$ [32].

Theorem 6. *([7]) Let the coefficients of the problem (2.2.31) satisfy the inequality (2.2.71) and $\kappa = 0$. Then, the problem (2.2.31) has a unique solution that can be found by uniformly convergent successive approximations.*

Therefore, the uniqueness takes place when (2.2.71) and $\kappa = 0$. The uniform convergence of successive approximations in this case also holds. The absolute convergence holds under the restriction (2.2.70). One can find further discussion devoted to convergence on page 71.

The term $\mathbb{R}$-linear problem was introduced by one of the authors (V.M.) and systematically used in [32] and following works on the subject. This term is associated to the $\mathbb{R}$-linear condition between two complex values $W = aZ + b\overline{Z}$.

–$\mathbb{R}$-linear problem [32]

2.2. Integral equations associated to the generalized alternating method of Schwarz for finitely connected domains

Method of potentials (single and double layers) is a method of integral equations applied to partial differential equations. In complex analysis, it is equivalent to the method of singular integral equations [14, 39, 47] when an unknown analytic function is represented as the singular integral with an unknown density on the curve. The *generalized alternating method of Schwarz*, shortly called *Schwarz's method*, can be considered as the method of integral equations of another type [27] when an unknown analytic function in each inclusion D_k is related to analytic functions in D_m ($m = 1, 2, \ldots, n$) by integral equations. Physically, these equations express interactions of the field in D_k with the fields in D_m ($m = 1, 2, \ldots, n$) and the external field.

Schwarz's method can be also applied to various partial differential equations. Schwarz's method in general form for linear equations in $\mathbb{R}^d$ ($d = 2, 3$) is outlined on page 59.

In the present section, we apply Schwarz's method to the $\mathbb{R}$-linear problem corresponding to the perfect contact between components of the composite with an external field modeled by a function $g(t)$

$$\varphi^-(t) = \varphi_k(t) - \rho_k \overline{\varphi_k(t)} - g(t), \quad t \in L_k \ (k = 1, 2, \ldots, n), \tag{2.2.72}$$

where the contrast parameter ρ_k has the form (2.2.48).

For fixed m introduce the operator

$$A_m g(z) = \frac{1}{2\pi i} \int_{L_m} \frac{g(t)dt}{t - z}, \ z \in G_m. \tag{2.2.73}$$

In accordance with the Sochocki formulae,

$$A_m g(\zeta) = \lim_{z \to \zeta} A_m g(z) = \frac{1}{2} g(\zeta) + \frac{1}{2\pi i} \int_{L_m} \frac{g(t)dt}{t - \zeta}, \ \zeta \in L_m. \tag{2.2.74}$$

Equations (2.2.73)–(2.2.74) determine the operator A_m in the space $\mathcal{H}(D_m)$.

Lemma 1. *The linear operator A_m is bounded in the space $\mathcal{H}(D_m)$.*

The proof is based on the definition of the bounded operator A_m for which $\|A_m g\| \leq C\|g\|$ and the fact that the norm in $\mathcal{H}(D_m)$ is equal to the norm of functions Hölder continuous on L_m. The estimation of the later norm follows from the boundness of the operator (2.2.74) in Hölder's space [14].

The lemma is proved.

The conjugation condition (2.2.72) can be written in the form

$$\varphi_k(t) - \varphi^-(t) = \rho_k \overline{\varphi_k(t)} + g(t), \quad t \in L_k \ (k = 1, 2, \ldots, n). \tag{2.2.75}$$

The difference between functions analytic in $D^+ = \cup_{k=1}^n D_k$ and in D appears in the left-hand part of the latter relation. Then, application of the Sochocki formulae (2.1.22) yield

$$\varphi_k(z) = \sum_{m=1}^n \frac{\rho_m}{2\pi i} \int_{L_m} \frac{\overline{\varphi_m(t)}}{t - z} dt + g_k(z), \quad z \in D_k \ (k = 1, 2, \ldots, n), \tag{2.2.76}$$

where the function

$$g_k(z) = \frac{1}{2\pi i} \sum_{m=1}^n \int_{L_m} \frac{g(t)}{t - z} dt$$

is analytic in D_k and Hölder continuous in its closure. The integral equations (2.2.76) can be continued to L_k as follows

$$\varphi_k(z) = \sum_{m=1}^n \rho_m \left[\frac{\overline{\varphi_k(z)}}{2} + \frac{1}{2\pi i} \int_{L_m} \frac{\overline{\varphi_m(t)}}{t - z} dt \right] + g_k(z), \ z \in L_k \ (k = 1, 2, \ldots, n). \tag{2.2.77}$$

One can consider equations (2.2.76), (2.2.77) as an equation with linear bounded operator in the space $\mathcal{H}(D^+)$ consisting of functions analytic in D^+ and Hölder continuous in the closure of D^+ (see page 10). The complex potential in the matrix surrounding inclusions is not directly presented in (2.2.76), (2.2.77) and the problem is reduced "only" to finding inclusions potentials.

Equations (2.2.76), (2.2.77) constitute the generalized method of Schwarz. Write, for instance, equation (2.2.76) in the form

$$\varphi_k(z) - \frac{\rho_k}{2\pi i} \int_{L_k} \frac{\overline{\varphi_k(t)}}{t - z} dt = \sum_{m \neq k} \frac{\rho_m}{2\pi i} \int_{L_m} \frac{\overline{\varphi_m(t)}}{t - z} dt + g_k(z), \ z \in D_k \ (k = 1, 2, \ldots, n).$$
$$\tag{2.2.78}$$

At the zeroth approximation we arrive at the problem for the single inclusion D_k ($k = 1, 2, \ldots, n$)

$$\varphi_k(z) - \frac{\rho_k}{2\pi i} \int_{L_k} \frac{\overline{\varphi_k(t)}}{t - z} dt = g_k(z), \ z \in D_k. \tag{2.2.79}$$

Let the problem (2.2.79) be solved. Further, its solution is substituted into the RHS of (2.2.78). Then, we arrive at the first-order problem and so forth. Therefore, the generalized method of Schwarz can be considered as the method of implicit iterations applied to integral equations (2.2.76), (2.2.77).

Remark 1. Another integral equation method was proposed in Chapter 4, [32] for the

Dirichlet problem. Direct iteration method applied to these equations can be viewed as another modification of Schwarz's method. However, emerging integral terms contain Green's functions of each simply connected domain D_k which should be constructed independently. One can obtain equations from Chapter 4,[32] by application of the operator S_k^{-1} to the both sides of (2.2.78), where the operator S_k solves equation (2.2.79).

Application of Theorem 1 (page 11) to equations (2.2.76), (2.2.77) yields

Theorem 7. *Let $|\rho_k| < 1$. Then the system of equations (2.2.76), (2.2.77) has a unique solution. This solution can be found by the method of successive approximations convergent in the space $\mathcal{H}(D^+)$.*

Proof. Let $|v| \leq 1$. Consider equations in $\mathcal{H}(D^+)$

$$\varphi_k(z) = v \sum_{m=1}^{n} \frac{\rho_m}{2\pi i} \int_{L_m} \frac{\overline{\varphi_m(t)}}{t - z} dt + g_k(z), \quad z \in D_k \ (k = 1, 2, \ldots, n). \qquad (2.2.80)$$

Equations on L_k has a form similar to (2.2.77).

Let $\varphi_k(z)$ be a solution of (2.2.80). Introduce the function $\varphi(z)$ analytic in D and Hölder continuous in its closure as follows

$$\varphi(z) = v \sum_{m=1}^{n} \frac{\rho_m}{2\pi i} \int_{L_m} \frac{\overline{\varphi_m(t)}}{t - z} dt, \quad z \in D. \qquad (2.2.81)$$

The expression in the right-hand part of (2.2.81) can be considered as Cauchy's integral

$$\Phi(z) = \frac{1}{2\pi i} \int_{L} \frac{\mu(t)}{t - z} dt$$

along $L = \cup_{m=1}^{n} L_m$ with the density $\mu(t) = \rho_m \overline{\varphi_k(t)}$ on L_k. Using the property of Cauchy's integral $\Phi^+(t) - \Phi^-(t) = \mu(t)$ on L and (2.2.80), we arrive at the $\mathbb{R}$-linear conjugation relation on each fixed curve L_k

$$\varphi_k(t) - g_k(t) - \varphi(t) = \rho_k \overline{\varphi_k(t)}, \quad t \in L_k. \qquad (2.2.82)$$

Here, $\Phi^+(t) = \varphi_k(t) - g_k(t)$, $\Phi^-(t) = \varphi(t)$. In accordance with Theorem 6 on page 25, the $\mathbb{R}$-linear problem (2.2.82) has a unique solution. This unique solution is the unique solution of the system (2.2.80).

Therefore, Theorem 1 yields the convergence of the method of successive approximations applied to the system (2.2.80).

This completes the proof of the theorem.   □

Remark 2. Though the method of integral equations discussed on page 31 is rather a numerical method, application of the residue theorem transforms the integral terms for special shapes of the inclusions to compositions of the functions. Therefore, at least for the boundaries expressed by algebraic functions one should arrive at the functional equations. The next section is devoted to circular inclusions. An example concerning elliptical inclusions is presented in [33]. This result [33] can be considered as an extension of Grave's method [18] to multiply connected domains.

3. Metod of functional equations

Boundary value problems can be reduced to functional equations. By *functional equations* we mean iterative functional equations [22, 32] including compositions of the unknown functions with shift inside the domain like $\varphi[\alpha(z)]$. Hence, we do not use traditional integral equations and infinite systems of linear algebraic equations. Operators expressing such compositions of functions are easily implemented in packages with symbolic computations yielding high-order approximate analytical formulae.

3.1. Functional equations

In the present section, we consider the ℝ-problem (2.2.56) for non-overlapping circular inclusions $|z - a_k| < r_k$ $(k = 1, 2, \ldots, n)$; $\mathbb{D}$ is the complement of the extended complex plane to all the closed disks $|z - a_k| \le r_k$ $(k = 1, 2, \ldots, n)$. We consider the case $\psi(\infty) = \varphi'(\infty) = 1$ most interesting in applications. Then, we have $n(t) = \frac{t - a_k}{r_k}$ and (2.2.56) becomes

$$\psi(t) = \psi_k(t) + \rho_k \left(\frac{r_k}{t - a_k} \right)^2 \overline{\psi_k(t)}, \quad |t - a_k| = r_k, \; k = 1, 2, \ldots, n. \tag{2.3.83}$$

The ℝ-linear problem (2.3.83) can be reduced to functional equations. Let

$$z_{(m)}^* = \frac{r_m^2}{\overline{z - a_m}} + a_m \tag{2.3.84}$$

denote the inversion with respect to the circle $|t - a_m| = r_m$. Following [32] introduce the function

$$\Phi(z) := \begin{cases} \psi_k(z) - \sum_{m \neq k} \rho_m \left(\frac{r_m}{z - a_m} \right)^2 \overline{\psi_m \left(z_{(m)}^* \right)}, & |z - a_k| \le r_k, \\ & \qquad\qquad\qquad\qquad k = 1, 2, \ldots, n, \\ \psi(z) - \sum_{m=1}^n \rho_m \left(\frac{r_m}{z - a_m} \right)^2 \overline{\psi_m \left(z_{(m)}^* \right)}, & z \in \mathbb{D}, \end{cases}$$

analytic in the disks $\mathbb{D}_k$ $(k = 1, 2, \ldots, n)$ and in $\mathbb{D}$. Calculate the jump across the circle $|t - a_k| = r_k$

$$\Delta_k := \Phi^+(t) - \Phi^-(t), \quad |t - a_k| = r_k,$$

where $\Phi^-(t) := \lim_{z \to t, z \in \mathbb{D}} \Phi(z)$, $\Phi^+(t) := \lim_{z \to t, z \in \mathbb{D}_k} \Phi(z)$. Using (2.3.83) we get $\Delta_k = 0$. It follows from the principle of analytic continuation[2] that $\Phi(z)$ is analytic in the extended complex plane. Moreover, $\psi(\infty) = 1$ yields $\Phi(\infty) = 1$. Then, Liouville's theorem (see Section A.2.6) implies that $\Phi(z) \equiv 1$. The definition of $\Phi(z)$ in $|z - a_k| \leq r_k$ yields the following system of functional equations

$$\psi_k(z) = \sum_{m \neq k} \rho_m \left(\frac{r_m}{z - a_m} \right)^2 \overline{\psi_m \left(z_{(m)}^* \right)} + 1, \quad |z - a_k| \leq r_k, \ k = 1, 2, \dots, n. \quad (2.3.85)$$

Let $\psi_k(z)$ $(k = 1, 2, \dots, n)$ be a solution of (2.3.85). Then the function $\psi(z)$ can be found from the definition of $\Phi(z)$ in $\mathbb{D}$

$$\psi(z) = \sum_{m=1}^{n} \rho_m \left(\frac{r_m}{z - a_m} \right)^2 \overline{\psi_m \left(z_{(m)}^* \right)} + 1, \quad z \in \mathbb{D} \cup \partial\mathbb{D}. \quad (2.3.86)$$

BOX A.2.6 Liouville's Theorem

Function $\psi(z)$ is called analytic at infinity, if $\psi \left(\frac{1}{z} \right)$ is analytic at the point $z = 0$ (see page 8).

Theorem 8 (Liouville). *Every function analytic in $\widehat{\mathbb{C}} = \mathbb{C} \cup \{\infty\}$ is constant.*

Physical proof. Consider an analytic in $\widehat{\mathbb{C}}$ function $\psi(z)$ as the complex flux $\psi = \frac{\partial u}{\partial x_1} - i \frac{\partial u}{\partial x_2}$ in $\widehat{\mathbb{C}}$. This flux has no singularity (interior motto) and no boundary condition (exterior motto) everywhere. Therefore, the flux vanishes, i.e., $\frac{\partial u}{\partial x_1}$ and $\frac{\partial u}{\partial x_2}$ equal to zero and the complex potential $\varphi(z)$ from (2.2.50) is constant. □

Theorem 9 (Traditional form of Liouville's theorem). *Every function analytic in $\mathbb{C}$ and bounded at infinity is constant.*

Theorems 8 and 9 are equivalent since any bounded isolated singularity including $z = \infty$ is removable.

Theorem 10. *Let a function $\psi(z)$ be analytic in $\widehat{\mathbb{C}} = \mathbb{C} \cup \{\infty\}$ except a finite set of points $z_1, z_2, \dots, z_m$ and infinity where it has poles with the principal parts, respectively,*

$$G_k(z) = \frac{c_{k1}}{z - a_k} + \frac{c_{k2}}{(z - a_k)^2} + \dots + \frac{c_{km_k}}{(z - a_k)^{m_k}} \ at \ z = a_k, \quad (2.3.87)$$

$$G_0(z) = c_{01}z + c_{02}z^2 + \dots + c_{0m_0}z^{m_0} \ at \ z = \infty.$$

---

[2] See analogous harmonic continuation on page 222.

Then, the function $\psi(z)$ is rational and has the form

$$\psi(z) = c + \sum_{k=0}^{m} G_m(z), \tag{2.3.88}$$

where c is a constant.

—Liouville's Theorem [13, p. 100; 14, 43]

3.2. Method of successive approximations

Solution to the system of functional equations (2.3.85) is based on Theorem 1 on page 11 and Corollary 1 to it. Consider the space $\mathcal{H}(\mathbb{D}^+)$ consisting of functions analytic in $\mathbb{D}^+ = \cup_{k=1}^{n} \mathbb{D}_k$ and Hölder continuous in the closure of $\mathbb{D}^+$ endowed with the norm (2.1.7).

Theorem 11. *Let $|\rho_m| \le 1$. The system (2.3.85) has a unique solution for any circular multiply connected domain $\mathbb{D}$. This solution can be found by the successive approximations convergent in the space $\mathcal{H}(\mathbb{D}^+)$, i.e., uniformly convergent in every disk $|z - a_k| \le r_k$.*

Proof. Let $|\nu| \le 1$. Consider equations in $\mathcal{H}(\mathbb{D}^+)$

$$\psi_k(z) = \nu \sum_{m \neq k} \rho_m \left(\frac{r_m}{z - a_m}\right)^2 \overline{\psi_m\left(z_{(m)}^*\right)}, \quad |z - a_k| \le r_k, \; k = 1, 2, \ldots, n. \tag{2.3.89}$$

Let $\psi_k(z)$ be a solution of (2.3.89). Introduce the function $\phi(z)$ analytic in $\mathbb{D}$ and Hölder continuous in its closure as follows

$$\psi(z) = \nu \sum_{m=1}^{n} \rho_m \left(\frac{r_m}{z - a_m}\right)^2 \overline{\psi_m\left(z_{(m)}^*\right)}, \quad z \in \mathbb{D} \cup \partial\mathbb{D}. \tag{2.3.90}$$

Calculating the difference $\psi(t) - \psi_k(t)$ on each $|t - a_k| = r_k$ we arrive at the $\mathbb{R}$-linear conjugation relations

$$\psi(t) = \psi_k(t) + \nu\rho_k \left(\frac{r_k}{t - a_k}\right)^2 \overline{\psi_k(t)}, \quad |t - a_k| = r_k. \tag{2.3.91}$$

Moreover, (2.3.90) implies that $\psi(\infty) = 0$. Let $|\nu\rho_k| < 1$ for all $k = 1, 2, \ldots, n$. This case corresponds to the elliptic case in Mikhailov's terminology. The winding number κ of the constant coefficient $a(t) = 1$ vanishes. Then, in accordance with Theorem 5 the $\mathbb{R}$-linear problem (2.3.91) has only zero solutions. Hence, the system (2.3.89) also has a unique solution. One can see that the zero solution satisfies it, hence, the problem

(2.3.91) has only zero solutions when $|v| < 1$.

Consider now the case $|v\rho_k| = 1$ for all $k = 1, 2, \ldots, n$. The latter is possible if only $v = -e^{2i\theta}$ for some θ and $\rho_k = \pm 1$ for all $k = 1, 2, \ldots, n$. Let sgn denote the signum function. Then, (2.3.91) can be written in the form

$$e^{-i\theta}\psi(t) = e^{-i\theta}\psi_k(t) - e^{i\theta}\text{sgn}\,\rho_k \left(\frac{r_k}{t - a_k}\right)^2 \overline{\psi_k(t)}, \quad |t - a_k| = r_k. \qquad (2.3.92)$$

Integration of (2.3.92) along $|t - a_k| = r_k$ yields

$$\varphi(t) = \varphi_k(t) + \text{sgn}\,\rho_k\,\overline{\varphi_k(t)} + d_k, \quad |t - a_k| = r_k, \qquad (2.3.93)$$

where $\varphi'(z) = e^{-i\theta}\psi(z)$, $\varphi'_k(z) = e^{-i\theta}\psi_k(z)$; d_k are constant of integration; $\varphi(z)$ is analytic in $\mathbb{D}$. Take for definiteness $\rho_k = 1$. Then, the imaginary part of (2.3.93) gives the problem

$$\text{Im}\,\varphi(t) = \text{Im}\,d_k, \quad |t - a_k| = r_k, \qquad (2.3.94)$$

which has only constant solutions [32]. Then, (2.3.93) yields

$$\text{Re}\,\varphi_k(t) = h_k, \quad |t - a_k| = r_k, \qquad (2.3.95)$$

for some constant h_k.[3] Therefore, each $\varphi_k(z)$ is also a constant. Then $\psi(z) \equiv 0$, $\psi_k(z) \equiv 0$. Consider the case when $\rho_k = 1$ if $k \in K$ and $|v\rho_k| < 1$ if $k \notin K$. Then, (2.3.94) for $k \in K$ yields the analytic continuation of $\varphi(z)$ through the circles $|t - a_k| = r_k$ by the relations $\varphi(z) = \overline{\varphi(z^*_{(k)})} + 2id_k$ into the domains $(\mathbb{D})^*_{(k)}$. The functions $\varphi_m(z)$ for $m \notin K$ produce the functions $\overline{\varphi_m(z^*_{(k)})}$ analytic in $(\mathbb{D}_m)^*_{(k)}$. These functions constitute the $\mathbb{R}$-linear problem with $|v\rho_m| < 1$.

Theorem 1 on page 11 yields the convergence of the method of successive approximations applied to the system (2.3.85).

The theorem is proved. □

Functional equations (2.3.85) and formula (2.3.86) are written for the complex flux. Following [28, 34] it is possible to write analogous functional equations for the complex potentials. Let $w \in \mathbb{D}$ be a fixed point not equal to infinity. It is convenient to put $\varphi(w) = 0$ instead of $\varphi(\infty) = 0$ that determines an additive constant of the complex

---

[3]In the case $\rho_k = -1$, the real and imaginary parts change places in (2.3.94) and (2.3.95).

potential $\varphi(z)$ in another way. Following the above scheme introduce the function

$$\Phi(z) := \begin{cases} \varphi_k(z) + \sum_{m \neq k} \rho_m \left[\overline{\varphi_m \left(z_{(m)}^* \right)} - \overline{\varphi_m \left(w_{(m)}^* \right)} \right] - \rho_k \overline{\varphi_m \left(w_{(k)}^* \right)} + z, \quad |z - a_k| \leq r_k, \\ \qquad\qquad\qquad\qquad\qquad\qquad\qquad\qquad\qquad\qquad\qquad k = 1, 2, \cdots, n, \\[2mm] \varphi(z) + \sum_{m=1}^n \rho_m \left[\overline{\varphi_m \left(z_{(m)}^* \right)} - \overline{\varphi_m \left(w_{(m)}^* \right)} \right], \quad z \in \mathbb{D} \end{cases}$$

One can check that $\Phi(z) \equiv \varphi(w) = 0$. This yields the functional equations

$$\varphi_k(z) = - \sum_{m \neq k} \rho_m \left[\overline{\varphi_m \left(z_{(m)}^* \right)} - \overline{\varphi_m \left(w_{(m)}^* \right)} \right] + z + \rho_k \overline{\varphi_m \left(w_{(k)}^* \right)}, \ |z - a_k| \leq r_k \ (k = 1, 2, \ldots, n)$$

$$(2.3.96)$$

and formula

$$\varphi(z) = - \sum_{m=1}^n \rho_m \left[\overline{\varphi_m \left(z_{(m)}^* \right)} - \overline{\varphi_m \left(w_{(m)}^* \right)} \right], \quad z \in \mathbb{D}. \qquad (2.3.97)$$

The functions $\varphi(z)$ and $\varphi_k(z)$ analytic in $\mathbb{D}$ and in $\mathbb{D}_k$, respectively, and continuously differentiable in the closures of the domains considered.

The main difference between the systems (2.3.85) and (2.3.96) is introduction of the additional constant terms with w in order to get equations for which the method of successive approximations converges. The proof of the convergence repeats the proof of Theorem 11 (see this proof in [28, 32]). If we delete the terms $\overline{\varphi_m \left(w_{(m)}^* \right)}$ in (2.3.96) we arrive at the functional equations

$$\varphi_k(z) = - \sum_{m \neq k} \rho_m \overline{\varphi_m \left(z_{(m)}^* \right)} + z, \quad |z - a_k| \leq r_k \ (k = 1, 2, \ldots, n) \qquad (2.3.98)$$

for which the convergence fails. It is easily seen, if we take a non-zero constant for the zero approximation of $\varphi_k(z)$.

In order to illustrate final formulae, consider the case $\rho_k = \rho$ for all $k = 1, 2, \ldots, n$. Application of the method of successive approximations to (2.3.85) or to (2.3.96) yields the exact formulae (for details see [32] and [34])

$$\varphi_k(z) = q_k + z + \rho \sum_{k_1 \neq k} \overline{(z_{(k_1)}^* - w_{(k_1)}^*)} + \rho^2 \sum_{k_1 \neq k} \sum_{k_2 \neq k_1} (z_{(k_2 k_1)}^* - w_{(k_2 k_1)}^*) + \qquad (2.3.99)$$

$$\rho^3 \sum_{k_1 \neq k} \sum_{k_2 \neq k_1} \sum_{k_3 \neq k_2} \overline{(z_{(k_3 k_2 k_1)}^* - w_{(k_3 k_2 k_1)}^*)} + \cdots, \quad |z - a_k| \leq r_k$$

and

$$\varphi(z) = z - w + \rho \sum_{k=1}^{n} \overline{(z^*_{(k)} - w^*_{(k)})} + \rho^2 \sum_{k=1}^{n} \sum_{k_1 \neq k} (z^*_{(k_1 k)} - w^*_{(k_1 k)}) + \qquad (2.3.100)$$

$$\rho^3 \sum_{k=1}^{n} \sum_{k_1 \neq k} \sum_{k_2 \neq k_1} \overline{(z^*_{(k_2 k_1 k)} - w^*_{(k_2 k_1 k)})} + \cdots, \quad z \in \mathbb{D},$$

where q_k are constants.

In the case $\rho = \pm 1$, we arrive at the famous Poincaré series (2.3.100) (see the auxiliary section below). This series converges uniformly in z in the closure of $\mathbb{D}$ [30]. This implies that the series (2.3.100) with $|\rho| < 1$ converge absolutely in ρ. Actually, this is the proof of the contrast parameter convergence for any $|\rho| < 1$. Convergence of the expansion in the parameter r^2, where r is the inclusion radius, can be obtained by expansion of the series (2.3.100). This approach will be described in Chapter 3 for periodic media when such an expansion in r^2 yields the concentration expansion.

Analogous series can be written for different ρ_k for general external sources modelled by the complex potential $g(z)$ [28].

B.2.2 Riemann–Hilbert problem for multiply connected domains
Reduction to the $\mathbb{R}$-linear problem

On page 24, we discuss the most general statement of the $\mathbb{R}$-linear problem (2.2.31) with non-constant coefficients. In the case $|a(t)| \equiv |b(t)|$, the problem (2.2.31) is reduced to the Riemann–Hilbert problem. This observation was first made by I.Sabitov and published in [26]. The Riemann–Hilbert problem has the form

$$\text{Re } \overline{\lambda(t)}\varphi(t) = g(t), \ t \in \partial D, \qquad (2.3.101)$$

where $\varphi(t) = \varphi^-(t)$ is the limit value of $\varphi(z)$ analytic in the multiply connectet domain D, the Hölder continuous functions $\lambda(t)$ and $g(t)$ are expressed in terms of the functions $a(t)$, $b(t)$ and $c(t)$ of the problem (2.2.31) by simple formulae.

The Riemann–Hilbert problem can be considered as a generalization of the classical Dirichlet and Neumann problems for harmonic functions. It includes as a particular case the mixed boundary value problem. The exact solution of the Dirichlet problem for annulus is due to Villat–Dini [11, 48] (see an independently established formula in [2], p.169). The computationally effective exact and asymptotic formulae for a doubly connected circular domain are derived in [42].

The complete solution of the Riemann–Hilbert problem (2.3.101) for an arbitrary multiply connected domain was first presented in [29, 31] and in Chapter 4, [32].[4] More

---

[4]The papers [29, 31] do not contain details of the complete solution. In the same time, a rather long presentation in the book [32] is based on the general introductory material, hence, it is long. A self-

precisely, we have to say that the problem (2.3.101) is solved in analytical form up to a conformal mapping of $\mathbb{D}$ onto a circular multiply connected domain $\mathbb{D}$.

We outline now the solution to the problem (2.3.101). First step is the factorization of the coefficient $\lambda(t)$ and reduction of the problem (2.3.103) to the problem with constant coefficients ν_k given at each boundary circle

$$\operatorname{Re} \nu_k \varphi(t) = h(t), \quad |t - a_k| = r_k \ (k = 1, 2, \ldots, n). \tag{2.3.102}$$

The coefficients ν_k are given explicitly in Chapter 4 [32]. The function $h(t)$ linearly depends on 2κ undetermined real constant in the case $\kappa \geq 0$. Here, $\kappa = wind\ \lambda(t)$ denotes the winding number (index) of $\lambda(t)$. In the case $\kappa < 0$, the function $\varphi(z)$ has zero of order $|\kappa|$ at infinity.

The next step is reduction of the problem (2.3.102) to the $\mathbb{R}$-linear problem

$$\nu_k \varphi^-(t) = \varphi^+(t) - \overline{\varphi^+(t)} + h(t), \ t \in \partial \mathbb{D}, \tag{2.3.103}$$

where $\varphi^-(t) \equiv \varphi(t)$, $\mathbb{D}^- \equiv \mathbb{D}$ and $\mathbb{D}^+$ is the complement of $\mathbb{D} \cup \partial \mathbb{D}$ to the extended complex plane. The unknown function $\varphi^+(z)$ is analytic in $\mathbb{D}^+$ and Hölder continuous in its closure. The proof that the problems (2.3.102) and (2.3.103) are equivalent, follows the lines of Theorem 4. Further, the problem (2.3.103) is reduced to functional equations and solved by successive approximations, see page 29.

Applications of functional equations to boundary value problems are described in literature. In 1934 Golusin [15, 17] reduced the Dirichlet problem for circular multiply connected domains to a system of functional equations and applied the method of successive approximations to obtain its solution under some geometrical restrictions. The restrictions ensure that each disk $\mathbb{D}_k$ lies sufficiently far away from all other disks $\mathbb{D}_m$ ($m \neq k$) (see, for instance, the restriction (2.3.106) below).

Golusin's approach was developed in [1, 12, 49]. Aleksandrov and Sorokin [1] extended Golusin's method to an arbitrary multiply connected circular domain. However, the analytic form of the Schwarz operator was not obtained and advantage of having a closed form solution was lost. More precisely, the Schwarz problem was reduced via functional equations to an infinite system of linear algebraic equations. Truncation of the infinite system was suggested and justified as well.[5]

We also reduce the problem to functional equations which are similar to Golusin's ones (see page 31). The main advantage of our modified functional equations is based on the possibility to solve them without any geometrical restriction by successive approximations.

---

consistent presentation of the complete solution is given in [35].

[5]Though the method of truncation can be effective in numeric computations, one can hardly accept that this method yields a closed form solution. Any way it depends on using of the term "closed form solution" discussed in Introduction. A regular infinite system [20] can be considered as an equation with compact operator, i.e., it is no more than a discreet form of a Fredholm integral equation. Therefore, the result [1] can be treated following item 3 of Introduction as an approximate numerical solution.

Schottky double

Though the solution is given by exact formulae (2.3.100), it is not an elementary function. More precisely, it is represented in the form of integrals involving the Abelian functions [4] (such as Poincaré series and Schottky–Klein prime function presented in the most general form as Poincaré α-series in [30, 32]). The reason, why the solution in general is not presented by integrals involving elementary kernels, has the topological nature. In order to explain this, we shortly recall the scheme of the solution to the Riemann–Hilbert problem

$$\varphi(t) + G(t)\overline{\varphi(t)} = g(t), \ t \in \partial\mathbb{C}^+, \tag{2.3.104}$$

for the upper half-plane $\mathbb{C}^+$ following [14, 39]. Define the function $\varphi^-(z) := \overline{\varphi(\bar{z})}$ analytic in the lower half-plane. Then the Riemann–Hilbert problem (2.3.104) becomes the $\mathbb{C}$-linear problem (Riemann problem in Gakhov's terminology [14])

$$\varphi^+(t) + G(t)\varphi^-(t) = g(t), \ t \in \partial\mathbb{C}^+. \tag{2.3.105}$$

The latter problem is solved in closed form in terms of the Cauchy type integrals by the factorization method (see details in [14, 39]).

Let us look at this scheme from another point of view [50]. Introduce a copy of the upper half-plane $\mathbb{C}^+$ with the local complex coordinate $\bar{z}$ and weld it with $\mathbb{C}^+$ along the real axis. Define the function $\varphi^-(\bar{z}) := \overline{\varphi(z)}$ analytic on the copy of $\mathbb{C}^+$. Then we again arrive at the $\mathbb{C}$-linear problem (2.3.105) but on the double of $\mathbb{C}^+$ which is conformally equivalent to the Riemann sphere $\hat{\mathbb{C}}$. The fundamental functionals of $\hat{\mathbb{C}}$ are expressed by means of meromorphic functions which produces the Cauchy type integrals. The same scheme holds for any n-connected domain $\mathbb{D}$. In result, we arrive at the problem (2.3.105) on the Schottky double of $\mathbb{D}$, the Riemann surface of genus $(n-1)$, where the life is more complicated than on the plane, i.e., on the Riemann sphere of zero genus. It does not described by simple meromorphic functions. Therefore, in order to solve the problem (2.3.105) on the double of $\mathbb{D}$, meromorphic analogies of the Cauchy kernel on Riemann surfaces have to be used, i.e., the Abelian functions [4]. In the case $n = 2$, the double of $\mathbb{D}$ becomes a torus in which meromorphic functions are replaced by the classical elliptic functions [2].

Poincaré series

Poincaré [41] introduced the θ_2-series associated to various types of the Kleinian groups. He did not study in depth the Schottky groups and just conjectured that the corresponding θ_2-series always diverges.[6] In 1891, Burnside [8] gave examples of convergent series for the Schottky groups (named by him the first class of groups) and studied their absolute convergence under some geometrical restrictions. In his study W. Burnside followed Poincaré's proof of the convergence of the θ_4-series. Burnside [8, p. 52] wrote "I have endeavoured to show that, in the case of the first class of groups, this series is convergent, but at present I have not obtained a general proof. I shall offer two partial proofs of the convergency; one of which applies only to the case of fuchsian

---

[6]H. Poincaré just said " Toujour dans le cas d'un groupe fuchsien, la série ... n'est pas convergent", see [41, p. 308] and [8, p. 51]

groups, and for that case in general, while the other will also apply to kleinian groups, but only when certain relations of inequality are satisfied." Furthermore, Burnside (see [8, p. 57]) gave a condition of the absolute convergence in terms of the coefficients of the Möbius transformations. He also noted that convergence holds if the radii of the circles $|z - a_k| = r_k$ are sufficiently less than the distances between the centers $|a_k - a_m|$ when $k \neq m$. Intrigued with [41] and [8] many mathematicians contributed to justification of the absolute convergence of the Poincaré series (2.3.100) under geometrical restrictions to the locations of the circles, for instance (2.3.106). In 1916, Myrberg [40] gave examples of absolutely divergent θ_2-series. It just seemed that the opposing conjectures of Poincaré and Burnside were the both wrong. However, it was proved in [30] that θ_2-series converges uniformly for any multiply connected domain $\mathbb{D}$ without any geometrical restriction that corresponds to Burnside's conjecture.

The uniform convergence does not directly imply the automorphy relation, i.e., invariance under the Schottky group of transformations, since it is forbidden to change the order of summation without absolute convergence. But this difficulty can be easily overcome by using functional equations. As a result, the Poincaré series satisfies the required automorphy relation and can be written in each fundamental domain with a prescribed summation depending on this domain [30].

Absolute and uniform convergence

Absolute convergence implies geometrical restrictions on the geometry which can be roughly presented as follows. Each inclusion $\mathbb{D}_k$ is sufficiently far away from other inclusions $\mathbb{D}_m$ ($m \neq k$). Here, we present such a typical restriction expressed in terms of the separation parameter Δ introduced by Henrici and used by DeLillo et al. [10]

$$\Delta = \max_{k \neq m} \frac{r_k + r_m}{|a_k - a_m|} < \frac{1}{(n-1)^{\frac{1}{4}}} \qquad (2.3.106)$$

for n-connected domain $\mathbb{D}$ bounded by the circles $|z - a_k| = r_k$ ($k = 1, 2, ..., n$).

The alternating Schwarz method for simply connected domains [20] was extended to multiply connected domains in [16, 27]. Mikhlin [27] called it *the generalized alternating Schwarz method*. Mikhlin [27] applied this method to the Dirichlet problem and proved its convergence under some geometrical restrictions similar to (2.3.106). Further modifying Michlin's work, we manage to lift any geometrical restriction and are in possession of the method convergent for any multiply connected domain, see page 26.

The tale of convergence repeats itself for the Schottky–Klein prime function $S(z)$. Baker [4] introduced this function in a form of absolutely convergent product under geometrical restrictions similar to (2.3.106). The function $S(z)$ was constructed for an arbitrary circular multiply connected domain in [29] (see also (4.4.28) from [32]) in the form of uniformly convergent product (see the proper terminology and extension to the α-functions in [37]). It is worth noting that the exact formulae for $S(z)$ from [9] do not hold in general case when the order of the summation (and product) is essential for convergence. Comparative analysis of absolute and uniform convergence can be found in [34, 36]. The θ_2-series and the infinite product $S(z)$ are related by formula $\theta_2(z) = (\ln S(z))'$. An analogous relation takes place for the α-functions [37].

This complicated situation concerning absolute and uniform convergences can be illustrated by a simple example. Let the almost uniformly convergent series

$$\sum_{n=1}^{\infty}(n-z)^{-2} \qquad (z \notin \mathbb{N})$$

be integrated term by term

$$\int_w^z \sum_{n=1}^{\infty} \frac{1}{(n-t)^2}\, dt = \sum_{n=1}^{\infty}\left(\frac{1}{n-z} - \frac{1}{n-w}\right). \qquad (2.3.107)$$

One can see that this series converges if and only if $w \neq \infty$. This unfortunate infinity is considered sometimes as a fixed point by specialists in Complex Analysis (see for instance Michlin's study [27] devoted to convergence of Schwarz's method).

Concluding remarks on the Riemann–Hilbert problem

Many important functions can be explicitly constructed as the solutions of the special Riemann–Hilbert problems. Harmonic measures, Green's function, the Schwarz operator [32], the Bergmann function [19], Schwarz–Cristoffel formula for multiply connected polygons [34, 36] were constructed via Riemann–Hilbert problems. These results cover an explicit construction of the objects for an arbitrary circular multiply connected domain.

We stress that the complete solution of the Riemann–Hilbert problem (analytical form of solution, necessary and sufficient conditions of solvability, investigation of the singular case, exactly written Bojarski's system [47]) is presented in [29, 32, 35]. All other methods deal with analytical solution to the particular cases (see, for instance, the condition (2.3.106)) or with the numerical solution as explained on page 1(item 3). Particular cases of the Riemann–Hilbert problems are discussed within the framework of various numerical methods on the level of item 3, for instance, a problem for a strip with circular holes, etc. But a straight line is a particular case of the circle on the extended complex plane $\widehat{\mathbb{C}}$. Hence, the scheme expounded in [29], Chapter 4 of [32], and [35] yields analytical solution including the case of touching circles [32].

Perhaps, one principal problem has not been solved yet, namely the Riemann–Hilbert problem with discontinuous coefficient $\lambda(t)$. It requires the factorization in Muskhelishvili's classes [14, 39].

–Scalar Riemann–Hilbert problem [29, 32, 35]

REFERENCE

1. Aleksandrov IA, Sorokin AS, The problem of Schwarz for multiply connected domains, Sib. Math. Zh. 1972; 13: 971–1001 [in Russian].
2. Akhiezer NI, Elements of the Theory of Elliptic Functions. Providence, Rhode Island: AMS; 1990.
3. Axler Sh, Bourdon P, Ramey W, Harmonic Function Theory. 2nd Berlin: Springer Verlag; 2001.
4. Baker HF, Abel's theorem and the allied theory, including the theory of the theta functions. Cambridge: Cambridge University Press; 1897.
5. Bojarski B, On a boundary value problem of the theory of analytic functions, Dokl. AN SSSR 1958; 119: 199–202 [in Russian].
6. Bojarski B, On generalized Hilbert boundary value problem, Soobsch. AN GruzSSR 1960; 25: 385–390 [in Russian].

7. Bojarski B, Mityushev V, ℝ-linear problem for multiply connected domains and alternating method of Schwarz, J. Math. Sci. 2013; 189: 68–77.

8. Burnside W, On a Class of Automorphic Functions, Proc. London Math. Soc. 1891; 23: 49–88.

9. Crowdy D, The Schwarz problem in multiply connected domains and the Schottky-Klein prime function, Complex Variables and Elliptic Equations 2008; 53: 221–236.

10. DeLillo TK, Driscoll TA, Elcrat AR, Pfaltzgraff JA, Radial and circular slit maps of unbounded multiply connected circle domains, Proc. R. Soc. London 2008; A 464: 1719–1737.

11. Dini U, Il problema di Dirichlet in unarea anulare, e nello spazio compreso fra due sfere concentriche, Rend. Circ. Mat. Palermo 1913; 36: 1–28 doi:10.1007/BF03016009.

12. Dunduchenko LE, On the Schwarz formula for an n–connected domain, Dopovedi AN URSR 1966; 5: 1386–1389 [in Ukrainian].

13. Freitag E, Busam R, Complex Analysis. 2nd Berlin: Springer Verlag; 2009.

14. Gakhov FD, Boundary Value Problems. 3rd Moscow: Nauka; 1977. [Engl. transl. of 1st ed.: Pergamon Press, Oxford, 1966.]

15. Golusin GM, Solution of basic plane problems of mathematical physics for the case of Laplace equation and multiply connected domains bounded by circles (method of functional equations), Math. zbornik 1934; 41: 246–276 [in Russian].

16. Golusin GM, Solution of spatial Dirichlet problem for Laplace equation and for domains enbounded by finite number of spheres, Math. zbornik 1934; 41: 277–283 [in Russian].

17. Golusin GM, Solution of plane heat conduction problem for multiply connected domains enclosed by circles in the case of isolated layer, Math. zbornik 1935; 42: 191–198 [in Russian].

18. Grave DA, On the main mathematical problems of construction of geographic maps. Sankt-Peterburg: Empire Academy of Sciences; 1896 [Chapter III], [in Russian].

19. Jeong Moonja, Mityushev V, The Bergman kernel for circular multiply connected domains, Pacific J. Math. 2007; 233: 145–157

20. Kantorovich LV, Krylov VI, Approximate methods of higher analysis. Groningen: Noordhoff; 1958.

21. Krasnosel'skii MA, Vainikko GM, Zabreiko PP, Rutickii JaB, Stecenko VJa, Approximate Methods for Solution of Operator Equations. Groningen: Wolters-Noordhoff Publ.; 1972.

22. Kuczma M, Chosewski B, Ger PP, Iterative functional equations. Encyclopedia Math.Appl., 32, Cambridge: Cambridge University Press; 1990.

23. Markushevich AI, On a boundary value problem of analytic function theory, Uch. zapiski MGU 1946; 1: 20–30 [in Russian].

24. Milne-Thompson LM, Theoretical Hydrodynamics. 5th New York: Dover Publications; 1968 [Chapter 8].

25. Mikhailov LG, On a boundary value problem, DAN SSSR 1961; 139: 294–297 [in Russian].

26. Mikhailov LG, New Class of Singular Integral Equations and its Applications to Differential Equations with Singular Coefficients. 2nd Berlin: Akademie Verlag; 1970.

27. Mikhlin SG, Integral Equations and Their Applications to Certain Problems in Mechanics, Mathematical Physics and Technology. 2nd New York: Macmillan; 1964.

28. Mityushev V, Plane problem for the steady heat conduction of material with circular inclusions, Arch. Mech. 1993; 45: 211–215.

29. Mityushev V, Solution of the Hilbert boundary value problem for a multiply connected domain, Slupskie Prace Mat.-Przyr. 1994; 9a: 37–69.

30. Mityushev VV, Convergence of the Poincaré series for classical Schottky groups, Proceedings Amer. Math. Soc. 1998; 126(8): 2399–2406.

31. Mityushev VV, Hilbert boundary value problem for multiply connected domains, Complex Variables 1998; 35: 283–295.

32. Mityushev VV, Rogosin SV, Constructive methods to linear and non-linear boundary value problems of the analytic function. Theory and applications. Boca Raton etc: Chapman & Hall / CRC; 1999/2000. [Chapter 4].

33. Mityushev V, Conductivity of a two-dimensional composite containing elliptical inclusions, Proc. R. Soc. London 2009; A465: 2991–3010.

34. Mityushev V, Riemann-Hilbert problems for multiply connected domains and circular slit maps, Computational Methods and Function Theory 2011; 11(2): 575–590.
35. Mityushev V, Scalar Riemann-Hilbert Problem for Multiply connected domains, ed. Th.M. Rassias and J. Brzdek, Functional Equations in Mathematical Analysis, Springer Optimization and Its Applications 52, Springer Science+Business Media, 599–632, 2012 doi:10.1007/978-1-4614-0055-4.
36. Mityushev V, Schwarz-Christoffel formula for multiply connected domains, Computational Methods and Function Theory 2011; 12(2): 449–463.
37. Mityushev V, Poincare α-series for classical Schottky groups and its applications, ed. Milovanović GV and Rassias MTh, Analytic Number Theory, Approximation Theory, and Special Functions, 827–852, 2014.
38. Muskhelishvili NI, To the problem of torsion and bending of beams constituted from different materials, Izv. AN SSSR 1932; 7: 907–945 [in Russian].
39. Muskhelishvili NI, Singular Integral Equations. 3rd Moscow: Nauka; 1968 [in Russian].
40. Myrberg PJ, Zur Theorie der Konvergenz der Poincaréschen Reihen, Ann. Acad. Sci. Fennicae 1916; A9(4): 1–75.
41. Poincaré H, Sur les Groupes des Équations Linéaires, Acta Math. 1884; 4: 201–312.
42. Rylko N, A pair of perfectly conducting disks in an external field, Mathematical Modelling and Analysis 2015; 20: 273–288.
43. Shabat BV, Introduction to complex analysis. 2nd Moscow: Nauka; 1985 [in Russian].
44. Tikhonov AN, Samarskii AA, Equations of Mathematical Physics. 2nd New York: Dover Publications; 1990.
45. Vekua IN, Rukhadze AK, The problem of the torsion of circular cylinder reinforced by transversal circular beam, Izv. AN SSSR 1933; 3: 373–386 [in Russian].
46. Vekua IN, Rukhadze AK, Torsion and transversal bending of the beam compounded by two materials restricted by confocal ellipses, Prikladnaya Matematika i Mechanika (Leningrad) 1933; 1(2): 167–178 [in Russian].
47. Vekua IN, Generalized Analytic Functions. 2nd Moscow: Nauka; 1988 [in Russian].
48. Villat H, Le probléme de Dirichlet dans une aire annulaire, Rend. Circ. Mat. Palermo 1912; 33: 134–174 doi:10.1007/BF03015296.
49. Zmorovich VA, On a generalization of the Schwarz integral formula on n-connected domains, Dopovedi URSR 1958; 5: 489–492 [in Ukrainian].
50. Zverovich EI, Boundary value problems of analytic functions in Hölder classes on Riemann surfaces, Uspekhi Mat. nauk 1971; 26: 113–179 [in Russian].

CHAPTER 3

Constructive homogenization

The infamous Colomban states that we have no proofs against Pyrot. He lies; we have them. I have in my archives seven hundred and thirty-two square yards of them which at five hundred pounds each make three hundred and sixty-six thousand pounds weight.

— Penguin Island, Anatole France

1. Introduction

The main purpose of this chapter is to work out a set of constructive analytical–numerical methods to calculate the effective constants of dispersed composites.

The local conductivity tensor $\sigma(\mathbf{x})$ of second order can be considered as a random function of the spatial variable $\mathbf{x} = (x_1, x_2, x_3)$. Similar approach is applicable to the local elastic tensor of fourth order and will be presented in Chapter 10. In the book, we study only dispersed composites made from non-overlapping inclusions. For different types of composites and porous media, for instance when $\sigma(\mathbf{x})$ obeys the Gaussian distribution, an inverse problem can be solved [3]. The reconstructions of such types of media from the correlation functions give excellent practical results (see [1] and works cited therein). Another type of composites, not discussed in the book, laminates and their polycrystalline compositions, was systematically investigated by Cherkaev [11]. His book also contains the general approach to structural optimization of composites.

In this chapter, we show that the results obtained in Chapter 2 lead to a constructive approach to the problems of classic theory of random composites for non-overlapping inclusions. The method of functional equations and Schwarz's method allow to go beyond the unrecognized myths discussed below, which permeate the theory of random composites.

- First of all, we solve in the analytical form a Laplacian and bi-harmonic problems for 2D composites with circular inclusions which are treated in literature as "impossible to solve in analytical form" problems. It is worth noting that any plane domain can be approximated by a disks packing. Moreover, the method of functional equations can be extended to algebraic form shapes [38, 41].

Computational Analysis of Structured Media
http://dx.doi.org/10.1016/B978-0-12-811046-1.50003-4

- Random composites have to be studied by application of the probabilistic and statistical methods. This assertion is beyond questions. However, such an investigation should not be based only on the correlation functions [50]. Immense computational restrictions in theory of random composites originate in a virtual impossibility to compute the correlation functions numerically. We obviate the need for correlation functions and calculate the mathematical expectation of the effective tensor after complete solving the deterministic problems.

- It was thought that the contrast parameter expansions are valid only for sufficiently small contrast parameters. However, it was first justified in [34] that such expansions for plane conductivity problems are valid for all physical values of the contrast parameters. We leave here aside a novel class of metamagnetic materials [45]. Proper expansions in concentration, or contrast parameter, will be developed in the present book.

- Heroic quest for the equivalent replacement of a random composite by a periodic regular composite (one inclusion per periodicity cell) with the same concentration invariably leads to a significantly lower effective conductivity than in the original random composites for high concentrations and high contrast parameters.

  It was noted in [7] that periodic arrangement of inclusions lowers the effective conductivity compared to non-periodic case, assuming that conductivity of individual inclusions is higher than conductivity of host. This observation can be expressed as a general *order principle in dispersed composites*: **stronger positional order yields deeper extremal values of the effective conductivity**. A similar effect was established in [30] and treated as the size effect of particles and layer thickness. The boundary conditions on the parallel plates yields the vertical periodicity, hence this size effect governed by the order principle in dispersed composites. For instance, the effective conductivities of square and hexagonal arrays of disks attain the local minima compared to perturbed regular locations [7]. The global minimum is attained for the hexagonal array [42].

- Self-consistent methods (SCM) for random composites are considered as a simple constructive alternative to the method of correlation functions. However, applications of Schwarz's method demonstrates essential limitations of these methods [43]. More precisely, SCM do not capture interactions among inclusions and give analytical formulae in general only up to $O(f^2)$ and up to $O(f^3)$ for macroscopically isotropic composites. Even solution to the n-particle problem in the infinite plane for a fixed n applies only to a diluted distributions of the clusters consisting of n inclusions (see illustration in Fig. 3.1). The concentration of a finite number of bounded inclusions in the whole space is equal to zero. A finite number of inclusions in a bounded domain Q serves to represent an infinite medium. Then, it is usually supposed that Q is a periodicity cell. Any doubly periodic structure in $\mathbb{R}^2$ can be represented by a parallelogram and any triply periodic structure in $\mathbb{R}^3$ by a parallelepiped.

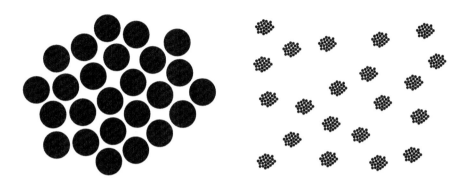

Figure 3.1 A cluster with a finite number of inclusions and an infinite number of clusters diluted on the plane.

- In addition to the previous item, we give the proper interpretation of the uncertainty arising in applications of the SCM. Consider a two-phase 3D composite with the normalized conductivities 1 and $\sigma > 1$ of the concentrations $1 - f$ and f, respectively. The celebrated Hashin–Shtrikman bounds [22] for the effective conductivity σ_e of possible macroscopically isotropic composites read as

$$\sigma_{HS}^-(\sigma, f) \leq \sigma_e \leq \sigma_{HS}^+(\sigma, f), \tag{3.1.1}$$

where

$$\sigma_{HS}^-(\sigma, f) = 1 + \frac{3f(\sigma - 1)}{3 + (1 - f)(\sigma - 1)}, \quad \sigma_{HS}^+(\sigma, f) = \sigma + \frac{3(1 - f)\sigma(1 - \sigma)}{3\sigma + f(1 - \sigma)}. \tag{3.1.2}$$

The Clausius–Mossotti approximation, a typical representative of SCM, coincides with the lower bound. It can be written in the standard form and further asymptotically transformed

$$\sigma_e \approx \frac{1 + 2\nu}{1 - \nu} = \frac{1}{1 - 3\nu} + O(\nu^2), \tag{3.1.3}$$

where $\nu = f\frac{\sigma - 1}{\sigma + 2}$. The graph $s(\sigma, f)$ in Fig. 3.2b is just a curve between the Hashin–Shtrikman bounds (3.1.2). Each point of the curve as well as each point

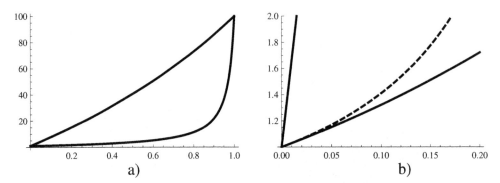

Figure 3.2 The Hashin–Shtrikman lower and upper bounds (3.1.2) (solid lines) and the graph of $\sigma_e \approx s(\sigma, f) = \left(1 - 3f\frac{\sigma-1}{\sigma+2}\right)^{-1}$ (dashed line) for $\sigma = 100$.

enclosed between (3.1.2) corresponds to composites with different geometry (inclusions location). The art of the SCM is based on manipulations like (3.1.3) within the accuracy $O(\nu)$ to write a universal formula. Of course, such a formula does not exist and the effective conductivity depends essentially on the locations of inclusions [37, 43].

Remark 3. This remark concerns designations used for conductivity. The conductivity of matrix is normalized to unity. In the previous chapter, the notation σ_k is used for the conductivity of kth inclusion. Hereafter, the main attention is paid to two-phase composites when all inclusions have the same constant conductivity, for shortness denoted by σ. In the same time, sometimes the notation $\sigma(\mathbf{x})$ is used for the local conductivity, i.e., $\sigma(\mathbf{x}) = 1$ in matrix and $\sigma(\mathbf{x}) = \sigma$ in inclusions.

2. Deterministic and stochastic approaches

We discuss the composites with non-overlapping inclusions when a collection of particles with fixed shapes and sizes is embedded into a matrix. More precisely, consider a set of particles as physical objects $\widetilde{D}_k$, $k = 1, 2, \ldots$ where each $\widetilde{D}_k$ has a fixed geometry. Let all the particles be randomly located in the space and each particle $\widetilde{D}_k$ occupy a domain D_k considered as a geometric object. Thus, the deterministic elements D_k are introduced independently but their joint set $\{D_1, D_2, \ldots\}$ is introduced randomly. The diversity of random locations is expressed by joint probabilistic distributions of the non-overlapping domains D_k.

In order to be precise in our theoretical study and to avoid misleading terminology and assertions, we must introduce the fundamental notations and axioms. Such a rigorous approach needs the advanced theory of homogenization based on the G-, H- and Γ-convergence; the measure theory; stochastic stationary processes, multiscale methods, PDE and others. Following Einstein's "Everything should be made as simple as possible, but not simpler", we emphasize what must be understood as *not simpler* in the theory of composites.

The theory of composites deals with infinite number of inclusions located in space. Dealing with infinity has been bringing many troubles in pure and applied mathematics. Composites are not an exception. First, one can note that the concentration of inclusions in $\mathbb{R}$ (1D layered composites) can be not properly defined, e.g., the concentration as the limit ratio measure of one phase $\lim_{L \to +\infty} \frac{f_L}{L}$ does not always exist. Let us restrict ourselves to such composites for which the concentration f exists. This assumption is sufficient to determine the effective conductivity σ_e as the harmonic mean of its constituents. However, it is not sufficient in higher dimensions.

The second-order term f^2 of $\sigma_e(f)$ for identical disks on the plane is not properly defined for some locations of the disks on the plane with certain prescribed concentration (see Remark 11 on page 88). An assumption of macroscopic isotropy determines the term f^2 in 2D, but does not help to to define the term f^3. This mathematical observation can be considered as a curious fact. But some estimations of σ_e for infinite number of inclusions assume the existence and dependence only on the concentration. The declaration that σ_e exists can be considered only as an incantation since one can repeat the presented arguments for a composite with non-existing σ_e and still obtain some formula for σ_e.

The classic asymptotic homogenization theory [4, 5, 24] deals with periodic composites. It is worth noting that we use the term "periodic" in the following context. *Periodic composite* consists of periodic sets of cells (doubly in $\mathbb{R}^2$ or triply periodic in $\mathbb{R}^3$). A finite number of inclusions is located in the same way in every periodicity cell. *Periodic regular composite* consists of periodic sets of cells with one inclusion per periodicity cell. The material effective constants are rigorously introduced for periodic composites in the framework of the asymptotic homogenization theory.

The homogenization theory was extended to non-periodic composites [16, 24, 49] when periodicity still holds but in the framework of probabilistic distribution. This is the main mathematical axiom of homogenzation. *Random composite* is a statistically homogeneous medium, i.e., the statistical properties of the medium are invariant under translations in space.[1] A lattice group generated by d fundamental translation vectors $\omega_j \in \mathbb{R}^d$ $(j = 1, \ldots, d)$ is assigned to any fixed random statistically homogeneous composite. The vectors ω_j $(j = 1, \ldots, d)$ form the fundamental cell which can be called by representative random cell. The homogenization theory rigorously justifies existence of the effective properties for statistically homogeneous random fields which constitute a subclass of heterogeneous fields discussed in [37, 43, 50].

---

[1]One may associate these spatial translations with translations in time of stationary signals having constant statistical parameters over time.

The gap between the proof of existence and actual computation of the effective properties can be bridged by means of the MMM principle and through rigorous introduction of the RVE in the next chapter. MMM is due to Hashin [20] and will be discussed in the next section.

2.1. MMM principle by Hashin

The effective constants of the deterministic and random composites can be estimated by two methods leading to the same result. The first method is based on the solution to doubly periodic problems discussed on page 48 and on page 63.

We now proceed to describe the second method consisting of two steps. Problem with a finite number of inclusions n is solved at the first step. Then, the limit of the obtained solution is investigated as $n \to \infty$. This formal mathematical approach is consistent with the MMM principle introduced by Hashin, who recognized the following three relevant scales [20]

$$\text{MICRO} \ll \text{MINI} \ll \text{MACRO} \tag{3.2.4}$$

First, following Hashin [20] microstructure of composites is analysed on the level MICRO. Further, a representative volume element (RVE) is introduced during the passage from MICRO to MINI. The macroscopic constants on the level MACRO are constructed by averaging over the RVE. We refer to the review [20] for the comprehensive physical discussion of the MMM principle.

Many discussions on the physical level can be found in literature devoted to MMM, in particular its application to a multilevel structures by repeated application of the scheme (3.2.4) with introduction of additional mesoscales. Some schemes skip the level MINI and treat homogenization as a "periodization", in fact approximating the structure of composite by a periodic material with one inclusion per cell. Such method leads to correct results only for regular composites as noted above.

The mathematical homogenization theory of random media [16, 24, 49] formalizes Hashin's ideas further as briefly discussed below. First, it is assumed that a random field which describes microstructure on the level MICRO is statistically homogeneous. This mathematical assumption is universally accepted as an axiom in the theory of random composites. It yields the existence of the RVE which represents the composite on the MINI level. The RVE is the pillar of homogenization. Further, it is proved that the composite can be homogenized, i.e., the corresponding PDE (partial differential equation) converges to the homogenized equation with constant coefficients called the effective constants.

It is worth noting that the real composite most likely is not periodic. But one can consider a periodic material on the level MINI when RVE-cells form a periodic structure. Though all the RVEs can be different internally, each RVE obeys the same statistical distribution. Hence, any RVE yields the same effective constants on the level MACRO. Therefore, for any statistically homogeneous field there exist a set of statistically equivalent RVEs which forms the class of equivalence. The investigation of the random (stochastic) composite can be simplified if we chose a single RVE from the class and consider a periodic composite constructed from this RVE-cell.

Thus, the homogenization theory justifies the physical MMM principle (3.2.4). As it is noted in the Introduction, the homogenization theory of random media deals with rigorous mathematical definition of the effective constants and refers to the mathematical quantitative methods of existence and uniqueness. How to compute these effective constants is a separate question of principal importance. On page 82, we describe such an analytical constructive method for 2D composites with circular inclusions.

In the present chapter, we apply a direct computational approach to the effective constants of random 2D composites. First, deterministic boundary value problems are solved for all locations of inclusions, i.e., for all events in the considered probabilistic space C. It is accomplished by applying Schwarz's method and solving a set of functional equations.

After the constructive solution to the deterministic problem is obtained, the ensemble average (mathematical expectation) for the effective constants is calculated in accordance with the given distribution. We manage to avoid computation of the correlation functions and do not require any preliminary knowledge of the complete system of correlation functions. Instead we actually have only to compute their weighted probabilistic moments, the basic sums introduced in the next chapter. The effective properties are completely expressed through these basic sums.

Analytical formulae and numerical simulations demonstrate the advantages of our approach in the case of non-overlapping inclusions. For instance for circular inclusions, our method yields amenable to computations formulae of order $O(f^{20})$ in concentration f explicitly written in the next chapter up to $O(f^7)$ in symbolic form (see equation (4.2.26) on page 86) and up to $O(f^{20})$ (see equation (6.2.4) on page 164).

Typical constructive formulae of the traditional approach [50] are deduced up to $O(f^5)$ for moderate contrast parameters in the range up to $\rho \leq 0.5$ (see discussion in [14] and Fig. 12 therein) and depend on the Torquato–Milton parameter ζ_1 [31], [50] for circular inclusions. Our approach yields the exact formula (4.2.42) on page 90 for the parameter ζ_1. It is worth noting that any plane domain can be approximated by disk packings .

Stochastic 2D problems are posed and solved in doubly periodic formulation in the plane. Theoretically, doubly periodic problems constitute the special class of problems in the plane with infinite number of inclusions [37]. However, the number of inclusion per periodicity cell, N is arbitrary, and the final formulae for the effective tensor contains N in symbolic form. Similar non-periodic statements can be put forward following [43]. Such an approach gives, for instance a proper treatment of the divergent sums (integrals) in applications of self-consistent methods. Such divergence is real (not spurious). It occurs for all non-zero concentrations of inclusions. The divergence and other similar effects do not arise for doubly periodic composites where homogenization always holds.

2.2. Effective conductivity of deterministic composites

The homogenization problems are frequently treated in the framework of Real Analysis [31, 50]. The electric conduction can be expressed in terms of harmonic functions arising from the local electric current $\mathbf{J}$ and the intensity field $\mathbf{E}$ and following equations

$$\nabla \cdot \mathbf{J} = 0, \quad \nabla \times \mathbf{E} = \mathbf{0}. \tag{3.2.5}$$

The linear constitutive relation is assumed

$$\mathbf{J}(\mathbf{x}) = \sigma(\mathbf{x})\mathbf{E}(\mathbf{x}). \tag{3.2.6}$$

Equations (3.2.5) imply existence of the potential u defined as

$$\mathbf{E} = -\nabla u. \tag{3.2.7}$$

The heat conduction is actually described by the same equations. The heat flux $\mathbf{q}(\mathbf{x})$ corresponding to $\mathbf{J}$ satisfies equation $\nabla \cdot \mathbf{q} = 0$. The temperature distribution $u(\mathbf{x})$ is related to the flux by the Fourier law $\mathbf{q}(\mathbf{x}) = -\sigma(\mathbf{x})\nabla u(\mathbf{x})$ where $\sigma(\mathbf{x})$ denotes the local thermal conductivity.

In the both cases, the potential $u(\mathbf{x})$ at the level MICRO satisfies Laplace's equation in a domain and the condition of perfect contact (2.2.37) on the curve dividing two different materials. Below, we describe 2D fields and constructive formulae for the effective conductivity using the deep relations between Real Analysis and Complex Analysis outlined in the previous chapter.

2.2.1. Statement of the double periodic conductivity problem

Let ω_1 and ω_2 be the fundamental pair of periods on the complex plane $\mathbb{C}$ such that $\omega_1 > 0$ and $\mathrm{Im}\,\omega_2 > 0$ where Im stands for the imaginary part. The fundamental parallelogram Q is defined by the vertices $\pm\frac{\omega_1}{2}$ and $\pm\frac{\omega_2}{2}$. Without loss of generality the area of Q can be normalized to one, hence,

$$\omega_1 \mathrm{Im}\,\omega_2 = 1. \tag{3.2.8}$$

For instance,

$$\omega_1 = 1, \quad \omega_2 = i$$

for the square array and

$$\omega_1 = \sqrt[4]{\frac{4}{3}}, \quad \omega_2 = \sqrt[4]{\frac{4}{3}}\left(\frac{1}{2} + i\frac{\sqrt{3}}{2}\right)$$

for the hexagonal array. The points $m_1\omega_1 + m_2\omega_2$ $(m_1, m_2 \in \mathbb{Z})$ generate a doubly periodic lattice Q where $\mathbb{Z}$ stands for the set of integer numbers. Introduce the zero-th cell

$$Q = Q_{(0,0)} = \left\{z = t_1\omega_1 + t_2\omega_2 \in \mathbb{C} : -\frac{1}{2} < t_1, t_2 < \frac{1}{2}\right\}.$$

The lattice Q consists of the cells $Q_{(m_1,m_2)} = Q_{(0,0)} + m_1\omega_1 + m_2\omega_2$.

Consider N non-overlapping simply connected domains D_k in the cell Q with Lyapunov's boundaries L_k and the multiply connected domain $D = Q\backslash \cup_{k=1}^N (D_k \cup L_k)$, the complement of all the closures of D_k to Q (see Fig. 3.3). Each curve L_k leaves D_k to the left. Let z denote a complex variable and a point a_k is arbitrary fixed in D_k $(k = 1, 2, \ldots, N)$.

We study conductivity of the doubly periodic composite when the host $D + m_1\omega_1 + m_2\omega_2$ and the inclusions $D_k + m_1\omega_1 + m_2\omega_2$ are occupied by conducting materials. Introduce the local conductivity as the function

$$\sigma(\mathbf{x}) = \begin{cases} 1, & \mathbf{x} \in D, \\ \sigma, & \mathbf{x} \in D_k, \ k = 1, 2, \ldots, N. \end{cases} \tag{3.2.9}$$

Here, $\mathbf{x} = (x_1, x_2)$ is related to the above introduced complex variable z by formula $z = x_1 + ix_2$. The potentials $u(\mathbf{x})$ and $u_k(\mathbf{x})$ are harmonic in D and D_k $(k = 1, 2, \ldots, N)$ and continuously differentiable in closures of the considered domains. The conjugation conditions express the perfect contact on the interface

$$u = u_k, \quad \frac{\partial u}{\partial \mathbf{n}} = \sigma\frac{\partial u_k}{\partial \mathbf{n}} \quad \text{on } L_k, \quad k = 1, 2, \ldots, N, \tag{3.2.10}$$

where $\frac{\partial}{\partial \mathbf{n}}$ denotes the outward normal derivative to L_k. The external field is modelled by the quasi-periodicity conditions

$$u(z + \omega_1) = u(z) + \xi_1, \ u(z + \omega_2) = u(z) + \xi_2, \tag{3.2.11}$$

where $\xi_{1,2}$ are constants. The external field applied to the considered doubly periodic composites is given by the vector

$$\mathbf{E}_0 \equiv (E_{01}, E_{02})^T = -\left(\frac{\xi_1}{\omega_1}, -\xi_1\text{Re }\omega_2 + \xi_2\omega_1\right)^T, \tag{3.2.12}$$

where ^{T} denotes the transposition. The vector $\mathbf{E}_0$ is isomorphic to the complex number $E_0 = -\frac{\xi_1}{\omega_1} + i(\xi_1\text{Re }\omega_2 - \xi_2\omega_1)$. It is obtained as the gradient of an $\mathbb{R}$-linear function,

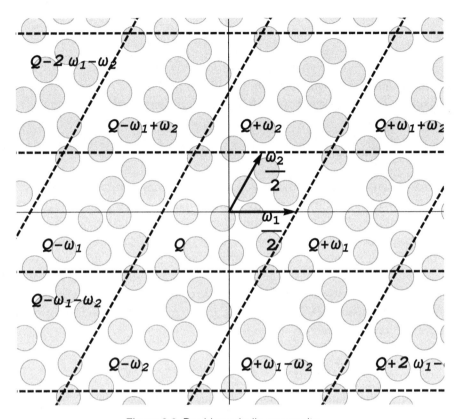

Figure 3.3 Doubly periodic composite.

i.e., linear in x_1 and x_2, satisfying (3.2.11) with the sign minus.

2.2.2. The effective conductivity tensor

The effective conductivity tensor of deterministic composites

$$\sigma_e = \begin{pmatrix} \sigma_{11} & \sigma_{12} \\ \sigma_{21} & \sigma_{22} \end{pmatrix} \qquad (3.2.13)$$

can be found in terms of the solutions of two independent problems (3.2.10), (3.2.11), i.e., for two linearly independent vectors (ξ_1, ξ_2). For instance, one can take

$$\xi_1 = \omega_1, \ \xi_2 = \mathrm{Re} \ \omega_2 \ \text{ and } \ \xi_1 = 0, \ \xi_2 = \mathrm{Im} \ \omega_2 \qquad (3.2.14)$$

that correspond to the complex macroscopic potentials z and $-iz$, respectively.

It follows from the theory of homogenization [24] that the effective conductivity tensor σ_e satisfies the relation

$$\langle \mathbf{J}(\mathbf{x}) \rangle = \sigma_e \mathbf{E}_0 \iff \langle \sigma(\mathbf{x}) \nabla u(\mathbf{x}) \rangle = -\sigma_e \mathbf{E}_0, \tag{3.2.15}$$

where the average over the cell Q of some quantity $F(\mathbf{x})$ is introduced as

$$\langle F(\mathbf{x}) \rangle := \int_Q F(\mathbf{x}) \, d\mathbf{x}. \tag{3.2.16}$$

We recall that the area $|Q| = 1$. It is worth noting that $u(\mathbf{x})$ and $\mathbf{E}_0$ linearly depend on $\xi_{1,2}$. In particular, for the first pair (3.2.14) the definition (3.2.15) gives

$$\langle \sigma(\mathbf{x}) \nabla u(\mathbf{x}) \rangle = \sigma_e \, (1, 0)^T. \tag{3.2.17}$$

The vector equality (3.2.15) can be written in the complex form

$$J \equiv \left\langle \sigma(\mathbf{x}) \left(\frac{\partial u}{\partial x_1}(\mathbf{x}) - i \frac{\partial u}{\partial x_2}(\mathbf{x}) \right) \right\rangle = \frac{\xi_1}{\omega_1}(\sigma_{11} - i\sigma_{12}) - i(\xi_1 \text{Re } \omega_2 - \xi_2 \omega_1)(\sigma_{22} + i\sigma_{12}). \tag{3.2.18}$$

Calculate Re J using (3.2.9)

$$\text{Re } J = \int_D \frac{\partial u}{\partial x_1} \, d\mathbf{x} + \sigma \sum_{k=1}^N \int_{D_k} \frac{\partial u_k}{\partial x_1} \, d\mathbf{x}. \tag{3.2.19}$$

Application of Green's theorem

$$\int_G \left(\frac{\partial g}{\partial x_1} - \frac{\partial h}{\partial x_2} \right) d\mathbf{x} = \int_{\partial G} h \, dx_1 + g \, dx_2 \tag{3.2.20}$$

to (3.2.19) yields

$$\text{Re } J = \int_{\partial Q} u \, dx_2 + (\sigma - 1) \sum_{k=1}^N \int_{L_k} u_k \, dx_2. \tag{3.2.21}$$

Here, we used the relation $\partial D = \partial Q - \cup_{k=1}^N L_k$. We have (see Fig.3.4)

$$\int_{\partial Q} u \, dx_2 = \int_{BC} u \, dx_2 - \int_{AD} u \, dx_2 = \xi_1 \text{Im } \omega_2,$$

where the first condition (3.2.11) is used. Applying again Green's theorem, we obtain

$$\text{Re } J = \xi_1 \text{Im } \omega_2 + (\sigma - 1) \sum_{k=1}^N \int_{D_k} \frac{\partial u_k}{\partial x_1} \, d\mathbf{x}. \tag{3.2.22}$$

Along the similar lines we have

$$\text{Im } J = \xi_2 \omega_1 - \xi_1 \text{Re } \omega_2 + (\sigma - 1) \sum_{k=1}^N \int_{D_k} \frac{\partial u_k}{\partial x_2} \, d\mathbf{x}. \tag{3.2.23}$$

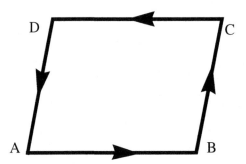

Figure 3.4 The cell $Q \equiv Q_{(0,0)}$ and its oriented boundary $\partial Q = ABCD$. Here, $AB \cong \omega_1$, $BC \cong \omega_2$.

As it is noted above it is sufficient to take two pairs of ξ_1 and ξ_2. Taking (3.2.14) we obtain

$$\sigma_{11} - i\sigma_{12} = 1 + (\sigma - 1) \sum_{k=1}^{N} \int_{D_k} \left(\frac{\partial u_k}{\partial x_1} - i \frac{\partial u_k}{\partial x_2} \right) d\mathbf{x} \qquad (3.2.24)$$

and

$$\sigma_{22} + i\sigma_{12} = 1 + (\sigma - 1) \sum_{k=1}^{N} \int_{D_k} \left(\frac{\partial u_k}{\partial x_1} - i \frac{\partial u_k}{\partial x_2} \right) d\mathbf{x}. \qquad (3.2.25)$$

It is worth noting that the functions u_k in (3.2.24) and (3.2.25) are different since they are solutions of the problem (3.2.10), (3.2.11) for the different pair (3.2.14).

In the case of macroscopically isotropic composites we have

$$\sigma_e = \sigma_e \mathbf{I}, \qquad (3.2.26)$$

where $\mathbf{I}$ stands for the unit matrix. Then, it is sufficient to solve only one problem and get

$$\sigma_e = 1 + (\sigma - 1) \sum_{k=1}^{N} \int_{D_k} \left(\frac{\partial u_k}{\partial x_1} - i \frac{\partial u_k}{\partial x_2} \right) d\mathbf{x}. \qquad (3.2.27)$$

In the theory of macroscopically isotropic composites [8, 24, 27] the following formula is deduced

$$\sigma_e = \int_Q \sigma(\mathbf{x})|\nabla u|^2 d\mathbf{x} = \int_D |\nabla u|^2 d\mathbf{x} + \sigma \sum_{k=1}^{N} \int_{D_k} |\nabla u_k|^2 d\mathbf{x}. \qquad (3.2.28)$$

We now proceed to prove that formulae (3.2.27) and (3.2.28) are equivalent.

We shall use Green's first identity

$$\int_G |\nabla F|^2 \mathbf{dx} = \int_{\partial G} F \frac{\partial F}{\partial \mathbf{n}} \mathrm{d}s, \qquad (3.2.29)$$

where $\mathbf{n}$ denotes the unit outward normal vector to ∂G. We have

$$\sigma \sum_{k=1}^{N} \int_{D_k} |\nabla u_k|^2 \mathbf{dx} = \sigma \sum_{k=1}^{N} \int_{L_k} u_k \frac{\partial u_k}{\partial \mathbf{n}} \mathrm{d}s. \qquad (3.2.30)$$

Application of (3.2.10) and again of Green's first identity yield

$$\sigma \sum_{k=1}^{N} \int_{D_k} |\nabla u_k|^2 \mathbf{dx} = - \int_D |\nabla u|^2 \mathbf{dx} + \int_{\partial Q} u \frac{\partial u}{\partial \mathbf{n}} \mathrm{d}s. \qquad (3.2.31)$$

It follows from (3.2.28) that

$$\sigma_e = \int_{\partial Q} u \frac{\partial u}{\partial \mathbf{n}} \mathrm{d}s. \qquad (3.2.32)$$

The equality of $\frac{\partial u}{\partial \mathbf{n}}$ on the opposite sides of Q yields

$$\sigma_e = \omega_1 \int_{BC} \frac{\partial u}{\partial \mathbf{n}} \mathrm{d}s - \mathrm{Re}\, \omega_2 \int_{DC} \frac{\partial u}{\partial \mathbf{n}} \mathrm{d}s \qquad (3.2.33)$$

for the first pair of (3.2.14). The meaning of the formula (3.2.33) is clear for the square array when $\omega_1 = 1$ and $\omega_2 = i$

$$\sigma_e = \int_{BC} \frac{\partial u}{\partial \mathbf{n}} \mathrm{d}s = \int_{-\frac{1}{2}}^{\frac{1}{2}} \frac{\partial u}{\partial x_1}\Big|_{\mathbf{x}=(\frac{1}{2},x_2)} \mathrm{d}x_2. \qquad (3.2.34)$$

The latter integral is equal to the normal flux going into the cell Q through the side CD. Recall that the flux direction is opposite to the gradient direction (see formula (3.2.7) and text after it on page 48). The same meaning has the integral (3.2.33) in the general case.

The following relation holds for a divergence-free vector function $\mathbf{q}$

$$\int_Q \mathbf{q}\, \mathrm{d}x_1 \mathrm{d}x_2 = \int_{\partial Q} (\mathbf{q} \cdot \mathbf{n})\mathbf{x}\, \mathrm{d}s. \qquad (3.2.35)$$

Here, we used the notation $\mathrm{d}x_1 \mathrm{d}x_2$ for the area differential form $\mathbf{dx}$ in order to avoid miscomprehensions in vector designations. In order to demonstrate (3.2.35), first, we use the identity

$$\mathbf{q} = \nabla \cdot (\mathbf{q} \otimes \mathbf{x}) \qquad (3.2.36)$$

valid for vector functions satisfying $\nabla \cdot \mathbf{q} = 0$. Here, the tensor product $\otimes$ of the vectors

$\mathbf{q} = (q_1, q_2)$ and $\mathbf{x} = (x_1, x_2)$ produces the dyadic

$$\mathbf{q} \otimes \mathbf{x} = \begin{pmatrix} q_1 x_1 & q_1 x_2 \\ q_2 x_1 & q_2 x_2 \end{pmatrix}$$

Using Ostrogradsky's (also known as Gauss's) theorem for a dyadic $\mathbf{D}$ (see Appendix B in [21])

$$\int_Q \nabla \cdot \mathbf{D} \, dx_1 dx_2 = \int_{\partial Q} (\mathbf{n} \cdot \mathbf{D}) \, ds \qquad (3.2.37)$$

we obtain

$$\int_{\partial Q} \nabla \cdot (\mathbf{q} \otimes \mathbf{x}) dx_1 dx_2 = \int_{\partial Q} \mathbf{n} \cdot (\mathbf{q} \otimes \mathbf{x}) \, ds. \qquad (3.2.38)$$

Using the directly checked identity $\mathbf{n} \cdot (\mathbf{q} \otimes \mathbf{x}) = (\mathbf{q} \cdot \mathbf{n})\mathbf{x}$, we arrive at the relation (3.2.35).

Put $\mathbf{q}(\mathbf{x}) = -\sigma(\mathbf{x})\nabla u(\mathbf{x})$ in (3.2.35) and take the first coordinate of the obtained relation

$$\int_Q \sigma(\mathbf{x}) \frac{\partial u}{\partial x_1}(\mathbf{x}) \, dx_1 dx_2 = \int_{\partial Q} \frac{\partial u}{\partial \mathbf{n}} x_1 \, ds. \qquad (3.2.39)$$

The latter integral is equal to the integral (3.2.32) since the functions $u(\mathbf{x})$ and x_1 satisfy the same jump conditions (3.2.11) in the first case (3.2.14). This proves the equivalence of formulae (3.2.27) and (3.2.33) which is, in turn, equivalent to (3.2.28).

Remark 4. The equivalence of the considered formulae actually includes the assumption that $\nabla \cdot (\sigma(\mathbf{x})\nabla u(\mathbf{x})) = 0$ in Q. This equation can be considered in the weak formulation in the corresponding Sobolev space [24].

Remark 5. Following [28] we check that (3.2.17) can be replaced by

$$\langle \sigma(\mathbf{x})\nabla u(\mathbf{x}) \rangle = \sigma_e \langle \nabla u(\mathbf{x}) \rangle, \qquad (3.2.40)$$

when the contact between materials is perfect. Using Ostrogradsky's theorem we have

$$\langle \nabla u(\mathbf{x}) \rangle = \int_D \nabla u(\mathbf{x}) \, d\mathbf{x} + \sum_{k=1}^N \int_{D_k} \nabla u_k(\mathbf{x}) \, d\mathbf{x} = \int_{\partial Q} u \mathbf{n} \, ds + \sum_{k=1}^N \int_{L_k} (u_k - u) \mathbf{n} \, ds. \qquad (3.2.41)$$

One can check by straight-forward calculations that $\int_{\partial Q} u \mathbf{n} \, ds = (1, 0)^T$. Then, equation $u = u_k$ on L_k yields $\langle \nabla u(\mathbf{x}) \rangle = (1, 0)^T$. The relation (3.2.40) is not true when the contact between materials is imperfect.

The same arguments hold for elastic problems [28].

Using the complex flux (2.2.51) defined in inclusions D_k we can write (3.2.24) in the form

$$\sigma_{11} - i\sigma_{12} = 1 + 2\rho \sum_{k=1}^{N} \int_{D_k} \psi_k(z) d\mathbf{x}, \qquad (3.2.42)$$

where $\rho = \frac{\sigma-1}{\sigma+1}$. Let D_k be a disk $\{z \in \mathbb{C} : |z - a_k| < r_k\}$. Then, the mean value theorem for harmonic functions yields

$$\sigma_{11} - i\sigma_{12} = 1 + 2\rho \sum_{k=1}^{N} \pi r_k^2 \psi_k(a_k). \qquad (3.2.43)$$

In the case of equal radii, formula (3.2.43) becomes

$$\sigma_{11} - i\sigma_{12} = 1 + 2\rho f \frac{1}{N} \sum_{k=1}^{N} \psi_k(a_k), \qquad (3.2.44)$$

where $f = N\pi r^2$ denotes the concentration of inclusions.

Remark 6. Formulae (3.2.42)–(3.2.44) have slightly different form for imperfect contact between phases. The method of functional equations was developed and applied to such problems in the papers [10, 15, 25]

2.3. Effective conductivity of random composites

We now proceed to discuss the problem (3.2.10), (3.2.11) in the probabilistic context when inclusions are randomly located within the host. For definiteness, consider a set of the domains D_k as a realization of the random locations of a collection of particles $\widetilde{D_k}$ considered as physical objects with fixed shapes and sizes (see page 44). Following the definition (3.4) from [50] we introduce the doubly periodic specific probability density $P(\mathbf{a})$ associated with finding a configuration of inclusions with position $\mathbf{a} := (a_1, a_2, \ldots, a_N)$. The point a_k determines the position of the particle $\widetilde{D_k}$ in the cell Q, i.e., determines the domain D_k.

The set of all the configurations is denoted by C. Let da_k denote a small area element about the point a_k and $d\mathbf{a} = da_1 da_2 \ldots da_N$. Then $P(\mathbf{a})d\mathbf{a}$ is equal to the probability of finding a_1 in da_1, a_2 in da_2, ..., a_N in da_N.

The ensemble average is introduced as the expectation in a probabilistic space

$$\widehat{F} = \int_C F(\mathbf{a}) P(\mathbf{a}) d\mathbf{a}. \qquad (3.2.45)$$

The effective conductivity tensor $\boldsymbol{\sigma}_e$ of random composites is determined by application of the operator (3.2.45) to the deterministic values (3.2.13)

$$\widehat{\sigma_e} = \int_C \sigma_e(\mathbf{a}) P(\mathbf{a}) \mathrm{d}\mathbf{a}. \tag{3.2.46}$$

Let the probabilistic distribution of the domains D_k $(k = 1, 2, \ldots, N)$ in Q be such that the corresponding two-dimensional composite is isotropic in macroscale, i.e., the expected effective conductivity of the composite is expressed by a scalar. It can be calculated as follows

$$\widehat{\sigma}_e = \int_C \sigma_e(\mathbf{a}) P(\mathbf{a}) \mathrm{d}\mathbf{a}, \tag{3.2.47}$$

where the deterministic value $\sigma_e(\mathbf{a})$ is calculated from (3.2.27). The tensor $\widehat{\sigma_e}$ and the scalar $\widehat{\sigma}_e$ are related by equation (3.2.26).

In the present chapter, we express the deterministic effective conductivity $\sigma_e(\mathbf{a})$ in terms of the series. Chapter 9 on random composites is devoted to averaging of the series in accordance with formula (3.2.47) by means of the Monte Carlo simulations and application of the asymptotic analysis to the averaged series for various probabilistic distributions of non-overlapping disks. The same approach can be applied to elastic problems discussed in Chapter 10.

3. Series expansions for the local fields and effective tensors. Traditional approach

In order to compare the results obtained by our method and by others we summarize in the present section the traditional approach. We follow the book [50] and do not present earlier classic results (see references and historical notes therein and also in [31]) in order to avoid multiple notations. For definiteness, only 2D two-phase dispersed composites with one phase embedded in the connected matrix are considered. Two main methods are distinguished in literature [31, 50] and discussed below: *cluster expansion* and *contrast expansion*. Below on page 62, we shall see that these traditional expansions correspond to two different iterative scheme for Schwarz's method.

3.1. Cluster expansions

This method gives a correction to the single-inclusion approximation by taking into account interactions between pairs of particles, triplets and so forth. It is worth repeating that only infinite number of inclusions on a plane gives a correct result [50], [43]. Study of the finite number of inclusions in the whole space yields the effective properties only of a dilute composites as it is demonstrated in [43] and noted in the item on page 42.

The intensity field $\mathbf{E}$ is considered as a function of $\mathbf{x}$ and denoted by $\mathbf{E}(\mathbf{x})$. Let the constant external field $\mathbf{E}_0$ be fixed. Torquato describes the following so-called cluster expansion (see (19.8) from [50])

$$\mathbf{E}(\mathbf{x}) = \mathbf{E}_0 + \sum_{k=1}^{N} \mathbf{M}_1(\mathbf{x}; k) \cdot \mathbf{E}_0 + \sum_{k=1}^{N} \sum_{k_1 \neq k}^{N} \mathbf{M}_2(\mathbf{x}; k, k_1) \cdot \mathbf{E}_0 + \cdots . \qquad (3.3.48)$$

Here, $\mathbf{M}_1(\mathbf{x}; k)$ is a single-inclusion operator which accounts for the first-order interactions over and above $\mathbf{E}_0$; $\mathbf{M}_2(\mathbf{x}; k, k_1)$ is the 2-inclusion operator which accounts for the *second*-order interactions.

The operator $\mathbf{M}_1(\mathbf{x}; k)$ can be derived by the following two methods. The first method corresponds to the Maxwell's approach applied to dilute composites. It consists of two steps. At the first step, a boundary value problem for a single inclusion D_k in the infinite plane is solved and the single-inclusion operator $\mathbf{K}_1(\mathbf{x}; k)$ in the plane is constructed (see for instance, formulae (17.4) and (19.6) from [50]). Next, $\mathbf{M}_1(\mathbf{x}; k)$ is constructed via $\mathbf{K}_1(\mathbf{x}; k)$ by a periodicity operator. The periodicity transformation $\mathcal{P} : \mathbf{K}_1(\mathbf{x}; k) \mapsto \mathbf{M}_1(\mathbf{x}; k)$ can be easily performed since $\mathbf{K}_1(\mathbf{x}; a_k)$ is expressed via the dipole tensor [44]. The periodicity operator transforms rational functions in $z = x_1 + ix_2$ into elliptic functions [2]; for instance, $\mathcal{P} : z^{-1} \mapsto \zeta(z)$ and $\mathcal{P} : z^{-2} \mapsto \wp(z)$. The second method to construct $\mathbf{M}_1(\mathbf{x}; k)$ is based on the same scheme realized in the torus topology.

The operator $\mathbf{K}_1(\mathbf{x}; k)$, hence $\mathbf{M}_1(\mathbf{x}; k)$, is available in closed form for the disk, ellipse and many other domains (see [50] and works cited therein). It can be constructed for any shape D_k in terms of the conformal mapping of D_k onto the unit disk. In general, $\mathbf{K}_p(\mathbf{x}; k, k_1, \ldots, k_p)$ and $\mathbf{M}_p(\mathbf{x}; k, k_1, \ldots, k_p)$ are the p-inclusion operator which accounts for the pth-order interactions.

The effective conductivity can be found through the polarization field defined by (19.3), (19.4) from [50]

$$\mathbf{P}(\mathbf{x}) = \begin{cases} \mathbf{0}, & z \in D, \\ (\sigma - 1)\mathbf{E}(\mathbf{x}), & z \in D_k, \ k = 1, 2, \ldots, N. \end{cases} \qquad (3.3.49)$$

Following the traditional approach [50] we introduce the "double" averaging operator over the unit cell Q and over the configuration space C. For instance in the case of equal disks, the probabilistic configuration space means the set of their centers $\mathbf{a} = \{a_1, a_2, \ldots, a_N\}$ in the periodicity cell. Introduce the average[2] of some quantity $g(\mathbf{x}, \mathbf{a})$

$$\langle g \rangle := \int_Q \int_C g(\mathbf{x}, \mathbf{a}) P(\mathbf{a}) \, \mathrm{d}\mathbf{a} \, \mathrm{d}\mathbf{x}, \qquad (3.3.50)$$

where $P(\mathbf{a})$ denotes the joint probabilistic density of disks, $\mathrm{d}\mathbf{x} = \mathrm{d}x_1 \mathrm{d}x_2$ and $\mathrm{d}\mathbf{a} =$

---

[2]This average may be considered as a generalization of the average (3.2.16) to random media.

$da_1 da_2 \cdots da_N$. The main computational problem of the traditional theory of random composites lies in the difficulty to estimate the integral (3.3.50). It can be written in terms of the correlation functions of all the orders, but we are still confronted with a hard computational problem. The effective tensor (3.2.13) can be defined by equation

$$\langle \mathbf{P} \rangle = (\sigma_e - \mathbf{I}) \cdot \langle \mathbf{E} \rangle, \qquad (3.3.51)$$

where $\mathbf{I}$ is the unit tensor. Having used (3.3.51) Torquato (see (19.33) from [50] and earlier studies cited therein) deduced formula for the macroscopically isotropic composites

$$\sigma_e = 1 + \sum_{n=1}^{\infty} \frac{1}{n!} \int_C W_n(\mathbf{a}) \, d\mathbf{a}, \qquad (3.3.52)$$

where $W_n(\mathbf{a})$ is a complicated functional of the n–inclusion cluster operators $\mathbf{M}_n(\mathbf{x}; k, k_1, \ldots, k_n)$. It is worth noting that $W_n(\mathbf{a})$ is derived through $\mathbf{K}_n(\mathbf{x}; k, k_1, \ldots, k_n)$ in [50]. Actually, our periodic modification yields the same final formula (3.3.52) by rearrangement of the terms in each $W_n(\mathbf{a})$.

Let $|D_k|$ denote the area of D_k. The concentration of inclusions is defined in the following way

$$f = \sum_{k=1}^{N} |D_k|. \qquad (3.3.53)$$

It is noted in [50] that (3.3.52) yields the volume-fraction expansion (see (19.34) from [50])

$$\sigma_e = 1 + \sum_{n=1}^{\infty} B_n f^n, \qquad (3.3.54)$$

where the coefficients B_n are multidimensional integrals on $\mathbf{M}_p(\mathbf{x}; k, k_1, \ldots, k_p)$ for $p = 1, 2, \ldots, n$. Direct applications of the correlation functions yield analytical formulae and numerical results for B_n $n = 2, 3, 4$ (see [50] and a discussion in [14]).

3.2. Contrast expansions

The so-called contrast expansions can be presented by formula (20.1) from [50] which after some modifications takes the following form

$$\sigma_e = 1 + \sum_{n=1}^{\infty} C_n \rho^n, \qquad (3.3.55)$$

where ρ is the contrast parameter (2.2.48). Many authors claimed that the series (3.3.55) converges only for sufficiently small ρ. Though convergence of the series for local fields had been proven in [34] for circular inclusions and in Section 4.9.2 of [38] for general shapes for all admissible $|\rho| \leq 1$. This of course implies the convergence of (3.3.55).

In order to estimate the first coefficients C_n ($n = 1, 2, 3, 4$), Torquato [50] derives an integral equation on $\mathbf{E}(\mathbf{x})$, which can be written in the form (compare to equation (20.17) from [50] for the cavity intensity field)

$$\mathbf{E}(\mathbf{x}) = \mathbf{E}_0 + \int_Q d\mathbf{x}' \ \mathbf{G}(\mathbf{x}, \mathbf{x}') \cdot [\sigma(\mathbf{x}) - 1]\mathbf{E}(\mathbf{x}'), \qquad (3.3.56)$$

where $\mathbf{G}(\mathbf{x}, \mathbf{x}')$ denotes the periodic Green's function. Substitution of (3.2.9) into (3.3.56) yields

$$\mathbf{E}(\mathbf{x}) = \mathbf{E}_0 + \rho(\sigma + 1) \sum_{k=1}^{N} \int_{D_k} d\mathbf{x}' \ \mathbf{G}(\mathbf{x}, \mathbf{x}') \cdot \mathbf{E}(\mathbf{x}'). \qquad (3.3.57)$$

The method of successive approximations applied to (3.3.57) yields the series on the contrast parameter ρ which can be compactly written in the form

$$\mathbf{E}(\mathbf{x}) = \sum_{n=0}^{\infty} \rho^n \mathbf{H}^n \cdot \mathbf{E}_0, \qquad (3.3.58)$$

where the operator $\mathbf{H}$ is defined by the right-hand part of (3.3.57). Use of (3.3.49) yields an analogous series for $\mathbf{P}(\mathbf{x})$. The tensor σ_e satisfies equation (3.3.51) which can be written in the form (20.37) from [50]. The latter formula for macroscopically isotropic media becomes (20.57) from [50]. In our designations it reads as follows

$$\rho^2 f^2 (\sigma_e - 1)^{-1}(\sigma_e + 1) = f\rho - \sum_{n=3}^{\infty} A_n^{(1)} \rho^n. \qquad (3.3.59)$$

The n-point coefficients $A_n^{(1)}$ are exactly expressed in terms of the p-point correlation functions ($p = 2, 3, \ldots, n$) (see formulae (20.38)–(20.41) and (20.59), (20.60) from [50]). Formula (3.3.59) can be written in the form (3.3.55).

The coefficients $A_n^{(1)}$ were calculated in [50] as a sums of multiple integrals containing the correlation functions which are presented also through multiple integrals. Calculation of these integrals is extremely difficult and blocks the practical application of (3.3.59). Only three to four starting terms were computed and it also turns out that the contrast expansion is limited to moderate ρ.

4. Schwarz's method

The generalized method of Schwarz (or just Schwarz's method) was proposed by Mikhlin [33] as a generalization of the classical alternating method of Schwarz to finitely connected domains. Schwarz's method belongs to decomposition methods [48] frequently used in numerical computations and realized in the form of alternating methods. Schwarz's method is associated with the method of integral equations discussed on page 26.

Schwarz's method is based on the decomposition of the domain with complex geometry onto simple domains and subsequent solution to the boundary value problems for simple domains. More precisely, let inclusions D_k ($k = 1, 2, \ldots, N$) be considered in some space domain. The problem is solved iteratively, based on the solution at each step p to the single-inclusion problems for each inclusion D_k ($k = 1, 2, \ldots, N$). The local field in the matrix $u^{(p)}$ is constructed from the pth-order approximation of fields inside the inclusions denoted as $u_k^{(p)}$. As the zeroth approximation $u^{(0)}$ one can take the external field applied at infinity, or at the boundary of the periodicity cell. In the first approximation, $u_k^{(1)}$ is calculated separately for each D_k in terms of the external field $u^{(0)}$. Then, $u^{(1)}$ is calculated as a field induced by the fields $u_k^{(1)}$ inside the inclusions. In the second approximation, $u_k^{(2)}$ is calculated in terms of the external field $u^{(1)}$ and $u^{(2)}$ and from the fields $u_k^{(1)}$ inside inclusions. Hence, $u_k^{(2)}$ is determined by the first-order interactions between inclusions, and so forth. Thus, $u_k^{(p)}$ takes into account interactions of order $(p - 1)$ between inclusions. The convergence of the method was established in [35, 38].

4.1. Scalar form of the method

As we shall see below application of Schwarz's method is equivalent to solving the integral equations by iterative schemes. One of the schemes corresponds to the modified cluster expansion when at each step of the algorithm the boundary conditions for the k-th simple domain are corrected with account to influence of all other simple domains with $m \neq k$. The second iterative scheme corresponds to the standard contrast expansion as already described above.

We now present the essentials of general scheme for Schwarz's method, first for the conductivity problem and then for a linear problem in general. The ideal contact condition (2.2.37) can be rewritten in the form

$$u_0 = u_k - u^{ext}, \quad \frac{\partial u_0}{\partial n} = \frac{\partial(u_k - u^{ext})}{\partial n} + \rho(\sigma + 1)\frac{\partial u_k}{\partial n} \quad \text{on } L_k, \quad k = 1, 2, \ldots, N,$$

(3.4.60)

where $u = u_0 + u^{ext}$ is the decomposition of the potential u onto the regular periodic part u_0 and the external (quasi-periodic) field u^{ext}. In the case of non-periodic problems the external field u^{ext} has a singularity at infinity.

Introduce the function

$$h(t) = (\sigma + 1)\frac{\partial u_k}{\partial n}(t), \quad t \in L_k \ (k = 1, 2, \ldots, N)$$

and the domains $D^+ := \cup_{k=1}^{n} D_k$, $D^- := D$. It is convenient to represent the harmonic functions in different domains as a single piece-wise harmonic function $u_0(\mathbf{x})$ introduced above in $D = D^-$ and equal to $u_k(\mathbf{x}) - u^{ext}(\mathbf{x})$ in $D_k \cup L_k$ $(k = 1, 2, \ldots, N)$. Then, (3.4.60) can be considered as the jump problem

$$u_0^+ = u_0^-, \quad \frac{\partial u_0^+}{\partial n} = \frac{\partial u_0^-}{\partial n} + \rho\, h(t) \quad \text{on } L = \cup_{k=1}^n L_k, \qquad (3.4.61)$$

where u_0^+ and u_0^- correspond to the limit values of $u(\mathbf{x})$ when $\mathbf{x}$ tends to t from D^+ and D^-, respectively. The jump problem (3.4.61) has a unique solution expressed in terms of the simple layer potential P for the curve L in the cell Q in the torus topology. We have

$$u_0(z) = \rho(Pf)(\mathbf{x}), \quad \mathbf{x} \in D^+ \cup D^-. \qquad (3.4.62)$$

The operator P is considered as an operator in an appropriate functional space [13, 17, 18, 19]. In the classic statement, Q should be replaced by the whole space. The operator P is decomposed onto the simple layer potentials P_k along the curves L_k $(k = 1, 2, \ldots, N)$ as $P = \frac{1}{\sigma+1} \sum_{k=1}^N P_k$. Then (3.4.62) implies that

$$u_k(\mathbf{x}) = \rho \sum_{m=1}^N \left(P_m \frac{\partial u_m^+}{\partial n} \right)(\mathbf{x}) + u^{ext}(\mathbf{x}), \quad \mathbf{x} \in D_k \ (k = 1, 2, \ldots, N), \qquad (3.4.63)$$

and

$$u(\mathbf{x}) = \rho \sum_{m=1}^N \left(P_m \frac{\partial u_m^+}{\partial n} \right)(\mathbf{x}) + u^{ext}(\mathbf{x}), \quad \mathbf{x} \in D. \qquad (3.4.64)$$

Equations (3.4.63) can be considered as a system of integral equations on the potentials $u_k(\mathbf{x})$ inside the inclusions D_k $(k = 1, 2, \ldots, N)$. If it is solved, the potential $u(\mathbf{x})$ is calculated in D by (3.4.64).

Equations (3.4.63) correspond to Schwarz's method in Mikhlin's form [33] when the method of successive approximations can diverge for closely spaced inclusions. The following slight modification yields a convergent algorithm for any configuration of inclusions and arbitrary contrast parameter satisfying $|\rho| \leq 1$

$$u_k(\mathbf{x}) = \rho \sum_{m=1}^N \left[\left(P_m \frac{\partial u_m^+}{\partial n} \right)(\mathbf{x}) - \left(P_m \frac{\partial u_m^+}{\partial n} \right)(\mathbf{w}) \right] + u^{ext}(\mathbf{x}) + \\ \rho \sum_{m=1}^N \left(P_m \frac{\partial u_m^+}{\partial n} \right)(\mathbf{w}), \quad \mathbf{x} \in D_k \cup L_k \quad (k = 1, 2, \ldots, N), \qquad (3.4.65)$$

where $\mathbf{w}$ is a fixed point in D.

Equations (3.4.62)–(3.4.65) are written for potentials. Their differentiation yields equations for the intensity field

$$\mathbf{E}_k(\mathbf{x}) = \rho \sum_{m=1}^N (Q_m \mathbf{E}_m)(\mathbf{x}) + \mathbf{E}_0(z), \quad \mathbf{x} \in D_k \cup L_k \quad (k = 1, 2, \ldots, N), \qquad (3.4.66)$$

where $\mathbf{E}_k(\mathbf{x}) = \mathbf{E}(\mathbf{x})$ in D_k and Q_m are appropriate operators (similar to double layer potentials). Equation (3.4.66) is similar to (3.3.56).

The method of successive approximations applied to equation (3.4.63) (and for the equivalent equations (3.4.65), (3.4.66)) leads to the contrast expansion, i.e., to power series in the parameter ρ.

Consider the following iterative scheme for equation (3.4.63) when the p-th approximation is calculated through the previous $(p-1)$th approximation by solution to the integral equation for each inclusion separately

$$u_k^{(p)}(\mathbf{x}) - \rho\left(P_k \frac{\partial u_k^{+(p)}}{\partial n}\right)(\mathbf{x}) = \rho \sum_{m \neq k}\left(P_m \frac{\partial u_m^{+(p-1)}}{\partial n}\right)(\mathbf{x}) + u^{ext}(\mathbf{x}), \ \mathbf{x} \in D_k \ (k = 1, 2, \dots, N).$$

(3.4.67)

Such scheme leads to the cluster expansion when for each $k = 1, 2, \dots, N$ at the step p the problem for kth inclusion is solved with accounting for the influence of the of the fields of all other inclusions calculated at the previous step $(p-1)$.

4.2. Vector form of the method

Let tensor fields $\mathbf{U}(z)$ in D and $\mathbf{U}_k(z)$ in D_k satisfy the contact conditions

$$\mathbf{T} \cdot \mathbf{U} = \mathbf{T}_k \cdot \mathbf{U}_k, \quad \text{on } L_k \quad (k = 1, 2, \dots, N),$$ (3.4.68)

where $\mathbf{T}$ and $\mathbf{T}_k$ are boundary operators involving physical parameters of the materials occupying the domains D and D_k, respectively. Let $\mathbf{T}_k$ be presented in the form $\mathbf{T}_k = \mathbf{T} - \boldsymbol{\rho}_k$, where $\boldsymbol{\rho}_k$ denotes a contrast parameter tensor. For instance, for elastic composites having isotropic components, $\boldsymbol{\rho}_k$ can be expressed through two scalar parameters $\frac{\mu_k - \mu}{\mu}$ and $\nu_k - \nu$ where μ and ν denote the shear modulus and Poisson's ratio (see Chapter 10). Then (3.4.68) becomes

$$\mathbf{U} = \mathbf{U}_k - \mathbf{T}^{-1} \cdot \boldsymbol{\rho}_k \cdot \mathbf{U}_k, \quad \text{on } L_k \quad (k = 1, 2, \dots, N),$$ (3.4.69)

Let $\mathbf{P}_k$ denote the simple layer potential corresponding to the considered linear equation for the domains D_k and D_k^- separated by L_k. Application of $\sum_{k=1}^N \mathbf{P}_k$ to (3.4.69) yields

$$\mathbf{U}_k(\mathbf{x}) = \sum_{m=1}^N (\mathbf{P}_m \cdot \mathbf{T}^{-1} \cdot \boldsymbol{\rho}_m \cdot \mathbf{U}_m)(\mathbf{x}) + \mathbf{U}^{ext}(\mathbf{x}), \ \mathbf{x} \in D_k \cup L_k \ (k = 1, 2, \dots, N).$$

(3.4.70)

There are two different methods to solve equations (3.4.70). The first method is based on direct iterations and corresponds to the contrast expansion in ρ (compare to the developments on page 58). The second method is based on implicit iterations applied to the same equations (3.4.70) written in the form

$$\mathbf{U}_k(\mathbf{x}) - (\mathbf{P}_k \cdot \mathbf{T}^{-1} \cdot \boldsymbol{\rho}_k \cdot \mathbf{U}_k)(\mathbf{x}) = \sum_{m \neq k}^{N} (\mathbf{P}_m \cdot \mathbf{T}^{-1} \cdot \boldsymbol{\rho}_m \cdot \mathbf{U}_m)(\mathbf{x}) + \mathbf{U}^{ext}(\mathbf{x}),$$
$$\mathbf{x} \in D_k \cup L_k \quad (k = 1, 2, \dots, N). \tag{3.4.71}$$

The single-inclusion problem for D_k ($k = 1, 2, \dots, N$) is solved at each step of iterations. This scheme corresponds to the cluster expansions discussed on page 56.

Remark 7. Equations (3.4.62) and (3.4.71) can't be reduced to the classic integral equations constructed in the framework of potenital's theory [13]. It is another type of equations.

Remark 8. Equations (3.4.62)–(3.4.64) and (3.4.70), (3.4.71) hold in the limit cases of soft and hard inclusions in terminology [44], by introduction of fictive potentials [38].

Remark 9. The 2D terminology is used in the present section. But the same consideration takes place in $\mathbb{R}^3$ by a simple reformulation.

4.3. Integral equations for doubly periodic problems

The problem (3.2.10), (3.2.11) given on page 49 can be written in terms of complex potentials by the method described on page 17. The corresponding $\mathbb{R}$-linear problem has the form

$$\varphi(t) = \varphi_k(t) - \rho \, \overline{\varphi_k(t)}, \ t \in L_k, \ k = 1, 2, \dots, N. \tag{3.4.72}$$

The unknown functions $\varphi(z)$ and $\varphi_k(z)$ are analytic in D and D_k, respectively, and continuously differentiable in the closures of the domains considered. The function $\varphi(z)$ is quasi-periodic, i.e., it has constant increments per periodicity cell

$$\varphi(z + \omega_j) - \varphi(z) = \xi_j + id_j \quad (j = 1, 2), \tag{3.4.73}$$

where the constants ξ_1 and ξ_2 are given in (3.2.11), d_1 and d_2 are undetermined real constants which should be found while solving the problem.

Introduce the functions $\psi(z)$ and $\psi_k(z)$ analytic in D and D_k, respectively, and continuous in the closures of D and D_k

$$\psi(z) = \frac{\partial u}{\partial x} - i \frac{\partial u}{\partial y} - 1, \quad z \in D,$$

$$\psi_k(z) = \frac{\sigma + 1}{2} \left(\frac{\partial u}{\partial x} - i \frac{\partial u}{\partial y} \right), \ z \in D_k. \tag{3.4.74}$$

Here, the complex flux has the form $\psi(z) + 1$ where 1 is the complex number which express the external vector field $(1, 0)$. The complex flux is double periodic, hence

$$\psi(z + \omega_1) = \psi(z), \quad \psi(z + \omega_2) = \psi(z). \tag{3.4.75}$$

Following (2.2.56) we have

$$\psi(t) = \psi_k(t) + \rho \overline{n_k^2(t)} \, \overline{\psi_k(t)}, \ t \in L_k, \ k = 1, 2, \ldots, N, \tag{3.4.76}$$

where $\psi(z) = \varphi'(z)$, $\psi_k(z) = \varphi_k'(z)$ and $n_k(t)$ stands for the unit normal outward vector to L_k expressed in terms of the complex values. The $\mathbb{R}$-linear problem (3.4.76) holds also in the limit cases for perfectly conducting inclusions ($\rho = 1$) and for perfect insulators ($\rho = -1$).

The $\mathbb{R}$-linear problem (3.4.72), (3.4.73) can be reduced to a system of integral equations in the following way. The cell Q in the torus topology is divided onto two domains $D^+ = \cup_{k=1}^n D_k$ (not connected) and $D^- = D$ (multiply connected). Let $L = \cup_{k=1}^n L_k$ denote the boundary of D^+ and $\mu(t)$ be a Hölder continuous function on L.

The Cauchy-type integral on torus was introduced with the ζ-function of Weierstrass $\zeta(z)$ in the kernel [12]. We introduce it in a slightly different form using the Eisenstein function $E_1(z) = \zeta(z) - S_2 z$ (S_2 is a constant; see details on page 71)

$$\Phi(z) = \frac{1}{2\pi i} \int_L \mu(t) E_1(t - z) dt + Cz + C_0. \quad z \in D^{\pm}, \tag{3.4.77}$$

where $\mu(t)$ is a Hölder continuous function on L, C and C_0 are complex constants. The function $\Phi(z)$ is analytic in $D^{\pm}$ and quasi-periodic, i.e., it has constant jumps per periodicity cell calculated with help of (3.4.130)

$$\Phi(z + \omega_1) - \Phi(z) = C\omega_1, \quad \Phi(z + \omega_2) - \Phi(z) = -\frac{1}{\omega_1} \int_L \mu(t) dt + C\omega_2. \tag{3.4.78}$$

The Sochocki formulae hold on torus in the form [12]

$$\Phi^{\pm}(t) = \pm \frac{1}{2} \mu(t) + \frac{1}{2\pi i} \int_L \mu(\tau) E_1(\tau - t) d\tau + Ct + C_0, \quad t \in L, \tag{3.4.79}$$

where $\Phi^{\pm}(t)$ denote the limit values of $\Phi(z)$ on L when $z \in D^{\pm}$ tends to $t \in L$, respectively. Formulae (3.4.79) imply that the function $\Phi(z)$ satisfies the jump condition

$$\Phi^+(t) - \Phi^-(t) = \mu(t), \ t \in L. \tag{3.4.80}$$

Equation (3.4.80) can be considered as a $\mathbb{C}$-linear conjugation problem [38] in a class of quasi-periodic functions with given jump $\mu(t)$. It follows from [12] that the general solution of (3.4.80) up to an additive arbitrary constant has the form (3.4.77).

Consider (3.4.72) as the problem (3.4.80) with $\mu(t) = \rho \overline{\varphi_k(t)}$ on L, $\Phi(z) = \varphi_k(z)$ in

$D_k \subset D^+$ and $\Phi(z) = \phi(z)$ in $D = D^-$. Then, (3.4.77) can be written in D^+ as follows

$$\varphi_k(z) = \rho \sum_{m=1}^{N} \frac{1}{2\pi i} \int_{L_m} \overline{\varphi_m(t)} E_1(t - z) dt + Cz + C_0, \ z \in D_k \ (k = 1, 2, \ldots, N). \quad (3.4.81)$$

(3.4.77) in D^- becomes

$$\varphi(z) = \rho \sum_{m=1}^{N} \frac{1}{2\pi i} \int_{L_m} \overline{\varphi_m(t)} E_1(t - z) dt + Cz + C_0, \quad z \in D. \quad (3.4.82)$$

Equations (3.4.81), (3.4.82) are deduced from the $\mathbb{R}$-linear problem (3.4.72). Using the same arguments one can derive analogous equations from the problem (3.4.76) for the complex flux

$$\psi_k(z) = -\rho \sum_{m=1}^{N} \frac{1}{2\pi i} \int_{L_m} \overline{n^2(t)} \ \overline{\psi_m(t)} E_1(t - z) dt + C, \ z \in D_k \ (k = 1, 2, \ldots, N), \quad (3.4.83)$$

$$\psi(z) = -\rho \sum_{m=1}^{N} \frac{1}{2\pi i} \int_{L_m} \overline{n^2(t)} \ \overline{\psi_m(t)} E_1(t - z) dt + C, \ z \in D. \quad (3.4.84)$$

It is also possible to obtain these equations by differentiation of (3.4.81)–(3.4.82).

The integral equations (3.4.83) are considered in the space $\mathcal{H}(D^+)$ introduced on page 10($D^+ = \cup_{k=1}^{N} D_k$). Equations (3.4.83) are written in the domains D_k and can be continued to the boundary by use of (3.4.79)

$$\psi_k(t) = -\rho \sum_{m=1}^{N} \left[\frac{1}{2} \overline{n_m^2(t)} \ \overline{\psi_m(t)} + \frac{1}{2\pi i} \int_{L_m} \overline{n_m^2(t)} \ \overline{\psi_m(t)} E_1(t - z) dt \right] + C, \\ t \in L_k \quad (k = 1, 2, \ldots, N). \quad (3.4.85)$$

Therefore, the integral equations in the space $\mathcal{H}(D^+)$ have the form (3.4.83), (3.4.85). We shall prove in the next section (see Theorem 12 on page 66) that equations (3.4.83), (3.4.85) have a unique solution. Solution to these equations corresponds to Schwarz's method which can be realized in different forms as noted on page 62.

The problem (3.4.72), (3.4.73) for the complex potentials has a unique solution up to an arbitrary additive constant C_0. The constants $\xi_1 + id_1$, $\xi_2 + id_2$ and C_0 disappear in the problem (3.4.76) after differentiation. The structure of the general solution of (3.4.76) has the form $\psi(z) = \xi_1 \psi^{(1)}(z) + \xi_2 \psi^{(2)}(z)$ [40], where the real constants ξ_1 and ξ_2 correspond to the external field (jumps of $u(\mathbf{x})$ per a cell). Hence, the $\mathbb{R}$-linear problem (3.4.76) has two $\mathbb{R}$-linear independent solutions. From the other side, these two independent solutions can be constructed from (3.4.83), (3.4.84) taking, for instance $C = 1$ and $C = i$. Therefore, the complex constant C can be expressed through the real constants ξ_1 and ξ_2. In order to calculate the effective properties of macroscopically isotropic composites it is sufficient to fix ξ_1, ξ_2 and find C. For simplicity, we put

$$\xi_1 = \omega_1, \quad \xi_2 = \text{Re } \omega_2. \tag{3.4.86}$$

Then the external field corresponds to the complex potential $\varphi^{(ext)}(z) = z$ and to the complex flux $\psi^{(ext)}(z) = 1$. This complex flux yields the physical external flux $(-1, 0)$. It follows from the first relation (3.4.78) that

$$C = 1 + \frac{id_1}{\omega_1}, \tag{3.4.87}$$

where $d_1 = \text{Im}[\varphi(z + \omega_1) - \phi(z)]$. Put $z = -\frac{1}{2}(\omega_1 + \omega_2)$ in this relation and calculate

$$d_1 = \text{Im} \int_{-\frac{1}{2}(\omega_1+\omega_2)}^{\frac{1}{2}(\omega_1-\omega_2)} \psi(t)dt = - \int_{-\frac{1}{2}(\omega_1+\text{Re}\omega_2)}^{\frac{1}{2}(\omega_1-\text{Re}\omega_2)} \frac{\partial u}{\partial x_2}\left(x_1 - \frac{i\text{Im } \omega_2}{2}\right) dx_1. \tag{3.4.88}$$

The integral from the left-hand side of (3.4.88) expresses the total flux through the lower side of the parallelogram Q parallel to the x_1-axis. It must be equal to zero because of the periodicity and that the external flux $(-1, 0)$ vanishes in the x_2-direction. Therefore, $d_1 = 0$, hence $C = 1$ by (3.4.87).

4.4. Convergence of Schwarz's method

This section is devoted to study of the integral equations (3.4.84), (3.4.85) in the space $\mathcal{H}(D^+)$.

Let $\Psi(z) = \psi_k(z)$ for $z \in D_k \cup L_k$ and A denote the operator from the right hand side of (3.4.84), (3.4.85). Then, equations (3.4.84), (3.4.85) can be written compactly as the following equation in the space $\mathcal{H}(D^+)$

$$\Psi = \rho A\Psi + 1, \tag{3.4.89}$$

where 1 is the unit function equal to the number 1 in D_k. The same notations are going to be applied below to the unit function and the number 1. We do not expect any confusion hereafter. The operator A is defined by the right-hand side of (3.4.84) and (3.4.85). In the present section, we demonstrate that equation (3.4.89) has a unique solution for $|\rho| < 1$ given by the series

$$\Psi = \sum_{n=0}^{\infty} \rho^n A^n 1, \tag{3.4.90}$$

where $A^n 1$ denotes the n-tuple application of A to the unity function.

We apply Theorem 1 (page 11) to equation (3.4.89) in the Banach space $\mathcal{H}$.

Theorem 12. *Let $|\rho| < 1$. Equation $\Psi = \rho A\Psi + 1$ has a unique solution. This solution can be found by the method of successive approximations convergent in the space*

$\mathcal{H}(D^+)$.

Proof. Let $|v| \le 1$. Consider equations

$$\psi_k(z) = -v\rho \sum_{m=1}^{N} \frac{1}{2\pi i} \int_{L_m} \overline{n_m^2(t)} \, \overline{\psi_m(t)} E_1(t-z) dt + g_k(z), \quad z \in D_k$$
$$(k = 1, 2, \dots, N),$$

$$\psi_k(t) = -v\rho \left[\frac{1}{2} \overline{n_k^2(t)} \, \overline{\psi_k(t)} + \sum_{m=1}^{N} \frac{1}{2\pi i} \int_{L_m} \overline{n_m^2(t)} \, \overline{\psi_m(t)} E_1(t-z) dt \right] + g_k(t),$$
$$t \in L_k \ (k = 1, 2, \dots, N),$$

$$(3.4.91)$$

where $g_k \in \mathcal{H}(D_k)$.

Let $\psi_k(z)$ be a solution of (3.4.91). Introduce the function $\psi(z)$ analytic in D and Hölder continuous in its closure as follows

$$\psi(z) = -v\rho \sum_{m=1}^{n} \frac{1}{2\pi i} \int_{L_m} \overline{n_m^2(t)} \, \overline{\psi_m(t)} E_1(t-z) dt, \quad z \in D. \qquad (3.4.92)$$

One can see that $\psi(z)$ is doubly periodic. The expression in the right-hand part of (3.4.92) can be considered as Cauchy's integral on torus

$$\Phi(z) = \frac{1}{2\pi i} \int_L \mu(t) E_1(t-z) dt \qquad (3.4.93)$$

for $z \in Q \backslash L$ with the density $\mu(t) = -v\rho \overline{n_m^2(t)} \, \overline{\psi_m(t)}$ on L_m. Using the property (3.4.80) of (3.4.93) and (3.4.91) on L_k we arrive at the $\mathbb{R}$-linear conjugation relation on each fixed curve L_k

$$\psi_k(t) - g_k(t) - \psi(t) = -v\rho \overline{n_k^2(t)} \, \overline{\psi_k(t)}, \quad t \in L_k. \qquad (3.4.94)$$

Here, $\Phi^+(t) = \psi_k(t) - g_k(t)$ and $\Phi^-(t) = \psi(t)$. Integration of (3.4.94) along L_k yields

$$\phi(t) = \varphi_k(t) - v\rho \, \overline{\varphi_k(t)} - h_k(t) + \gamma_k, \quad t \in L_k \ (k = 1, 2, \dots, N), \qquad (3.4.95)$$

where the functions $\phi(z)$, $\varphi_k(z)$ are analytic in D, D_k and belong to $C^{(1,\alpha)}$ in the closures of the domains considered; $\phi'(z) = \psi(z)$, $\varphi_k'(z) = \psi_k(z)$; $h_k(z)$ is a primitive function of $g_k(z)$; γ_k is an arbitrary constant. Moreover, the function $\phi(z)$ is quasi-periodic, hence

$$\phi(z) = \varphi(z) + cz, \qquad (3.4.96)$$

where $\varphi(z)$ is periodic and c is a constant. The constant γ_k can be included into $\varphi_k(z)$ by the representation $\gamma_k = \alpha_k - v\rho \, \overline{\alpha_k}$. Then, (3.4.95) becomes

$$\varphi(t) = \varphi_k(t) - v\rho \, \overline{\varphi_k(t)} - h_k(t) - ct, \quad t \in L_k, \ (k = 1, 2, \dots, N). \qquad (3.4.97)$$

It follows from [38] and Appendix B of [37] that the general solution of the problem

(3.4.97) has the following structure

$$\varphi(z) = \varphi^{(1)}(z) + \varphi^{(2)}(z), \quad \varphi_k(z) = \varphi_k^{(1)}(z) + \varphi_k^{(2)}(z), \tag{3.4.98}$$

where

$$\varphi^{(1)}(t) = \varphi_k^{(1)}(t) - v\rho \, \overline{\varphi_k^{(1)}(t)} - h_k(t), \quad t \in L_k, \ (k = 1, 2, \ldots, N) \tag{3.4.99}$$

and

$$\varphi^{(2)}(t) = \varphi_k^{(2)}(t) - v\rho \, \overline{\varphi_k^{(2)}(t)} - ct, \quad t \in L_k, \ (k = 1, 2, \ldots, N). \tag{3.4.100}$$

Each of the problems (3.4.99) and (3.4.100) (with fixed c) has a unique solution since $|v\rho| < 1$ (the elliptic case in Mikhailov's terminology [32]). The uniqueness also follows from [9]. This unique solution of the problem (3.4.100) can be easily found as $\varphi^{(2)}(z) = cz$ and $\varphi_k^{(2)}(z) = 0$. Then, differentiation of (3.4.98)–(3.4.100) implies that the general solution of the problem (3.4.94) has the form

$$\psi(z) = \psi^{(1)}(z) + c, \tag{3.4.101}$$

where the function $\psi_k^{(1)}(z)$ is uniquely determined from the problem

$$\psi^{(1)}(t) = \psi_k^{(1)}(t) + v\rho \overline{n_k^2(t)} \, \overline{\psi_k^{(1)}(t)} - g_k(t), \ t \in L_k \ (k = 1, 2, \ldots, N). \tag{3.4.102}$$

Comparison of (3.4.94) and (3.4.102) proves the theorem. $\square$

The series

$$\Psi = \sum_{k=0}^{\infty} \rho^k A^k 1 \tag{3.4.103}$$

converges absolutely for $|\rho| < 1$, since it can be represented in the form $\Psi = \sum_{k=0}^{\infty} s^k \rho_0^k A^k 1$ with $s\rho_0 = \rho$ and $|\rho| < |\rho_0| < 1$. The latter series converges absolutely with the rate $|s|$ because the series $\sum_{k=0}^{\infty} \rho_0^k A^k 1$ converges uniformly. Hence, its general term tends to zero as $k \to \infty$.

Uniform convergence of (3.4.103) for $|\rho| = 1$ can be justified with the arguments from [38].

4.5. Contrast expansions from Schwarz's method

It was proved in the previous section that the unique solution of (3.4.84), (3.4.85) with $C = 1$ can be found by successive approximations uniformly convergent in $\cup_{k=1}^{N}(D_k \cup L_k)$. The method of successive approximations can be also presented as the following iterative scheme

$$\psi_k^{(0)}(z) = 1,$$

$$\psi_k^{(p)}(z) = -\rho \sum_{m=1}^{N} \frac{1}{2\pi i} \int_{L_m} \overline{n_m^2(t)}\ \overline{\psi_m^{(p-1)}(t)} E_1(t-z)\mathrm{d}t + 1,\ \ p = 1, 2, \ldots, \quad (3.4.104)$$

$$z \in D_k \quad (k = 1, 2, \ldots, N).$$

When the functions $\psi_k(z)$ $(k = 1, 2, \ldots, N)$ are found, the effective conductivity is calculated from (3.2.42).

In order to compare equations (3.3.57) and (3.4.85), we introduce the complex intensity field $E(z) = E_1(z) + iE_2(z)$ isomorphic to the vector field $\mathbf{E}(\mathbf{x}) = (E_1(\mathbf{x}), E_2(\mathbf{x}))$. Then, the vector equation (3.3.57) can be written as a complex equation on $E(z)$. For definiteness, we consider the complex potential $\psi_k(z)$ in D_k. It follows from (2.2.51) and $\psi_k(z) = \phi_k'(z)$ that

$$-E(z) = \frac{\partial u_k}{\partial x_1} - i\frac{\partial u_k}{\partial x_2} = \frac{2}{\sigma + 1}\overline{\psi_k(z)} \quad (3.4.105)$$

for $\mathbf{E}_0 = (-\frac{2}{\sigma+1}, 0)$. Therefore, the complex equation on $E(z)$ in D_k can be written as the equation (3.4.84) on $\psi_k(z)$ (equation (3.4.85) on the boundary). We do not directly prove that equations (3.3.57) and (3.4.84) are the same. But it is easy to show that the forms of their solutions (3.3.58) and (3.4.90) coincide. Since the solutions $\mathbf{E}$ and Ψ coincide (more precisely, isomorphic), the power series (3.3.58) and (3.4.90) in ρ can be considered as the presentations of the same function analytic in ρ for sufficiently small $|\rho|$. Therefore, the coefficients of (3.3.58) and (3.4.90) coincide. In particular, this fact implies that substitution of (3.4.90) into (3.2.42) yields a series which coincides with the series (3.3.55) obtained by the traditional approach. It follows from [34, 38] that the contrast expansion (3.4.90) converges for all admissible $|\rho| \leq 1$.

4.6. Cluster expansions from Schwarz's method

Schwarz's method can be realized by means of the different iterative scheme corresponding to the traditional cluster expansions summarized on page 56. First, we rewrite (3.4.81) in the form

$$\psi_k(z) + \frac{\rho}{2\pi i} \int_{L_k} \overline{n_k^2(t)}\ \overline{\psi_k(t)} E_1(t-z)\mathrm{d}t =$$

$$-\rho \sum_{m \neq k} \frac{1}{2\pi i} \int_{L_m} \overline{n_m^2(t)}\ \overline{\psi_m(t)} E_1(t-z)\mathrm{d}t + 1, \quad z \in D_k\ (k = 1, 2, \ldots, n). \quad (3.4.106)$$

Let $g_m(z) \in \mathcal{H}(D_m)$. For each fixed $k = 1, 2, \ldots, N$ introduce the operators

$$(P_{km}g_m)(z) = \frac{1}{2\pi i} \int_{L_m} \overline{n_m^2(t)}\ \overline{g_m(t)} E_1(t-z)\mathrm{d}t,\ z \in D_k\ (m = 1, 2, \ldots, n). \quad (3.4.107)$$

One can check that the operator P_{kk} is singular and P_{km} $(m \neq k)$ are compact in $\mathcal{H}(D_k)$. Equations (3.4.106) can be shortly written in the form

$$(I + \rho P_{kk})\psi_k = -\rho \sum_{m \neq k} P_{km}\psi_m + 1, \quad k = 1, 2, \ldots, n. \tag{3.4.108}$$

The zeroth approximation in the concentration f for (3.4.106) can be written as the following integral equation

$$\psi_k^{(0)}(z) + \frac{\rho}{2\pi i} \int_{L_k} \overline{n_k^2(t)} \, \overline{\psi_k^{(0)}(t)} E_1(t - z)\mathrm{d}t = 1, \quad z \in D_k. \tag{3.4.109}$$

Equation (3.4.109) corresponds to the $\mathbb{R}$-linear problem (3.4.76) for one inclusion D_k in the cell Q. Solution to the one-inclusion problem (3.4.109) defines the inverse operator[3] $(I + \rho P_{kk})^{-1}$ bounded in $\mathcal{H}(D_k)$, i.e.,

$$\psi_k^{(0)} = (I + \rho P_{kk})^{-1}1. \tag{3.4.110}$$

Then, (3.4.108) can be written in the equivalent form

$$\psi_k = -\rho \sum_{m \neq k}(I + \rho P_{kk})^{-1} P_{km}\psi_m + \psi_k^{(0)}, \quad k = 1, 2, \ldots, n. \tag{3.4.111}$$

As established in Section 4.5 the vector-function $\mathbf{E}(\mathbf{x})$ and the complex functions $\psi_k(z)$ express the same vector field in D_k up to an isomorphism for $\mathbf{E}_0 = (-\frac{2}{\sigma+1}, 0)$. Therefore, application of the successive approximations to (3.4.111) yields the same series as (3.3.48) in D_k. A detailed comparison can be performed. It shows, for instance, that the term $\mathbf{E}_0 + \sum_{k=1}^{N} \mathbf{M}_1(\mathbf{x}; k) \cdot \mathbf{E}_0$ from (3.3.48) corresponds to $\psi_k^{(0)} = (I + \rho P_{kk})^{-1}1$.

The solution of the problem (3.4.109) depends on the shape of D_k. Therefore, the first-order approximation in f for the effective conductivity also depends on the shape. If all the inclusions have the same form, the functions $\psi_k^{(0)}(z)$ do not depend on k. Then, we arrive at the formula valid up to $O(f^2)$

$$\sigma_e \approx 1 + 2\rho f\alpha, \tag{3.4.112}$$

where the shape factor α is introduced following [29], [43]

$$\alpha = \frac{1}{|D_1|} \int_{D_1} \psi_1^{(0)}(x_1 + ix_2)\mathrm{d}x_1\mathrm{d}x_2.$$

The Clausius–Mossotti formula equivalent to (3.4.112) up to $O(f^2)$ is widely applied

---

[3]Here, we want to stress that (3.4.110) is not a formula for the solution. Actually it is the same integral equation (3.4.109). The notation (3.4.110) just shows that the integral equation has a unique solution, allowing to use the notion $(I + \rho P_{kk})^{-1}$ (see discussion in Introduction).

[31]

$$\sigma_e \approx \frac{1 + \rho f \alpha}{1 - \rho f \alpha}. \tag{3.4.113}$$

One can find the discussion of limitations of the formulae (3.4.112), (3.4.113) and a comparison of their accuracy in [43]. The phase-interchange symmetry, exact property of the conductivity in 2D, follows from (3.4.113) as well [31]. Equation (3.4.113) can be also viewed as the simple Padé approximant applied to the expansion in concentration f, available only up to first-order terms inclusively. The Clausius–Mossotti formula is very intuitive and symmetric. It will be further discussed and extended in Chapters 6,7, and 9.

The higher-order terms in the cluster expansion can be obtained from higher-order iterations to the integral equation on $\psi_k^{(p)}(z)$

$$\psi_k^{(p)}(z) + \frac{\rho}{2\pi i} \int_{L_k} \overline{n_k^2(t)} \; \overline{\psi_k^{(p)}(t)} E_1(t - z) dt =$$

$$-\rho \sum_{m \neq k} \frac{1}{2\pi i} \int_{L_m} \overline{n_m^2(t)} \; \overline{\psi_m^{(p-1)}(t)} E_1(t - z) dt + 1, \quad z \in D_k \tag{3.4.114}$$

$$(k = 1, 2, \ldots, n); \quad p = 1, 2, \ldots.$$

The uniform convergence of the iterative scheme (3.4.114) was proved in the case of perfectly conducting inclusions $\rho = 1$ in Section 4.9.2 of [38]. The higher-order iterations (3.4.114) yield approximate analytical formulae for σ_e for arbitrary non-overlapping locations of different disks [7, 39].

Using the notations of the equation (3.4.111), we rewrite the iterative scheme in the form

$$\psi_k^{(p)} = -\rho \sum_{m \neq k} (I + \rho P_{kk})^{-1} P_{km} \psi_m^{(p-1)} + \psi_k^{(0)}, \quad k = 1, 2, \ldots, n; \; p = 1, 2, \ldots. \tag{3.4.115}$$

Introduce the function $\Psi^{(p)}(z) = \psi_k^{(p)}(z)$ for $z \in D_k$ ($k = 1, 2, \ldots, n$) and shortly write (3.4.115) with the convolution operator (see the explicit form (3.4.114))

$$\Psi^{(p)} = \Psi^{(p-1)} * E_1 + \Psi^{(0)}, \quad p = 1, 2, \ldots. \tag{3.4.116}$$

Formal application of the successive approximations to (3.4.116) yields the series

$$\Psi = \Psi^{(0)} + \Psi^{(0)} * E_1 + \Psi^{(0)} * \Psi^{(0)} * E_1 + \cdots. \tag{3.4.117}$$

Here, the footnote on page 70 on the form of solution can be applied to (3.4.117).

B.3.3 Eisenstein's series

Following [14], we present constructive formulae for the Eisenstein–Rayleigh sums S_m and the Eisenstein functions $E_m(z)$ corresponding to the lattice Q.

The Eisenstein–Rayleigh lattice sums S_m are defined as

$$S_m = \sum_{m_1,m_2} (m_1\omega_1 + m_2\omega_2)^{-m}, \quad m = 2, 3, \ldots, \tag{3.4.118}$$

where m_1, m_2 run over all integers except $m_1 = m_2 = 0$. The series (3.4.118) is conditionally convergent, hence, its value depends on the order of summation. Here, the order is fixed by using of the Eisenstein summation [51]

$$\sum_{p,q}^{e} := \lim_{M_2 \to +\infty} \lim_{M_1 \to +\infty} \sum_{q=-M_2}^{M_2} \sum_{p=-M_1}^{M_1}. \tag{3.4.119}$$

The sums (3.4.118) are slowly convergent if computed directly. But they can be easily calculated through the rapidly convergent series

$$S_2 = \left(\frac{\pi}{\omega_1}\right)^2 \left(\frac{1}{3} - 8\sum_{m=1}^{\infty} \frac{mq^{2m}}{1 - q^{2m}}\right), \quad q = \exp\left(\pi i \frac{\omega_2}{\omega_1}\right), \tag{3.4.120}$$

$$S_4 = \left(\frac{\pi}{\omega_1}\right)^4 \left(\frac{1}{45} + \frac{16}{3}\sum_{m=1}^{\infty} \frac{m^3 q^{2m}}{1 - q^{2m}}\right), \tag{3.4.121}$$

$$S_6 = \left(\frac{\pi}{\omega_1}\right)^6 \left(\frac{2}{945} - \frac{16}{15}\sum_{m=1}^{\infty} \frac{m^5 q^{2m}}{1 - q^{2m}}\right). \tag{3.4.122}$$

S_{2n} ($n \geq 4$) can be calculated by the recurrence formula

$$S_{2n} = \frac{3}{(2n + 1)(2n - 1)(n - 3)} \sum_{m=2}^{n-2} (2m - 1)(2n - 2m - 1) S_{2m} S_{2(n-m)}. \tag{3.4.123}$$

The rest of sums vanish. The following formula was proved in [36]

$$S_2 = \frac{2}{\omega_1} \zeta\left(\frac{\omega_1}{2}\right). \tag{3.4.124}$$

Application of the formula (5) from [23, p. 210] for $\zeta\left(\frac{\omega_1}{2}\right)$ yields (3.4.120). It is worth noting that Akhiezer (see [2, p. 204]) deduced a similar formula which after substitution into (3.4.124) yields

$$S_2 = \left(\frac{\pi}{\omega_1}\right)^2 \left(\frac{1}{3} - 8\sum_{m=1}^{\infty} \frac{q^{2m}}{(1 - q^{2m})^2}\right). \tag{3.4.125}$$

Formulae (3.4.121) and (3.4.122) follow from the formulae (5) from [23, p. 210], for the invariants of the Weierstrass functions $g_2 = 60 S_4$ and $g_3 = 140 S_6$. The review of other formulae for S_2 can be found in [52].

For the hexagonal array,

$$S_2 = \pi, \quad S_4 = 0, \quad S_6 \approx 3.80815, \quad S_8 = 0, \tag{3.4.126}$$

while for the square array,

$$S_2 = \pi, \quad S_4 \approx 3.151211, \quad S_6 = 0, \quad S_8 \approx 4.2557732. \tag{3.4.127}$$

Let $\zeta(z)$ denote the Weierstrass ζ-function for which [2]

$$\zeta(z + \omega_j) - \zeta(z) = \delta_j \quad (j = 1, 2). \tag{3.4.128}$$

where $\delta_j = 2\zeta\left(\frac{\omega_j}{2}\right)$. Using the Eisenstein summation we introduce the Eisenstein function

$$E_1(z) = \sum_{m_1, m_2 \in \mathbb{Z}} \frac{1}{z + m_1\omega_1 + m_2\omega_2}, \tag{3.4.129}$$

The Weierstrass function $\zeta(z)$ and Eisenstein function are related by formula $E_1(z) = \zeta(z) - S_2 z$. It follows from Legendre's identity $\delta_1\omega_2 - \delta_2\omega_1 = 2\pi i$ (see [2, 51]) and (3.4.128) that the jumps of $E_1(z)$ have the form

$$E_1(z + \omega_1) - E_1(z) = 0, \quad E_1(z + \omega_2) - E_1(z) = -\frac{2\pi i}{\omega_1}. \tag{3.4.130}$$

The high-order Eisenstein functions are related to the Weierstrass function $\wp(z)$ [2] by the identities [51]

$$E_2(z) = \wp(z) + S_2, \quad E_m(z) = \frac{(-1)^m}{(m-1)!} \frac{d^{m-2}\wp(z)}{dz^{m-2}}, \quad m = 3, 4, \ldots. \tag{3.4.131}$$

Every function (3.4.131) is doubly periodic and has a pole of order m at $z = 0$. The Eisenstein functions of the even-order $E_{2m}(z)$ can be presented in the form of the series [51]

$$E_{2m}(z) = \frac{1}{z^{2m}} + \sum_{k=1}^{\infty} s_k^{(m)} z^{2(k-1)}, \tag{3.4.132}$$

where

$$s_k^{(m)} = \frac{(2m + 2k - 3)!}{(2m - 1)!(2k - 2)!} S_{2(m+k-1)}. \tag{3.4.133}$$

The function $E_1(z)$ is expanded into Laurent's series

$$E_1(z) = \frac{1}{z} - \sum_{k=1}^{\infty} S_{2k} z^{2k-1}. \tag{3.4.134}$$

The Eisenstein functions are related by the formula

$$E'_m(z) = -m E_{m+1}(z). \tag{3.4.135}$$

The relation (3.4.135) implies the following form of the nth derivative of E_k

$$E_k^{(n)}(z) = (-1)^n \frac{(k+n-1)!}{(k-1)!} E_{k+n}(z). \tag{3.4.136}$$

To make the formulae for the effective conductivity more transparent it is convenient to adopt the following definition

$$E_n(0) := S_n. \tag{3.4.137}$$

Such an expression appears often in the sums involving the terms $E_n(a_k - a_m)$ which has to be calculated by (3.4.137) for $a_k = a_m$.

5. Remark on asymptotic methods

Let us proceed to discuss the essentials of an asymptotic methods applied to the different limit of closely spaced disks by other authors. Consider the case of a square array and introduce the dimensionless gap parameter $\delta = \frac{1-2r}{r}$ where $1 - 2r$ is the distance between the neighboring disks of radius r. The gap parameter can be expressed through the concentration f as $\delta = \sqrt{\frac{\pi}{f}} - 2$. The limit concentration $f = f_c = \frac{\pi}{4}$ holds. Keller [26] obtained the following expression for the effective conductivity

$$\sigma_e = \frac{\pi^{\frac{3}{2}}}{2\sqrt{\frac{\pi}{4}-f}} + O(\delta^0) = \frac{\pi}{\sqrt{\delta}} + O(\delta^0), \quad \text{as } \delta \to 0. \tag{3.5.138}$$

For the hexagonal regular array, the effective conductivity (3.5.138) as a function of δ must be multiplied by $\sqrt{3}$ because of the following arguments. Consider the conductivity of the both arrays when the external flux is applied along the x_1-axis. For sufficiently small δ the main flux is concentrated in the necks between the disks, hence, it is determined by the graphs displayed in Fig. 3.5.

Let $I(\delta)$ denote the total flux between the disks in the square array, i.e., the flux in the edge $\overrightarrow{AB}$ in Fig. 3.5a. According to Keller's formula $I(\delta) = \frac{\pi}{\sqrt{\delta}}$. The effective conductivity (3.5.138) is obtained by multiplication of $I(\delta)$ by the area of the dashed cell in Fig. 3.5a equal to unity. Consider now the hexagonal array displayed in Fig. 3.5b. The flux in every edge is equal to $\frac{\sqrt{3}}{2}I(\delta)$ since the edge length of the hexagonal array is greater in $\frac{\sqrt{2}}{\sqrt[4]{3}}$ times than the edge length of the square array.[4] The total flux between A and B is combined with the flux along $\overrightarrow{AB}$ and two fluxes $\frac{\sqrt{3}}{4}I(\delta)$ along $\overrightarrow{AC}, \overrightarrow{CB}$ and $\overrightarrow{AD}, \overrightarrow{DB}$. Hence, $\sqrt{3}I(\delta)$ has to be multiplied by the area of the rhombus $ACBD$ equal

---

[4]Recall that the areas of periodicity cells in the both lattices are normalized to unity.

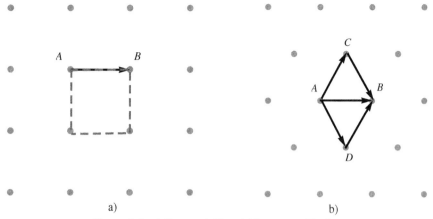

Figure 3.5 a) Square lattice, b) hexagonal lattice.

to unity. This yields the effective conductivity of the hexagonal array

$$\sigma_e = \frac{\sqrt{3}\pi}{\sqrt{\delta}} + O(\delta^0) = \sqrt[4]{\frac{3}{4}}\frac{\pi^{\frac{3}{2}}}{\sqrt{f - \frac{\pi}{\sqrt{12}}}} + O(\delta^0), \quad \text{as } \delta \to 0, \qquad (3.5.139)$$

where $\delta = \frac{\sqrt[4]{\frac{4}{3}}-2r}{r} = \sqrt[4]{\frac{4}{3}}\sqrt{\frac{\pi}{f}} - 2$. Formulae (3.5.138) and (3.5.139) were rigorously justified by variational methods in the books [6, 27], i.e., the asymptotic Keller formula for the pair interactions was proved.

Analysis of pair interactions of inclusions not always can be applied to estimations of the effective properties as explained in the item devoted to the self-consistent methods on page 42. We now discuss two different formulae concerning the same asymptotic problem. O'Brien (1977 unpublished, see discussion in Section 5 of [46]) considered pair interactions for the total flux I in the unit cell of the square array for highly conducting closed inclusions when $\sigma \gg 1$ and $\delta \ll 1$

$$I \approx \left[\frac{2\ln\sigma}{\pi}\frac{\ln\sigma}{\sigma} + \frac{\sqrt{\delta}}{\pi}\right]^{-1}. \qquad (3.5.140)$$

Rylko (see formula (4.21), [47]) deduced another relation for the local flux in complex form between two disks

$$I \approx e^{i\theta}\frac{2\text{Re } ce^{i\theta} - \varepsilon(ce^{i\theta} + 2\bar{c}e^{i\theta}) - 2\bar{c}e^{i\theta}\sqrt{\delta}}{4\sqrt{\delta}(1-2\varepsilon) + 2\varepsilon}, \qquad (3.5.141)$$

where $\varepsilon = 1 - \rho = \frac{2}{\sigma+1}$. The parameter θ is the inclination of the edge in a fixed co-ordinate system. The constant c is determined by locations of surrounding disks, in particular, c depends on the considered lattice. Though finite clusters of inclusions were investigated in [47] the derivation of the asymptotic equation (3.5.141) holds for doubly periodic sets of inclusions as well as Keller's formula.

It is difficult to compare directly (3.5.140) and (3.5.141) since the constant c for the square array was not computed. But we can accomplish qualitative study when ε and δ tend to zero. The structure of the complex flux Ψ is described by the operator series (3.4.103). The operator A in (3.4.103) for $\delta \to 0$ depends analytically on $\sqrt{\delta}$ for the circular inclusions [47]. The series (3.4.103) converges uniformly for $|\rho| \leq |\rho_0| < 1$, i.e., for $\varepsilon \geq \varepsilon_0 > 0$ where ρ_0 and ε_0 are arbitrarily fixed in the prescribed domains. Therefore, the complex flux Ψ is an analytic function on $\sqrt{\delta}$ and $\sigma = \frac{2}{\varepsilon} - 1$. This contradicts to appearance of the term $\log \sigma = \log(2 - \varepsilon) - \log \varepsilon$ in (3.5.140). The above arguments confirm that Rylko's formula (3.5.141) is the unique known acceptable formula for the case when $\delta \to 0$ and $\sigma \to \infty$. Further progress could be based on the computations of the constant c.

Finally we want to comment on the limit case of densely packed random composites. It is expected that Keller-type formula for the effective conductivity will hold if average inter-particle distance parameter $\langle \delta \rangle$ is introduced. But its presence is formal since it is not known how to get back to the variable $f_c - f$ and extract the critical index for random composite. Therefore, random case at high-concentration requires different methodology suggested in Chapter 9.

REFERENCE

1. Adler PM, Thovert JF, Mourzenko VV, Fractured porous media. 2rd Oxford: Oxford University Press; 2012.
2. Akhiezer NI, Elements of the Theory of Elliptic Functions. RI, American Mathematical Society, Providence, 1990.
3. Ammari H, Kang H, Polarization and Moment Tensors. NY: Springer Science+Business Media; 2007.
4. Bakhvalov NS, Panasenko GP, Averaging processes in periodic media. Mathematical problems in mechanics of composite materials. Dordrecht: Kluwer; 1989.
5. Bensoussan A, Lions JL, Papanicolaou G, Asymptotic Analysis for Periodic Structures. Amsterdam: North-Holland; 1978.
6. Berlyand L, Kolpakov AG, Novikov A, Introduction to the Network Approximation Method for Materials Modeling. Cambridge: Cambridge University Press; 2012.
7. Berlyand L, Mityushev V, Generalized Clauisius-Mossotti Formula for Random Composite with Circular Fibers, J. Stat. Phys. 2001; 102: 115–145.
8. Berlyand L, Kolpakov AG, Novikov A, Introduction to the Network Approximation Method for Materials Modeling. Cambridge: Cambridge University Press; 2012.
9. Bojarski B, Mityushev V, $\mathbb{R}$-linear problem for multiply connected domains and alternating method of Schwarz, J. Math. Sci. 2013; 189: 68–77.
10. Castro LP, Pesetskaya E, Rogosin SV, Effective conductivity of a composite material with non-ideal contact conditions, Complex Variables and Elliptic Equations 2009; 54: 1085–1100.
11. Cherkaev A, Variational methods for Structural Optimization. New York etc: Springer Verlag; 2000.
12. Chibrikova LI, Fundamental boundary value problems for the analytic functions. Kazan Univ. Publ., Kazan, 1977 [in Russian].

13. Colton D, Kress R, Integral Equation Methods in Scattering Theory. New York etc: John Wiley & Sons; 1983.
14. Czapla R, Nawalaniec W, Mityushev V, Effective conductivity of random two-dimensional composites with circular non-overlapping inclusions, Computational Materials Science 2012; 63: 118–126.
15. Drygaś P, Mityushev V, Effective conductivity of arrays of unidirectional cylinders with interfacial resistance, Q J Mechanics Appl Math 2009; 62: 235–262.
16. Golden K, Papanicolaou G, Bounds for effective parameters of heterogeneous media by analytic continuation, Communications in Mathematical Physics 1983; 90: 473–491.
17. Grigolyuk EI, Filshtinsky LA, Perforated Plates and Shells. Moscow: Nauka; 1970 [in Russian].
18. Grigolyuk EI, Filshtinsky LA, Periodical Piece–Homogeneous Elastic Structures. Moscow: Nauka; 1991 [in Russian].
19. Grigolyuk EI, Filshtinsky LA, Regular Piece-Homogeneous Structures with defects. Moscow: Fiziko-Matematicheskaja Literatura; 1994 [in Russian].
20. Hashin Z, Analysis of composite materials a survey, Journal of Applied Mechanics 1983; 50: 481–505.
21. Happel J, Brenner H, Low Reynolds Number Hydrodynamics. Prentice-Hall, 1965.
22. Hashin Z, Shtrikman S, A variational approach to the theory of the effective magnetic permeability of multiphase materials, J. Appl. Phys. 1962; 33: 3125–3131.
23. Hurwitz A, Courant R, Vorlesungen Über allgemeine Funktionentheorie und elliptische Funktionen. Berlin: Springer-Verlag; 1964.
24. Jikov VV, Kozlov SM, Oleinik OA, Homogenization of differential operators and integral functionals. Berlin, New York: Springer-Verlag, 1994.
25. Kapanadze D, Mishuris G, Pesetskaya E, Improved algorithm for analytical solution of the heat conduction problem in doubly periodic 2D composite materials, Complex Variables and Elliptic Equations 2015; 60: 1–23.
26. Keller JB, Conductivity of a Medium Containing a Dense Array of Perfectly Conducting Spheres or Cylinders or Nonconducting Cylinders, Journal of Applied Physics 1963; 34: 991–993.
27. Kolpakov AA, Kolpakov AG, Capacity and Transport in Contrast Composite Structures: Asymptotic Analysis and Applications. Boca Raton etc: CRC Press Inc.; 2009.
28. Kushch I, Micromechanics of Composites. Butterworth-Heinemann: Elsevier; 2013.
29. Landau LD, Lifshitz EM, Pitaevskii LP, Electrodynamics of Continuous Media. Oxford: Pergamon Press; 1960.
30. Le Quang H, Bonnet G, Pham DC, Bounds and correlation approximation for the effective conductivity of heterogeneous plates, Phys. Rev. E 2011; 84: 061153.
31. Milton GW, The theory of composites. Cambridge University Press, Cambridge, 2002.
32. Mikhailov LG, New Class of Singular Integral Equations and its Applications to Differential Equations with Singular Coefficients. 2nd Berlin: Akademie Verlag; 1970.
33. Mikhlin SG, Integral Equations and Their Applications to Certain Problems in Mechanics, Mathematical Physics and Technology. 2nd New York: Macmillan; 1964.
34. Mityushev VV, Plane problem for the steady heat conduction of material with circular inclusions, Arch. Mech. 1993; 45: 211–215.
35. Mityushev VV, Generalized method of Schwarz and addition theorems in mechanics of materials containing cavities, Arch. Mech. 1995; 47, 6: 1169–1181.
36. Mityushev V, Transport properties of regular array of cylinders, ZAMM 1997; 77, 2: 115–120.
37. Mityushev V, Transport properties of two-dimensional composite materials with circular inclusions, Proc. R. Soc. London 1999; A455: 2513–2528.
38. Mityushev VV, Rogosin SV, Constructive methods to linear and non-linear boundary value problems of the analytic function. Theory and applications. Boca Raton etc: Chapman & Hall / CRC; 1999/2000. [Chapter 4].
39. Mityushev V, Transport properties of doubly periodic arrays of circular cylinders and optimal design problems, Appl. Math. Optimization, 2001; 44: 17–31.
40. Mityushev V, R-linear problem on torus and its application to composites, Complex Variables

2005; 50, 7-10: 621–630.

41. Mityushev V Conductivity of a two-dimensional composite containing elliptical inclusions, Proc. Roy. Soc. London A, 2009; 465: 2991–3010.

42. Mityushev V, Rylko N, Optimal distribution of the non-overlapping conducting disks, Multiscale Model. Simul. 2012; 10: 180–190.

43. Mityushev V, Rylko N, Maxwell's approach to effective conductivity and its limitations, The Quarterly Journal of Mechanics and Applied Mathematics 2013.

44. Movchan AB, Movchan NV, Poulton CG, Asymptotic Models of Fields in Dilute and Densely Packed Composites. Imperial College Press, London, 2002.

45. Nicorovici NA, McPhedran RC, Milton GW, Optical and dielectric properties of partially resonant composites, Phys. Rev. B 1994; 49: 8479–8482.

46. Perrins WT, McKenzie DR, McPhedran RC, Transport properties of regular array of cylinders, Proc. R. Soc.A 1979; 369: 207–225.

47. Rylko N, Structure of the scalar field around unidirectional circular cylinders, Proc. R. Soc. A 2008; 464: 391–407.

48. Smith B, Björstad P, Gropp W, Domain decomposition. Parallel multilevel methods for elliptic partial differential equations. Cambridge: Cambridge University Press; 1996.

49. Telega JJ, Stochastic homogenization: convexity and nonconvexity, eds. Castañeda PP, Telega JJ, Gambin B. Nonlinear Homogenization and its Applications to Composites, Polycrystals and Smart Materials, NATO Science Series, 305–346: Dordrecht, Kluwer Academic Publishers, 2004.

50. Torquato S, Random Heterogeneous Materials: Microstructure and Macroscopic Properties. New York. Springer-Verlag: (2002).

51. Weil A, Elliptic Functions according to Eisenstein and Kronecker. Springer-Verlag, Berlin etc, 1999.

52. Yakubovich S, Drygas P, Mityushev V, Closed-form evaluation of 2D static lattice sums, Proc Roy. Soc. London 2016; A472, 2195: 20160662. DOI: 10.1098/rspa.2016.0510.

CHAPTER 4

From Basic Sums to effective conductivity and RVE)

The main purpose of this chapter is to suggest a constructive analytical-numerical approach to construct an RVE (representative volume element) theory.

A periodicity cell can be treated as a representative cell and vice versa. From practical point of view, each sample is finite, hence, it can be considered as an RVE with many inclusions. Application of the RVE theory can reduce the number of inclusions per periodicity cell but not necessarily down to just a few inclusions. "Bold periodization" to a small number of inclusions per cell can considerably distort the effective properties of the original sample. Correct introduction of RVE becomes possible with notion of Basic Sums.

1. Basic Sums

In the present section, we introduce an important for our studies geometrical characteristics called *the basic sums* or *the e-sums* first arisen in the papers [3, 20] and systematically discussed in [5, 21, 23].

At the beginning, we consider mono-dispersed composites with identical circular non-overlapping inclusions arbitrary distributed in the periodicity cell. Let q be a natural number; k_s runs over 1 to N, $m_q = 2, 3, \ldots$. Let $\mathbf{C}$ denote the operator of complex conjugation. Introduce the multiple convolution sums

$$e_{m_1,\ldots,m_q} = \frac{1}{N^{1+\frac{1}{2}(m_1+\cdots+m_q)}} \sum_{k_0,k_1,\ldots,k_n} E_{m_1}(a_{k_0} - a_{k_1})\overline{E_{m_2}(a_{k_1} - a_{k_2})}\ldots \mathbf{C}^{q+1}E_{m_q}(a_{k_{q-1}} - a_{k_q}),$$

(4.1.1)

where the Eisenstein functions $E_m(z)$ are introduced on page 71. We pay attention that definition (3.4.137) is used for shortness.

For instance, basic sums e_2 and e_{22} take the following form

$$e_2 = \frac{1}{N^2} \sum_{k_0=1}^{N} \sum_{k_1=1}^{N} E_2(a_{k_0} - a_{k_1}),$$

(4.1.2)

$$e_{22} = \frac{1}{N^3} \sum_{k_0=1}^{N} \sum_{k_1=1}^{N} \sum_{k_2=1}^{N} E_2(a_{k_0} - a_{k_1})\overline{E_2(a_{k_1} - a_{k_2})}.$$

It is worth noting that the triple sum e_{22} can be written as the double sum (4.2.39).

We now proceed to discuss polydispersed composites. Assume inclusions are modeled by N non-overlapping disks of different radii r_j ($j = 1, 2, 3, \ldots, N$) (see Fig. 4.1). Thus, the total concentration of inclusions equals

$$f = \pi \sum_{j=1}^{N} r_j^2.$$

Let $r = \max_{1 \leq j \leq N} r_j$. Following Berlyand and Mityushev [4] introduce the constants

$$f_j = \left(\frac{r_j}{r}\right)^2, \quad j = 1, 2, 3, \ldots, N,$$

(4.1.3)

describing polydispersity of inclusions. Then, the corresponding basic sums of the multi-order $\mathbf{p} = (p_1, \ldots, p_n)$ are defined by the following formulae

$$e_{p_1,p_2,p_3,\ldots,p_n}^{f_0,f_1,f_2,\ldots,f_n} = \frac{1}{\eta^{1+\frac{1}{2}(p_1+\cdots+p_n)}} \sum_{k_0,k_1,\ldots,k_n} f_{k_0}^{t_0} f_{k_1}^{t_1} f_{k_2}^{t_2} \cdots f_{k_n}^{t_n} E_{p_1}(a_{k_0} - a_{k_1})\overline{E_{p_2}(a_{k_1} - a_{k_2})}$$
$$\times E_{p_3}(a_{k_2} - a_{k_3}) \cdots C^{n+1} E_{p_n}(a_{k_{n-1}} - a_{k_n}),$$

(4.1.4)

where $\eta = \sum_{j=1}^{N} f_j$. The superscripts t_j ($j = 0, 1, 2 \ldots, n$) are given by the recurrence relations

$$t_0 = 1,$$
$$t_j = p_j - t_{j-1}, \quad j = 1, 2, \ldots, n.$$

(4.1.5)

$$t_n = 1.$$

(4.1.6)

In the case of equal disks, (4.1.4) becomes (4.1.1).

The basic sums can be introduced for an arbitrary shaped inclusions after substitution of the local field (3.4.117) into (3.2.42). In order to demonstrate how to do it, consider the simplest case of identical inclusions. Arrange them in such a way that each inclusion D_k is obtained from a fixed sufficiently smooth domain G by the translation $z = a_k + \zeta$ where $z \in D_k$ and $\zeta \in G$. Introduce the functions $\omega_k(\zeta) = \psi_k(z)$

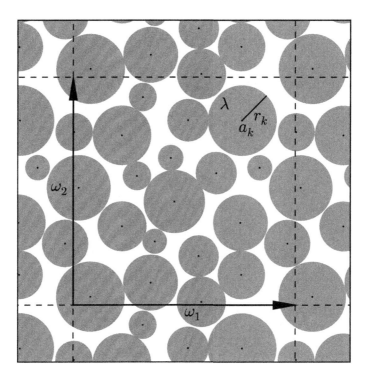

Figure 4.1 Polydispersed two-dimensional composite modelled as a two-periodic cell $Q_{(0,0)}$

analytic in G and Hölder continuous in $G \cup \partial G$. Then, (3.4.114) becomes

$$\omega_k^{(p)}(\zeta) + \frac{\rho}{2\pi i} \int_{\partial G} \overline{n^2(\tau)} \; \overline{\omega_k^{(p)}(\tau)} E_1(\tau - \zeta) d\tau =$$

$$-\rho \sum_{m \neq k} \frac{1}{2\pi i} \int_{\partial G} \overline{n^2(\tau)} \; \overline{\omega_m^{(p-1)}(\tau)} E_1(a_m - a_k + \tau - \zeta) d\tau + 1, \qquad (4.1.7)$$

$$\zeta \in G \quad (k = 1, 2, \ldots, n); \quad p = 1, 2, \ldots,$$

where $\tau = t - a_m \in \partial G$ when $t \in L_m$, $n(\tau)$ is the unit outward normal vector to ∂G. Using (3.4.135), we have

$$E_1(a_m - a_k + \tau - \zeta) = E_1(a_m - a_k) - E_2(a_m - a_k)(\tau - \zeta) + \cdots . \qquad (4.1.8)$$

Substitute (4.1.8) into (4.1.7)

$$\omega_k^{(p)}(\zeta) + \frac{\rho}{2\pi i} \int_{\partial G} \overline{n^2(\tau)} \; \overline{\omega_k^{(p)}(\tau)} E_1(\tau - \zeta) d\tau =$$

$$1 - \rho \sum_{m \neq k} \left[E_1(a_m - a_k) \frac{1}{2\pi i} \int_{\partial G} \overline{n^2(\tau)} \; \overline{\omega_m^{(p-1)}(\tau)} d\tau - \right.$$

$$\left. E_2(a_m - a_k) \frac{1}{2\pi i} \int_{\partial G} \overline{n^2(\tau)} \; \overline{\omega_m^{(p-1)}(\tau)}(\tau - \zeta) d\tau + \cdots \right]$$

$$\zeta \in G \quad (k = 1, 2, \ldots, n); \quad p = 1, 2, \ldots.$$

(4.1.9)

The integral $\frac{1}{2\pi i} \int_{\partial G} \overline{n^2(\tau)} \; \overline{\omega_m^{(p-1)}(\tau)} \, d\tau$ is of order $|G| \max_{\partial G} |\omega_m^{(p-1)}(\tau)|$. Taking into account that $f = N|G|$ one can see that the expression (4.1.9) can be presented as a linear combinations of the power terms $\rho^q f^s$. The normalized coefficients with these terms depend only on geometry and can be considered as the basic sums. This point will remain here just mentioned in passing but can be developed further in special study.

2. Identical circular inclusions.

Below, we consider the case of identical circular inclusions $D_k = \{z \in \mathbb{C} : |z - a_k| < r\}$, i.e., all $r_k = r$ and $\rho_k = \rho$. The concentration of the inclusions is given by formula

$$f = N\pi r^2.$$

(4.2.10)

Deterministic case is solved below completely and explicit expressions for local fields are obtained below. This step is vital since it allows to obtain an explicit expression for the effective conductivity. Based on it we develop in subsequent chapters an efficient computer algorithm for regular and random composites.

2.1. Local fields

Following the developments on page 29, we can write the condition of perfect contact as the $\mathbb{R}$-linear condition

$$\psi(t) = \psi_k(t) + \rho \left(\frac{r_k}{t - a_k} \right)^2 \overline{\psi_k(t)} - 1, \quad |t - a_k| = r_k, \; k = 1, 2, \ldots, N. \quad (4.2.11)$$

The $\mathbb{R}$-linear problem (4.2.11) can be reduced to a system of functional equations as demonstrated below. First, consider the integral operator from (3.4.81)

$$(J_{km}F)(z) = \frac{1}{2\pi i} \int_{L_m} \overline{n_m^2(t)} \; \overline{F(t)} E_1(t - z) dt, \quad z \in D_k, \quad (4.2.12)$$

where $F(z)$ is analytic in the disk D_m. We have $t = t^*_{(m)}$ on L_m. Hence, the operator (4.2.12) can be written in the form

$$(J_{km}F)(z) = \frac{1}{2\pi i} \int_{L_m} \left(\frac{r}{t - a_m}\right)^2 \overline{F\left(t^*_{(m)}\right)} E_1(t - z)dt, \quad z \in D_k, \tag{4.2.13}$$

where the function $\overline{F\left(t^*_{(m)}\right)}$ is analytically continued into $|z - a_k| > r$.

The function $E_1(z)$ has poles of first order at the points $z = m_1\omega_1 + m_2\omega_2$ (see [30]). The function $\left(\frac{r}{z - a_m}\right)^2 \overline{F\left(z^*_{(m)}\right)}$ is analytic in the domain $|z - a_m| > r$ and the integral $(J_{km}F)(z)$ can be calculated by method of residues at $|z - a_m| > r$. Introduce the operator

$$\left(W^{(m)}_{m_1,m_2}F\right)(z) = \left(\frac{r}{z + m_1\omega_1 + m_2\omega_2}\right)^2 \overline{f\left(a_m + \frac{r^2}{z - a_m + m_1\omega_1 + m_2\omega_2}\right)}. \tag{4.2.14}$$

We have for $m = k$

$$(J_{kk}F)(z) = -\sum'_{m_1,m_2\in\mathbb{Z}} \left(W^{(k)}_{m_1,m_2}F\right)(z) \tag{4.2.15}$$

and for $m \neq k$

$$(J_{km}F)(z) = -\sum_{m_1,m_2\in\mathbb{Z}} \left(W^{(m)}_{m_1,m_2}F\right)(z). \tag{4.2.16}$$

The double series (4.2.14)–(4.2.16) are defined by the Eisenstein summation that correctly determines their convergence [3].

Substitution of (4.2.14)–(4.2.16) into (3.4.81) yields the system of functional equations

$$\psi_k(z) = \rho \sum_{m=1}^{N} \sum^*_{m_1,m_2\in\mathbb{Z}} \left(W^{(m)}_{m_1,m_2}\psi_m\right)(z) + 1, \quad |z - a_k| \leq r \ (k = 1, 2, \ldots, N), \tag{4.2.17}$$

where m_1, m_2 run over integers in the sum $\sum^*$ with the excluded term $m_1 = m_2 = 0$ for $m = k$.

Remark 10. Functional equations (4.2.17) can be obtained by the method described on page 29 for inclusions with the centers $a_k + m_1\omega_1 + m_2\omega_2$.

Each term $\left(W^{(m)}_{m_1,m_2}\psi_m\right)(z)$ can be treated as the shift operator written in the form of functional compositions. Such equations can be easily solved by use of symbolic computations. The difficulty related to the infinite double summations in m_1, m_2 can be overcome by application of the Eisenstein functions $E_n(z)$. Equations (4.2.17) can be solved by the contrast expansions in ρ (see page 68) and by the cluster expansions in f (see page 69). We refer to [3, 20] where this approach is explained in details. Computational efficiency of the method can be seen from the analytical formulae shown below. Consider the cluster expansion in the concentration of inclusions $f = N\pi r^2$ which is equivalent to the expansion in r^2

$$\psi_m(z) = \sum_{n=0}^{\infty} \psi_m^{(n)}(z) r^{2n}, \quad |z - a_k| \le r \ (k = 1, 2, \dots, N). \qquad (4.2.18)$$

Few first functions in the expansion (4.2.18) are given by the following exact formulae

$$\psi_m^{(0)}(z) = 1, \quad \psi_m^{(1)}(z) = \rho \sum_{k=1}^{N} E_2(z - a_k),$$

$$\psi_m^{(2)}(z) = \rho^2 \sum_{k,k_1=1}^{N} \overline{E_2(a_k - a_{k_1})} E_2(z - a_k),$$

$$\psi_m^{(3)}(z) = \rho^3 \sum_{k,k_1,k_2=1}^{N} E_2(a_k - a_{k_1}) \overline{E_2(a_{k_1} - a_{k_2})} E_2(z - a_{k_2}) - \qquad (4.2.19)$$

$$2\rho^2 \sum_{k,k_1=1}^{N} \overline{E_3(a_k - a_{k_1})} E_3(z - a_k),$$

where $E_n(z - a_k)$ has to be replaced by

$$\widetilde{E}_n(z - a_k) = E_n(z - a_k) - (z - a_k)^{-n} \text{ for } k = m \qquad (4.2.20)$$

and $E_n(0)$ is defined by (3.4.137). It is possible to proceed with (4.2.19) employed as an input and calculate the next approximations.

The above scheme is formal. Rigorous justification is based on the following

Theorem 13 ([18]). *i) Let k be fixed. The series $\sum_{m_1,m_2} (W_{m_1,m_2}^{(k)} \psi_k)(z)$ converges absolutely and uniformly in the perforated cell $\mathbb{D} \cup \partial\mathbb{D}$. It defines a function which is analytic in $\mathbb{D}$, continuous in $\mathbb{D} \cup \partial\mathbb{D}$ and doubly periodic. This function can be written in the form*

$$\sum_{m_1,m_2} (W_{m_1,m_2}^{(k)} \psi_k)(z) = \sum_{l=0}^{\infty} \overline{\psi_{lk}} r^{2(l+1)} E_{l+2}(z - a_k), \quad z \in \mathbb{D}, \qquad (4.2.21)$$

where $E_l(z)$ is the Eisenstein function of order l.
ii) The series

$$\sum_{m_1,m_2}{}' (W_{m_1,m_2}^{(k)} \psi_k)(z) := \sum_{m_1,m_2} (W_{m_1,m_2}^{(k)} \psi_k)(z) - \left(\frac{r}{z - a_k}\right)^2 \overline{\psi_k\left(\frac{r^2}{\overline{z - a_k}} + a_k\right)}$$

defines a function analytic in the unit cell Q and continuous in $Q \cup \partial Q$. This function can be written in the form

$$\sum_{m_1,m_2}{}' (W_{m_1,m_2}^{(k)} \psi_k)(z) = \sum_{l=0}^{\infty} \overline{\psi_{lk}} r^{2(l+1)} E_{l+2}(z - a_k), \qquad (4.2.22)$$

where in the last formula the modified Eisenstein function (4.2.20) is used.

iii)   The linear operator $\sum_j{}' W_{jk}\psi_k(z)$ is compact in the space $C(\mathbb{D}_k)$ of functions continuous in $\mathbb{D}_k \cup L_k$ and analytic in the disk $\mathbb{D}_k$.

The above theorem yields [3, 20] the following

Theorem 14. *Equation (4.2.17) has the unique solution in $C\left(\cup_{k=1}^N \mathbb{D}_k\right)$. It can be found by the method of successive approximations, which gives the following series*

$$\psi_m(z) = \rho \sum_{k_1=1}^N \sum_{m_1',m_2'}{}' W_{m_1',m_2'}^{(k_1)} 1(z) + \rho^2 \sum_{k_1} \sum_{k_2} \sum_{m_1',m_2'} \sum_{m_1'',m_2''}{}' W_{m_1',m_2'}^{(k_1)} W_{m_1'',m_2''}^{(k_2)} 1(z) + \cdots,$$

$$(4.2.23)$$

$$|z - a_m| \le r, \quad m = 1, 2, \cdots, N.$$

The complex flux has the form $\Psi(z) = 1 + \psi(z)$ and

$$\Psi(z) = \rho \sum_{k=1}^N \sum_{m_1,m_2}{}' (W_{m_1,m_2}^{(k)} \psi_k)(z), \quad z \in \mathbb{D} \cup \partial\mathbb{D}. \qquad (4.2.24)$$

The series (4.2.23), (4.2.24) give exact solution to the conductivity problem of 2D composites with circular inclusions. One could possibly think that these series are only formal construction like the Neumann series used in the theory of operators. However, it is not so. As we see below, though the series (4.2.23), (4.2.24) are complicated, it is possible to deduce formulae in very high orders, thus reaching much higher orders that could be reached by the traditional expansion methods (see page 56, the book [29] and works cited therein). Below we demonstrate how to work effectively with the series (4.2.23), (4.2.24).

Example 1. Consider a regular array with one circular inclusion ($N = 1$) per periodicity cell as displayed in Fig. 4.2. Then, the series (4.2.23), (4.2.24) are simplified and the local flux is given by the following formula

$$\Psi(z) = 1 + \sum_{k=1}^\infty (\rho r^2)^k \sum_{n_1,n_2,\ldots n_k} s_{n_1}^{(1)} s_{n_2}^{(n_1)} \ldots s_{n_{k-1}}^{(n_{k-2})} E_{n_k}(z) r^{4(n_1+n_2+\cdots+n_k-k)}, \qquad (4.2.25)$$

where the coefficients $s_{n_{k-1}}^{(n_{k-2})}$ are calculated by formula (3.4.133). It is worth noting that this formula presents the flux exactly.

The flux depends on the lattice through the lattice sums S_m in the coefficients $s_{n_{k-1}}^{(n_{k-2})}$ calculated with (3.4.133). The lattice sums are calculated by the fast formu-

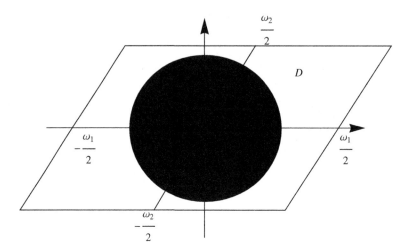

Figure 4.2 General periodic cell with one disk per cell.

lae (3.4.120)–(3.4.123).

2.2. Effective conductivity.

Substitution of (4.2.19) and high-order terms into (3.2.44) yields

$$\sigma_{11} - i\sigma_{12} = 1 + 2\rho f(1 + A_1 f + A_2 f^2 + \cdots), \qquad (4.2.26)$$

where

$$A_1 = \frac{\rho}{\pi} e_2, \quad A_2 = \frac{\rho^2}{\pi^2} e_{22}, \quad A_3 = \frac{1}{\pi^3}\left[-2\rho^2 e_{33} + \rho^3 e_{222}\right],$$

$$A_4 = \frac{1}{\pi^4}\left[3\rho^2 e_{44} - 2\rho^3(e_{332} + e_{233}) + \rho^4 e_{2222}\right],$$

$$A_5 = \frac{1}{\pi^5}\left[-4\rho^2 e_{55} + \rho^3(3e_{442} + 6e_{343} + 3e_{244}) - \right.$$
$$\left. -2\rho^4(e_{3322} + e_{2332} + e_{2233}) + \rho^5 e_{22222}\right], \qquad (4.2.27)$$

$$A_6 = \frac{1}{\pi^6}\left[5\rho^2 e_{66} - 4\rho^3(e_{255} + 3e_{354} + 3e_{453} + e_{552}) + \right.$$
$$+\rho^4(3e_{2244} + 6e_{2343} + 4e_{3333} + 3e_{2442} + 6e_{3432} + 3e_{4422}) -$$
$$\left. -2\rho^5(e_{22233} + e_{22332} + e_{23322} + e_{33222}) + \rho^6 e_{222222}\right].$$

A symbolic-numerical algorithm to compute the coefficients A_n is described in [5]. The next coefficients A_n can be written in a closed form by application of this algorithm.

Example 2. In the case of the regular array, the coefficients (4.2.27) are simplified by substitution $e_{m_1 \ldots m_q} = S_{m_1} \cdots S_{m_q}$ where the lattice sums are computed with (3.4.120)–(3.4.123). The effective conductivity can be computed directly from (4.2.25)

$$\sigma_{11} - i\sigma_{12} = 1 + 2\rho f + 2\rho^2 f^2 \frac{S_2}{\pi} +$$

$$\frac{2\rho^2 f^2}{\pi} \sum_{k=1}^{\infty} \rho^k \sum_{m_1=1}^{\infty} \sum_{m_2=1}^{\infty} \cdots \sum_{m_k=1}^{\infty} S_{m_1}^{(1)} S_{m_2}^{(m_1)} \cdots S_{m_k}^{(m_{k-1})} S_1^{(m_k)} \left(\frac{f}{\pi}\right)^{2(m_1+m_2+\cdots+m_k)-k} \tag{4.2.28}$$

Extract the terms with $m_1 = m_2 = \cdots = 1$ in the last sum. Then, (4.2.28) becomes

$$\sigma_{11} - i\sigma_{12} = 1 + 2\rho f + \frac{2\rho^2 f^2 \frac{S_2}{\pi}}{1 - \rho f \frac{S_2}{\pi}} +$$

$$\frac{2\rho^2 f^2}{\pi} \sum_{k=1}^{\infty} \rho^k \sum_{m_1,m_2\cdots m_k}' S_{m_1}^{(1)} S_{m_2}^{(m_1)} \cdots S_{m_k}^{(m_{k-1})} S_1^{(m_k)} \left(\frac{f}{\pi}\right)^{2(m_1+m_2+\cdots+m_k)-k}, \tag{4.2.29}$$

where the term with $m_1 = m_2 = \cdots = 1$ is omitted in the last sum. For the square and hexagonal arrays, $S_2 = \pi$ and the first three terms in (4.2.29) yield exactly the Clausius–Mossotti approximation. The starting terms of (4.2.29) are explicitly written for the hexagonal array by formula (7.2.6) on page 195 for $\rho = 1$ and by (7.9.73) on page 215.

Example 3. Formulae (4.2.26)-(4.2.27) up to $O((|\rho|f)^3)$ yield

$$\sigma_{11} - i\sigma_{12} = 1 + 2\rho f + 2\rho^2 f^2 \frac{e_2}{\pi} + O((|\rho|f)^3), \tag{4.2.30}$$

$$\sigma_{22} + i\sigma_{21} = 1 + 2\rho f + 2\rho^2 f^2 \left(2 - \frac{e_2}{\pi}\right) + O((|\rho|f)^3), \tag{4.2.31}$$

For a macroscopically isotropic medium $\sigma_e = \sigma_{11} = \sigma_{22}$, $\sigma_{12} = \sigma_{21} = 0$. Hence, e_2 must be equal to π, in particular, for the square and hexagonal arrays $e_2 = S_2 = \pi$. Then, (4.2.31) becomes the Clausius–Mossotti approximation

$$\sigma_e = 1 + 2\rho f + 2\rho^2 f^2 + O((|\rho|f)^3) = \frac{1 + \rho f}{1 - \rho f} + O((|\rho|f)^3). \tag{4.2.32}$$

The approximation in the right part of (4.2.32) for the square array holds up to $O((|\rho|f)^5)$ (see formula (28) from [27] for the square array of perfectly conducting inclusions when $\rho = 1$)

$$\sigma_e = \frac{1 + f}{1 - f} + 6S_4^2 \pi^{-2} \frac{f^5}{(1-f)^2} + 2(9S_4^2 + 7S_8^2)\pi^{-8} f^9 + O(f^{10}). \tag{4.2.33}$$

Formula (4.2.30) becomes (4.2.31) under the clockwise rotation of structure about 90^0. This implies that the original and rotated e-sums, e_2 and e_2^*, respectively, are related by the identity $e_2 + e_2^* = 2\pi$ which is equivalent to $S_2 + S_2^* = 2\pi$. This identity relates the results of the Eisenstein summation in (3.4.118) under this rotation.

Remark 11. Consider a general distribution of disks on the plane with well-defined concentration f. The numeration $|a_1| \leq |a_2| \leq |a_3| \leq \cdots$ is prescribed.

Let γ be a smooth curve in D_0 and to connect point z_0 with infinity. In the periodic case, the value e_2 has the form (4.1.2). For general non-periodic composites e_2 is expressed through the limit [19]

$$e_2(\gamma) = \lim_{n \to \infty} \frac{1}{n} \sum_{k=1}^{n} \sum_{m \neq k} \left[\frac{1}{(a_k - a_m)^2} - \frac{1}{a_m^2} \right] + \lim_{\gamma \ni z \to \infty} \sum_{k=1}^{\infty} \left(\frac{1}{a_k^2} - \frac{1}{(z_0 - a_k)(z - a_k)} \right).$$

(4.2.34)

This limit is not always well-defined. It was proven in [19] that both limits in (4.2.34) exist but the second limit depends on the curve γ. Therefore, the homogenization in the whole plane is impossible if the second limit changes with γ. This theoretical example is useful as a test of the non-rigorous arguments applied for deduction of a high concentration formula for the effective constants (see discussion on page 45).

Example 4. Consider the effective conductivity of macroscopically isotropic 2D composites as a function of the contrast parameter $\sigma_e(\rho)$. The quantity $\sigma_e(-\rho)$ is nothing else but the ratio of the effective conductivity to the matrix conductivity when the materials in the matrix and in the inclusions are interchanged. Let us substitute (4.2.26)-(4.2.27) into Keller's identity [13] $\sigma_e(\rho)\sigma_e(-\rho) = 1$ and extract the coefficients in every order f^m. Equations obtained for the coefficients of the low even orders f^0, f^1 and f^3, appear to be identities. The coefficients in the odd orders f^2, f^4, and f^5 yield non-trivial expressions[1]

$$e_2 = \pi, \quad e_{222} = 2\pi e_{22} - \pi^3, \quad e_{233} + e_{332} = 2\pi e_{23}.$$

(4.2.35)

These series of relations can be easily continued. Equations (4.2.35) yield the criterion of macroscopically isotropic composite with accuracy up to $O(f^6)$ [23]. It is worth noting that the standard autocorrelation analysis stops on the first-order equation (4.2.35) since it is based on the 2-point correlation functions.

Following [23] consider three different composites displayed in Fig. 4.3 with the

---

[1]It is a mathematical, unsolved problem to prove (4.2.35) not using Keller's identity based on physical considerations.

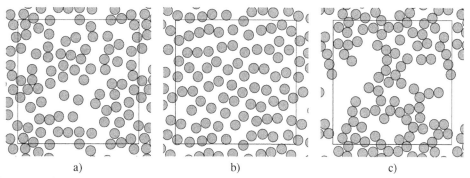

Figure 4.3 Try to guess which structure is isotropic. The answer is given via the application of basic sums.

same concentration $f = 0.4$. Let us analyse them with the accuracy $O(f^3)$. For the structure displayed in Fig. 4.3a, $e_2 = 3.13306 - 0.0942445i$, $e_{22} = 12.4064$ and $e_{222} = 46.6899 - 1.0006i$ that satisfies the first two equations of (4.2.35) with the error 3%. Hence, one may affirm that the structure in Fig. 4.3a is isotropic. Since $e_2 = 3.8454 - 0.12758i$ for the structure displayed in Fig. 4.3b, this composite is not isotropic. The coefficient of anisotropy introduced in [20] as the error is equal to 0.23. For the structure displayed in Fig. 4.3c, $e_2 = 3.14417 - 0.096571i$ that gives the error 3% in the first equation. Therefore, one may think that this structure is isotropic. The same conclusion should follow from the analysis of the 2-point correlation function. However, the values $e_{22} = 16.2179$ and $e_{222} = 65.2098 - 3.28886i$ imply that the second equation has the error 10%. This means that the composite in Fig. 4.3c is rather anisotropic.

The deterministic formula (4.2.27) is of essence to compute the effective conductivity of random composites. Following the Monte Carlo method one can take a random representative sets of $\mathbf{a} \in C$ to statistically calculate the expectations

$$\widehat{e}_{m_1 \ldots m_q} = \int_C e_{m_1 \ldots m_q}(\mathbf{a}) P(\mathbf{a}) \mathrm{d}\mathbf{a}. \tag{4.2.36}$$

The infinite set $\{\widehat{e}_{m_1 \ldots m_q}, m_j = 2, 3, \ldots\}$ completely determines the random geometric structure of the considered class of composites and can be taken as the basic set in the RVE theory presented in Section 4.

2.3. Contrast parameter expansion and Torquato–Milton parameter ζ_1

Formulae (4.2.26), (4.2.27) correspond to the expansion in concentration and can be compared with cluster expansion. Application of the explicit iterations to the functional equations (4.2.17) yields the expansion in contrast parameter ρ. One can see that the following third-order formula in ρ takes place

$$\sigma_{11} - i\sigma_{12} = 1 + 2\rho f + 2\rho^2 f^2 \frac{e_2}{\pi} + 2\rho^3 f^3 \sum_{n=2}^{\infty}(-1)^n(n-1)e_{nn}\frac{f^{n-2}}{\pi^n} + O(\rho^4). \quad (4.2.37)$$

For macroscopically isotropic composites, the expectation of (4.2.37) has the form

$$\sigma_e = 1 + 2\rho f + 2\rho^2 f^2 + 2\rho^3 f^3 \sum_{n=2}^{\infty}(-1)^n(n-1)\widehat{e}_{nn}\frac{f^{n-2}}{\pi^n} + O(\rho^4), \quad (4.2.38)$$

where the relation $\widehat{e}_2 = \pi$ from [22] is used. Another formula deduced in [22] can be useful in simulations and estimations of the third-order term of (4.2.37), (4.2.38)

$$e_{nn} = \frac{(-1)^n}{N^{n+1}} \sum_{m=1}^{N}\left|\sum_{k=1}^{N} E_n(a_m - a_k)\right|^2. \quad (4.2.39)$$

One can find the numerical values of $\widehat{e}_{nn}$ for uniform non-overlapping distributions in [5].

In order to compare the contrast expansion formula (4.2.38) and formula (3.3.59), we write (3.3.59) up to $O(\rho^4)$

$$\rho^2 f^2(\sigma_e - 1)^{-1}(\sigma_e + 1) = f\rho - A_3^{(1)}\rho^3 + O(\rho^4). \quad (4.2.40)$$

Substitution of (4.2.38) into (4.2.40) yields

$$A_3^{(1)} = f^3\left(\sum_{n=2}^{\infty}(-1)^n(n-1)\widehat{e}_{nn}\frac{f^{n-2}}{\pi^n} - 1\right). \quad (4.2.41)$$

The latter formula gives the Torquato–Milton parameter ζ_1 (see (20.59) and (20.66) from [29])

$$\zeta_1 = \frac{f^2}{1-f}\left[\sum_{n=2}^{\infty}(-1)^n(n-1)\widehat{e}_{nn}\frac{f^{n-2}}{\pi^n} - 1\right]. \quad (4.2.42)$$

The three- and four-point contrast bounds on the effective conductivity include this parameter (see (21.33)–(21.35) and (21.42)–(21.44) from [29]).

Remark 12. Formula (4.2.42) is valid for arbitrary distributed identical circular inclusions. Any other shape of inclusions can be approximated by a disk packing.

Remark 13. In accordance with (3.4.104) computation of the Torquato–Milton parameter ζ_1 for general shapes of inclusions can be reduced to a sum of the triple integrals

$$\int_{L_k} \int_{L_{k_1}} \int_{L_{k_2}} t_2 \overline{E_1(t_2 - t_1)} E_1(t_1 - t) \mathrm{d}\bar{t} \mathrm{d}t_1 \mathrm{d}\bar{t_2} \qquad (4.2.43)$$

that should be easier than computation of the corresponding triple integral (20.66) from [29] on the 3-point correlation function which in its turn is another triple integral.

3. Representative volume element

One of the most important notions for composite materials is the *representative volume element* (RVE). It is also important for the reconstruction of porous media from statistical data [1]. In the present section, we propose a mathematically rigorous and computationally effective RVE theory following [21].

Hill [11] introduced the fundamental physical postulates of the RVE as

> ... a sample that (a) structurally entirely typical of the whole mixture on average, and (b) contains sufficiently number of inclusions for the apparent overall moduli to be effectively independent of the surface values of traction and displacement, so long, as these values are 'macroscopically uniform'. That is, they fluctuate about a mean with a wavelength small compared with the dimensions of the sample, and the effects of such fluctuations become insignificant with a few wavelengths of the surface. The contribution of these surface layer to any average can be made negligible by taking the sample large enough.

One can consider Hill's postulates as the physical interpretation of the mathematical RVE theory. Application of Hill's postulates leads to the statistical approach to RVE problem when the effective constants are sought by means of simulations. The main parameters of such simulations are the number of particles per cell N and the number of Monte Carlo realizations M for each fixed N. The number N increases and computations are stopped when the scatter among the effective constants averaged over M experiments is small. Such a statistical direct approach to RVE construction was summaraized in [12, 32].

The most advanced 3D simulations can be found in [32, p. 93], for $N = 2, 4, 8, 16,$ $32, 64$ and $M = 100$. Such an approach could be classified as a statistical analysis of the computational experiments and of the corresponding physical measures. The main drawback of this approach is computational restrictions on sample sizes and impossibility to clearly recognize and separate the influence of geometry and physics onto the effective properties. Moreover, some authors present a single " random picture" for simulations, hence take $M = 1$, assuming that there is only one random structure and the one is displayed at the given picture.

Our 2D simulations based on the *e*-sums demonstrate that for the uniform non-

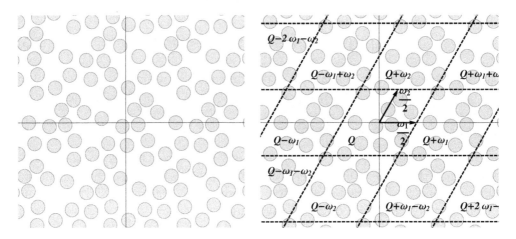

Figure 4.4 From 100 inclusions in large cell to 12 inclusions per representative cell by instant computer computations.

overlapping distribution N should be at least 81 and $M = 1500$ to get the precision 3% in the series coefficients. In the present section, we describe a fast method to construct RVE based on the e-sums. The method works almost instantly. Thus, it does not require long, time-consuming computer simulations.

First, we give a rigorous definition of the representative element following [21] and determine its minimal size, i.e., construct the RVE. The geometrical interpretation of the problem is shown in Fig. 4.4 where the large cell $Q'_{(0,0)}$ with 100 inclusions is replaced by the smaller one, $Q_{(0,0)}$, with 12 inclusions per cell. The method is developed for 2D structures and can be used for the following tasks:

- to construct an RVE as a pre-requisite for long numerical computations in a single class of structures. Principal application is for the effective properties of the considered class;

- to compare two cells or two classes of cells to detect their subtle "hidden" differences.

Remark 14. In this section, the area of the cell is not normalized to unity, since the size of cells is going to vary during determination of the appropriate minimal size cell.

It is established above that the effective conductivity tensor σ_e of two-phase dispersed composites has the form of a double series on the concentration of inclusions and on the quantified basic elements.

Definition 1. We say that two composites are equivalent within a given precision if the expansions of their σ_e have equal basic sums up to the given precision.

Therefore, we divide the set of the composites onto classes of equivalence determined only by geometrical structure of the composite. In particular, composites with the same locations of inclusions but with different contrast parameters ρ belong to the same class of equivalence. Note that composites belonging to a class of equivalence can have different σ_e; and composites from different classes can have the same σ_e. Each composite is represented by a periodicity cell.

Definition 2. Let a class of equivalence be fixed. The cell having the minimal size within the considered class is called the representative cell of the considered class.

Using the basic sums, we construct the (minimal) representative cell, i.e., we calculate its fundamental translation vectors and determine the positions of inclusions within this cell. Rigorous definition of the representative cell in turn requires well-defined notion of precision. It is given below in Definition 3.

Consider a large fundamental region $Q'_{(0,0)}$ constructed from the fundamental translation vectors ω'_1 and ω'_2. Let $Q'_{(0,0)}$ to contain N' non-overlapping disks D'_k of the radius r with the centers $a'_k \in Q_{(0,0)}$ ($k = 1, 2, \ldots, N'$). Let σ_e' be the effective conductivity tensor of the composite represented by the region $Q'_{(0,0)}$ with inclusions D'_k.

We are interested in the following problem: How to replace $Q'_{(0,0)}$ by another smaller cell $Q_{(0,0)}$ which contains inclusions $D_k = \{z \in \mathbb{C} : |z - a_k| < r\}$ ($k = 1, 2, \ldots, N$) and which has an effective conductivity tensor σ_e close to σ_e'. We assume that the concentration f of the inclusions in both materials is the same. Closeness is defined by the precision $O(f^{L+1})$ for the difference $\Delta\sigma_e = \|\sigma_e - \sigma_e'\|$ with a prescribed norm and the number L.

Definition 3. Let $Q_{(0,0)}$ and $Q'_{(0,0)}$ be representative cells within the accuracy $O(f^{L+1})$, i.e., $\Delta\sigma_e = O(f^{L+1})$. We say that $Q_{(0,0)}$ is the minimal representative cell for the region $Q'_{(0,0)}$, if $Q_{(0,0)}$ is a representative cell with minimal possible area $|Q_{(0,0)}|$.

For brevity we are not going to distinguish between the minimal representative cell and simply representative cell. The existence of the representative cell is evident, since in the worst case one can take $Q_{(0,0)} = Q'_{(0,0)}$.

It follows from (4.2.26) that the effective conductivity tensor σ'_e has the following structure

$$\sigma'_e = (1 + 2\rho f)\mathbf{I} + 2\rho f \sum_{k=1}^{\infty} \mathbf{A}'_k f^k. \tag{4.3.44}$$

For instance, $(1, 1)$-component of $\mathbf{A}'_k$ is given as follows

$$A'_k = \left| Q'_{(0,0)} \right|^k \sum_{n_1 \cdots n_p} b^{(k)}_{n_1 \ldots n_p} e_{n_1 \cdots n_p} (\omega'_1, \omega'_2) \tag{4.3.45}$$

with appropriate constants $b^{(k)}_{n_1 \ldots n_p}$. The tensor σ_e has an analogous form. $\Delta \sigma_e$ is of order $O(f^{L+1})$ if $A'_k = A_k$ for $k = 1, 2, \ldots, L - 1$. Therefore, $\Delta \sigma_e$ is of order $O(f^{L+1})$ if and only if the following RVE conditions

$$\left| Q_{(0,0)} \right|^k e_{n_1 \cdots n_p} (\omega_1, \omega_2) = \left| Q'_{(0,0)} \right|^k e_{n_1 \cdots n_p} (\omega'_1, \omega'_2) \tag{4.3.46}$$

are satisfied for $k = 1, 2, \ldots, L - 1$ and corresponding sets of the numbers $n_1, \ldots, n_p$. According to our definition $Q_{(0,0)}$ is a representative cell for the region $Q'_{(0,0)}$ within the accuracy $O(f^{L+1})$ if and only if, the relations (4.3.46) are fulfilled.

One can consider (4.3.46) as a system of equations with respect to $\omega_1, \omega_2, a_1, a_2, \ldots, a_N$ including the unknown number N with the restriction $\left| a_j - a_m \right| \leq 2r \, (j \neq m)$. One can assume that one of the centers, say a_N, lies at the origin, since geometrically any cell is determined up to a translation. The fundamental region $Q_{(0,0)}$ as well as the translation vectors ω_1, ω_2 can be chosen in infinitely many ways [2]. For any doubly periodic structure on the plane it is always possible to construct such a pair ω_1, ω_2 that $\omega_1 > 0$ and Im $\tau > 0$. Here, $\tau = \frac{\omega_2}{\omega_1}$.

The area of $Q_{(0,0)}$ is calculated through ω_1 and ω_2

$$\left| Q_{(0,0)} \right| = \omega_1^2 \, \text{Im} \, \tau. \tag{4.3.47}$$

On the other hand, we also have

$$\left| Q_{(0,0)} \right| = \frac{N \pi r^2}{f} \tag{4.3.48}$$

that yields the formula

$$\omega_1 = \sqrt{\frac{N \pi r^2}{f \, \text{Im} \, \tau}}. \tag{4.3.49}$$

In order to construct the representative cell with given accuracy $O(f^{L+1})$, we prescribe to solve the system (4.3.46) with fixed L, while increasing the number of inclusions in the cell N from 1 to N'. Then N is fixed at each step of the solution to the system (4.3.46).

In terms of τ and applying (4.3.47), we rewrite (4.3.46) in the form

$$(\text{Im} \, \tau)^k e_{n_1 \ldots n_p} (1, \tau) = \left| Q'_{(0,0)} \right|^k e_{n_1 \ldots n_p} (\omega'_1, \omega'_2), \quad k = 1, 2, \ldots, L - 1. \tag{4.3.50}$$

We can consider (4.3.50) as a system with respect to $\tau, a_1, a_2, \ldots, a_{N-1} \, (a_N = 0)$ with the restriction $|a_j - a_m| \geq 2r$ ($j \neq m$). RHS of (4.3.50) is known. If we know a solution of (4.3.50), we can calculate ω_1 from (4.3.49).

It is also possible to state the problem of finding the representative cell with prescribed form of the cell $Q_{(0,0)}$. Let us consider the case when $Q_{(0,0)}$ is a rectangle. Then $\tau = i\alpha$, where α is positive and (4.3.49) implies

$$\omega_1 = \sqrt{\frac{N\pi r^2}{\alpha f}}.$$ (4.3.51)

Equations (4.3.50) become

$$\alpha^k e_{n_1 \dots n_p}(1, i\alpha) = \left|Q'_{(0,0)}\right|^k e_{n_1 \dots n_p}(\omega'_1, \omega'_2), \quad k = 1, 2, \dots, L - 1.$$ (4.3.52)

It is hard to investigate analytically the systems (4.3.50), (4.3.52), since it is difficult to extract independent equations from the sets (4.3.50), (4.3.52). However, the numerical solution to the system is practically instant. Below several examples and solutions to (4.3.52) are considered.

Example 5. Consider case of two inclusions in a rectangular cell ($N = 2$). Then the positions of the inclusions are determined by one complex parameter $a = a_2 - a_1$. One can check that

$$e_n(1, \tau) = \begin{cases} \frac{1}{2}(S_n(1, \tau) + E_n(a; 1, \tau)) & \text{if } n \text{ is even,} \\ 0 & \text{if } n \text{ is odd,} \end{cases}$$ (4.3.53)

$$e_{mn}(1, \tau) = e_m(1, \tau)\overline{e_n(1, \tau)}, \quad e_{mnp}(1, \tau) = e_m(1, \tau)\overline{e_n(1, \tau)}e_p(1, \tau),$$ (4.3.54)

and so on. Therefore, instead of the general basic sums in (4.3.52), it is sufficient to consider equations with even sums

$$\alpha^k e_k(1, i\alpha) = \left|Q'_{(0,0)}\right|^k e_k(\omega'_1, \omega'_2), \quad k = 2, 4, \dots$$ (4.3.55)

Substitution of (4.3.53) into (4.3.55) in the case of $\tau = i\alpha$ yields

$$\alpha^k[S_k(1, i\alpha) + E_k(a; 1, i\alpha)] = p_k, \quad k = 2, 4, \dots,$$ (4.3.56)

where

$$p_k = 2\left|Q'_{(0,0)}\right|^k e_k(\omega'_1, \omega'_2)$$

are known constants. Equation (4.3.56) with $k = 2, 4$ becomes

$$\alpha^2[S_2(1, i\alpha) + E_2(a; 1, i\alpha)] = p_2, \quad \alpha^4[S_4(1, i\alpha) + E_4(a; 1, i\alpha)] = p_4.$$ (4.3.57)

Using the relations [2, 30]

$$E_4(z; 1, i\alpha) = \frac{1}{6} \frac{d^2}{dz^2} E_2(z; 1, i\alpha),$$

$$\frac{d^2}{dz^2} E_2(z; 1, i\alpha) = 6(E_2(a; 1, i\alpha) - S_2(1, i\alpha))^2 - \frac{1}{2} g_2(1, i\alpha)$$

and (3.4.72), we obtain

$$E_4(z; 1, i\alpha) = (E_2(a; 1, i\alpha) - S_2(1, i\alpha))^2 - 5S_4(1, i\alpha). \qquad (4.3.58)$$

Then the second equation (4.3.57) is transformed to the following form

$$\alpha^4[(E_2(a; 1, i\alpha) - S_2(1, i\alpha))^2 - 4S_4(1, i\alpha)] = p_4. \qquad (4.3.59)$$

We express $E_2(a; 1, i\alpha)$ from the first equation (4.3.57)

$$E_2(a; 1, i\alpha) = \frac{p_2}{\alpha^2} - S_2(1, i\alpha) \qquad (4.3.60)$$

and substitute it in (4.3.59)

$$\alpha^4\left[\left(\frac{p_2}{\alpha^2} - 2S_2(1, i\alpha)\right)^2 - 4(S_4(1, i\alpha))\right] = p_4. \qquad (4.3.61)$$

The latter equation is real, since S_2 and S_4 are real for rectangular arrays. Hence, we have obtained the real number equation (4.3.61) with respect to real unknown α.

Let us consider a numerical example with $p_2 = 10$, $p_4 = 50$, $r = 0.15$, $f = 0.3$. Equation (4.3.61) has the solution $\alpha = 0.820$. Substituting α in (4.3.60) and solving the obtained equation with respect to a we get $a = 0.331$. Then (4.3.51) implies $\omega_1 = 0.758$. Therefore, the representative cell is described by the fundamental vectors $\omega_1 = 0.934$, $\omega_2 = i0.820$ with two inclusions with the centers $a_1 = 0$ and $a_2 = 0.331$.

Example 6. Consider now three inclusion in the cell ($N = 3$). In this case the positions of the inclusions are determined by two complex parameters a_1 and a_2 ($a_3 = 0$). It follows from equation (4.3.49) that

$$\omega_1^2 = \frac{3\pi r^2}{f \operatorname{Im} \tau}. \qquad (4.3.62)$$

Consider the following equations (4.3.50)

$$(\operatorname{Im} \tau)^k e_{2k}(1, \tau) = p_{2k}, \quad k = 1, 2, 3. \qquad (4.3.63)$$

For numerical computations, it is convenient to use formulae (3.4.131) involving the Weierstrass function $\wp(z)$ and its derivatives. Then $e_{2k}(1, \tau)$ becomes

$$e_2(1, \tau) = S_2(1, \tau) + \frac{2}{9}(\wp(a_1) + \wp(a_2) + \wp(a_1 - a_2)), \qquad (4.3.64)$$

$$e_4(1,\tau) = -3S_4(1,\tau) + \frac{2}{9}\left(\wp^2(a_1) + \wp^2(a_2) + \wp^2(a_1 - a_2)\right), \qquad (4.3.65)$$

$$e_6(1,\tau) = \frac{1121}{3}S_6(1,\tau) + 6S_4(1,\tau)\left(\wp(a_1) + \wp(a_2) + \wp(a_1 - a_2)\right)$$
$$-\frac{22}{45}\left(\wp^3(a_1) + \wp^3(a_2) + \wp^3(a_1 - a_2)\right). \qquad (4.3.66)$$

Here we use the relations (3.4.131) and the following formulae from [2], [30]

$$E_4(z) = \frac{1}{6}\wp''(z) = \wp^2(z) - 5S_4(1,\tau), \qquad (4.3.67)$$

$$E_6(z) = -\frac{11}{5}\wp^3(z) + 27S_4(1,\tau)\wp(z) + 560S_6(1,\tau). \qquad (4.3.68)$$

Therefore, we arrive at three equations (4.3.63) where e_{2k} have the form (4.3.64)–(4.3.66) with respect to three unknowns a_1, a_2, and τ.

Consider a numerical example in which the large cell $Q'_{(0,0)}$ with $N' = 60$ inclusions of the radius $r = 0.12$ and the concentration $f = 0.1$ is determined by the translation vector $\omega'_1 = 4$, $\omega'_2 = 4i$. In this case $p_2 = 0.78 - 0.66i$, $p_4 = -2.15 + 2.27i$, $p_6 = -6.28 - 51i$. The cell $Q'_{(0,0)}$ is replaced by a smaller cell $Q_{(0,0)}$ with $N = 3$ inclusions. In order to find parameters of $Q_{(0,0)}$, we solve the system (4.3.63). One of the solutions has the form $a_1 = -0.92$, $a_2 = -0.36 + 0.36i$, $\tau = 0.06 + 0.39i$. Then (4.3.49) yields $\omega_1 = 1.08$.

In **summary** a rigorous constructive theory of the RVE in periodic composites is proposed. It is applicable to the conductivity of a two-dimensional, two-component composite materials made from a collection of non-overlapping, identical circular disks, arbitrary embedded in a matrix. The definition of the RVE is based on the representation of the effective conductivity tensor in terms of the basic sums. We say that the cells $Q_{(0,0)}$ and $Q'_{(0,0)}$ are equivalent if they have the same basic elements in the representation (4.3.44)–(4.3.45). Thus, all composites are divided onto classes of equivalence. The minimal size cell in each class is called the RVE. The basic elements of (4.3.44), (4.3.45) are expressed in terms of the basic sums.

The problem of determination of the RVE is reduced to the finite system of equation (4.3.50) with respect to its parameters. We would like to make some remarks about the general system (4.3.50). First, it is evident that it has infinite number of solutions, since any doubly periodic structure is determined by an infinite number of the pairs of the fundamental translation vectors. Moreover, as it follows from relations (4.3.53), (4.3.54), the system (4.3.50) can contain redundant equations.

Finally we briefly explain how the method can be developed and applied to other problems from the theory of structured media. First, let us consider the same problem but with inclusions having the shape S different from circular. Any plane domain S can be approximated by disk (of radius r) packings. This approximation can be expressed by appropriate conditions on the centers of the disks $b_1, b_2,..., b_P$. We write them in the form of constrains on b_j

$$b_j - b_1 = B_j e^{i\psi}, \quad j = 3, 4, ..., P, \quad |B_2| = |b_2 - b_1|. \quad (4.3.69)$$

Here, the constants $B_2, B_3,..., B_P$ are given, $\psi = \arg \frac{B_2}{b_2 - b_1}$. The constrains (4.3.69) mean that the points $b_1, b_2,..., b_P$ are tied and may only translate and rotate as a rigid body. We can replace all inclusions (say M inclusions per cell) by a set of points $a_1, a_2,..., a_N$ divided onto M subsets each of them contains P points. We assume that points of each subset satisfy the constrains (4.3.69), i.e., each subset of the disks approximates an inclusion of the form S. These constrains on $a_1, a_2,..., a_N$ should be added to equations (4.3.50) in order to obtain a system of equations corresponding to the representative cell with M inclusions of the shape S.

Remark 15. The proposed method could be also applied to elastic problems. Generally speaking, one should complement the series already expressed through the classical Eisenstein functions with some other series presented on page 295. However it turns that the latter series could be expressed in terms of the classic Eisenstein series using the algebraic equations established in [8]. Thus, a direct application of the method presented in this section to elastic composites holds.

Remark 16. The method was extended to a 3D composite with spherical inclusions in [28] by investigation of the 2D cross sections obtained by cutting through the 3D composite.

Systematic applications of the RVE method in the material science can be found in [14, 15]. The RVE for real composites obtained through friction stir process was computed from typical experimental sample shown in Fig. 4.5.

4. Method of Rayleigh

The first approximate solution solution to the problem (3.2.10)–(3.2.11) in a class of doubly periodic functions was obtained in 1892 by Lord Rayleigh [26]. The problem was reduced to an infinite system of linear algebraic equations. It was the first mathematical treatment of the boundary value problem on the Riemann surface. Rayleigh's approach was further elaborated by McPhedran with coworkers [16, 17, 24, 25]. They developed the method of Lord Rayleigh and applied it to various problems of the theory of composites. The papers [26] and [10, 16, 17, 24, 25, 31] contain detailed description of the method and an exhaustive review of other works. Perhaps only the works [6, 7, 8] should be added for completeness. The method of Rayleigh was extended there to a 2D elastic doubly periodic problems with the necessary mathematical justification, but without reference to Rayleigh (see the review [9]).

It is worthwhile to mention the famous results by Eisenstein (1848) outlined in [30] (see page 71 of the present book devoted to Eisenstein's series). Rayleigh (1892) did not refer to Eisenstein (1847) and credited Weierstrass (1856). Perhaps it happened because Eisenstein treated his series formally, without considering their uniform convergence introduced by Weierstrass. It is curious that Rayleigh used the Eisenstein summation and proved the fascinating formula $S_2 = \pi$ for the square array (see [31]).

The method of Rayleigh is closely related to the method of functional equations applied in the present book. Roughly speaking, Rayleigh's infinite system is a discrete form of the functional equations. We obtain below Rayleigh's system following [27] starting from the equations (4.2.17) for $N = 1$

$$\psi_1(z) = \rho \sum_{m_1,m_2\in\mathbb{Z}}{}' \frac{r^2}{(z - \omega_1 m_1 - \omega_2 m_2)^2} \overline{\psi_1\left(\frac{r^2}{z - \omega_1 m_1 - \omega_2 m_2}\right)} + 1, \ |z| \le r,$$
(4.4.70)

where m_1, m_2 run over integers in the sum $\sum^*$ with the excluded term $m_1 = m_2 = 0$ for $m = k$. Let us look for a function $\psi_1(z)$ in the form of the Taylor expansion

$$\psi_1(z) = \sum_{m=0}^{\infty} \alpha_m z^m, \ |z| \le r,$$
(4.4.71)

Substituting this expansion in (4.4.70), we obtain

$$\sum_{m=0}^{\infty} \alpha_m z^m = \rho \sum_{m=0}^{\infty} r^{2(m+1)} \overline{\alpha_m} \sum_{m_1,m_2\in\mathbb{Z}}{}' \frac{1}{(z - \omega_1 m_1 - \omega_2 m_2)^{m+2}} + 1, \ |z| \le r, \quad (4.4.72)$$

where the Eisenstein summation is used (see page 71). Applying the expansion (3.4.132) and selecting the coefficients in the same powers of z, we arrive at the infinite $\mathbb{R}$-linear algebraic system

$$\alpha_l = \rho \sum_{m=0}^{\infty} (-1)^m \frac{(l + m + 1)!}{l!(m + 1)!} S_{l+m+2} r^{2(m+1)} \overline{\alpha_m} + 1, \ |z| \le r,$$
(4.4.73)

where S_{l+m+2} are the Eisenstein–Rayleigh sums. The real part of (4.4.73) is Rayleigh's system.

The infinite system (4.4.73) is truncated to get approximate formulae for the coefficient $\alpha_0 = \psi_1(0)$, needed to calculate the effective conductivity by (3.2.44) for $N = 1$

$$\sigma_{11} - i\sigma_{12} = 1 + 2\rho f \ \psi_1(0).$$
(4.4.74)

Various analytical and numerical formulae were deduced. For instance, for regular arrays formulae (4.4.73) are asymptotically equivalent to the truncated series (4.2.28). In

Chapter 6, we deduce higher-order 2D formulae by the method of functional equations for circular inclusions which include the known 2D formula obtained by the method of Rayleigh.

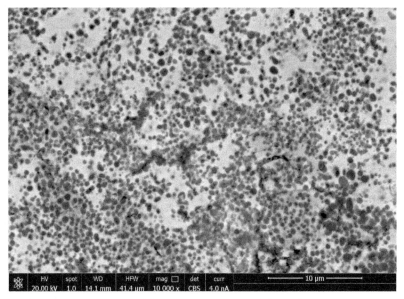

Figure 4.5 Typical microstructure of the composite zone, TiC–FeCr type, obtained in situ from steel casting. The figure was kindly supplied by Ewa Olejnik and Paweł Kurtyka.

REFERENCE

1. Adler PM, Thovert JF, Mourzenko VV, Fractured porous media. 2rd Oxford: Oxford University Press; 2012.
2. Akhiezer NI, Elements of the Theory of Elliptic Functions. RI, American Mathematical Society, Providence, 1990.
3. Berlyand L, Mityushev V, Generalized Clauisius-Mossotti Formula for Random Composite with Circular Fibers, J. Stat. Phys. 2001; 102: 115–145.
4. Berlyand L, Mityushev V, Increase and decrease of the effective conductivity of a two phase composites due to polydispersity, J. Stat. Phys. 2005; 118: 481–509.
5. Czapla R, Nawalaniec W, Mityushev V, Effective conductivity of random two-dimensional composites with circular non-overlapping inclusions, Computational Materials Science 2012; 63: 118–126.
6. Chibrikova LI, Fundamental boundary value problems for the analytic functions. Kazan Univ. Publ., Kazan, 1977 [in Russian].
7. Grigolyuk EI, Filshtinsky LA, Perforated Plates and Shells. Moscow: Nauka; 1970 [in Russian].
8. Grigolyuk EI, Filshtinsky LA, Periodical Piece–Homogeneous Elastic Structures. Moscow: Nauka; 1991 [in Russian].
9. Filshtinsky L, Mityushev V, Mathematical models of elastic and piezoelectric fields in two-dimensional composites, ed. P.M. Pardalos , T.M. Rassias VIII, 648 p., Mathematics Without Boundaries. Surveys in Interdisciplinary Research, 217–262, 2014.

10. Godin YuA, Effective complex permittivity tensor of a periodic array of cylinders, Journal of Mathematical Physics 2013; 54: 053505-1.
11. Hill R, Elastic properties of reinforced solids: some theoretical principles, Journal of the Mechanics and Physics of Solids 1963; 11: 357–372. doi:10.1016/0022-5096(63)90036-x.
12. Kanit T, Forest S, Galliet I, Mounoury V, Jeulin D, Determination of the size of the representative volume element for random composites: statistical and numerical approach, Int. J. Solids and Structures, 2003; 40: 3647–3679.
13. Keller JB, Conductivity of a Medium Containing a Dense Array of Perfectly Conducting Spheres or Cylinders or Nonconducting Cylinders, Journal of Applied Physics 1963; 34: 991–993.
14. Kurtyka P, Rylko N, Structure analysis of the modified cast metal matrix composites by use of the RVE theory, Archives of Metallurgy and Materials, 2013; 58: 357–360.
15. Kurtyka P, Rylko N, Tokarski T, Wojicicka A, Pietras A, Cast aluminium matrix composite modified with using FSP process a "Changing of the structure and mechanical properties", Composite Structure 2015; 133 (C): 959–967.
16. McKenzie DR, McPhedran RC, Derrick GH, The conductivity of lattices of spheres II. The body centred and face centred cubic lattices, Proceedings of the Royal Society of London A 1978; 362: 211–232.
17. McPhedran RC, McKenzie DR, The conductivity of lattices of spheres I. The simple cubic lattice, Proceedings of the Royal Society of London A 1978; 359: 45–63.
18. Mityushev V, A functional equation in a class of analytic functions and composite materials, Demostration Math. 1997; 30: 63–70.
19. Mityushev V, Transport properties of two-dimensional composite materials with circular inclusions, Proc. R. Soc. London 1999; A455: 2513–2528.
20. Mityushev V, Transport properties of doubly periodic arrays of circular cylinders and optimal design problems, Appl. Math. Optimization, 2001; 44: 17–31.
21. Mityushev V, Representative cell in mechanics of composites and generalized Eisenstein–Rayleigh sums, Complex Variables 2006; 51: 1033–1045.
22. Mityushev V, Rylko N, Optimal distribution of the non-overlapping conducting disks, Multiscale Model. Simul. 2012; 10: 180–190.
23. Mityushev V, Nawalaniec W, Basic sums and their random dynamic changes in description of microstructure of 2D composites, Comput. Mater. Sci. 2015; 97: 64–74.
24. Poladian L, McPhedran RC, Effective Transport Properties of Periodic Composite Materials, Proc. R. Soc. Lond. A 1986; 408: 45–59.
25. Poladian L, Effective Transport Properties of Periodic Composite Materials Constants of Composite Materials, Proc. R. Soc. Lond. A 1989; 426: 343–359.
26. Rayleigh Lord, On the influence of obstacles arranged in rectangular order upon the properties of medium. Phil. Mag. 1892; 34, 481–502.
27. Rylko N, Transport properties of the regular array of highly conducting cylinders, J Engrg Math 2000; 38: 1–12.
28. Rylko N, Representative volume element in 2D for disks and in 3D for balls, Journal of Mechanics of Materials and Structures 2014; 9: 427–439.
29. Torquato S, Random Heterogeneous Materials: Microstructure and Macroscopic Properties. New York. Springer-Verlag: (2002).
30. Weil A, Elliptic Functions according to Eisenstein and Kronecker. Springer-Verlag, Berlin etc, 1999.
31. Yakubovich S, Drygas P, Mityushev V, Closed-form evaluation of 2D static lattice sums, Proc Roy. Soc. London 2016; A472, 2195: 20160662. DOI: 10.1098/rspa.2016.0510.
32. Zohdi TI, Wriggers P, An introduction to Computational mechanics. Berlin etc: Springer-Verlag; 2008. [Chapter 6].

CHAPTER 5

Introduction to the method of self-similar approximants

In Reimann, Hilbert or in Banach space
Let superscripts and subscripts go their ways.
Our asymptotes no longer out of phase,
We shall encounter, counting, face to face.

—Stanislaw Lem, The Cyberiad

1. Brief introduction to extrapolation

There exists a very old problem of extrapolating asymptotic perturbation-theory expansions in powers of a small variable to large values of the variable tending to infinity. It is constantly met in various aspects of applied mathematics. Very often realistic problems are so much complicated that do not allow for exact solutions. The natural solution is the use for such problems of some kind of perturbation theory [12, 46]. The answers are given in terms of expansions in powers of a small positive parameter, say x, as $x \to 0$.

However, the problem of interest corresponds not to a small variable, but to large values of this variable; very often it is the infinite limit, as $x \to \infty$ that is of the most interest [36]. One could find this limit, provided the general formula of expansion terms would be given and the derived expansion would produce convergent series. None of these conditions is usually valid. As a rule, only several expansion terms can be derived. And the resulting series are divergent, being only asymptotic [15, 31]. Then the question arises: how from the knowledge of several terms of an asymptotic expansion as $x \to 0$ one could find the limit corresponding to $x \to \infty$? The variable $x > 0$ can represent, e.g., a coupling constant or concentration properly transformed.

One often extrapolates small-variable expansions by means of Padé approximants $P_{M,N}(x)$ (see Section A.5.7 below based on [5]). However, the straightforward use of these approximants yields

$$P_{M,N}(x) \sim x^{M-N} \qquad (x \to \infty),$$

which, depending on the relation between M and N, can tend to either infinity (when $M > N$), to zero (when $M < N$), or to a constant (if $M = N$). In that sense, the limit $x \to \infty$ is ill-defined. In what follows, in order to avoid the uncertainty, we would treat the asymptotic behavior by means of different self-similar approximants. Then, only

Computational Analysis of Structured Media
http://dx.doi.org/10.1016/B978-0-12-811046-1.50005-8

diagonal Padé approximants remain to be employed.

BOX A.5.7 Padé approximants

Box 1 "One-point" Padé

Padé approximant $P_{M,N}$ to an analytical function $\Phi(x)$ at a regular point $x_0 = 0$ is a rational function

$$P_{M,N} = \frac{p(x)}{q(x)},$$

the ratio of two polynomials. Namely,

$$p(x) = p_0 + p_1 x + \cdots + p_M x^M$$

is of the degree M and

$$q(x) = 1 + q_1 x + \cdots + q_N x^N$$

is of the degree N. The first $N + M + 1$ terms in the Taylor expansion of $P_{M,N}(x)$ match the first $N + M + 1$ terms of the Taylor series of $\Phi(x)$. If $M = N$ the approximant is called a diagonal Padé approximant of order N.

Box 2 "Two-point" Padé

In addition to the expansion about $x_0 = 0$,

$$\Phi(x) \sim \sum_{n=0}^{\infty} a_n x^n,$$

an additional information is available and contained in the expansion about $x = \infty$,

$$\Phi(x) \sim \sum_{n=0}^{\infty} b_n x^{-n}. \tag{5.1.1}$$

A two-point Padé approximant to $\Phi(x)$ is a rational function

$$F(x) = \frac{P_M(x)}{Q_N(x)}, \tag{5.1.2}$$

and $Q_N(0) = 1$. The polynomials of degrees M and N have their coefficients chosen arbitrarily to make the first J terms of the Taylor series of $F(x)$ around $x = 0$ agree with the expansion (5.1.1), and the first L terms of the expansion of $F(x)$ about $x = \infty$ to agree with the given expansion (5.1.2), so that also $J + L = N + M + 1$.

[5, 8]

When the character of the large-variable limit is known, one can invoke the two-point Padé approximants [5]. However the accuracy of the latter is not high and confronts several difficulties. First, when constructing these approximants, one often obtains spurious poles yielding unphysical singularities [5], sometimes a large number of poles [48]. Second, there are the cases when Padé approximants are not able to sum perturbation series even for small values of an expansion parameter [51]. Third, in the majority of cases, to reach a reasonable accuracy, one needs to have tens of terms in the expansions [5], while often interesting problems provide only a few terms. Fourth, defining the two-point Padé approximants, one always confronts the ambiguity in distributing the coefficients for deciding which of these must reproduce the left-side expansion and which the right-side series. This ambiguity aggravates with the increase of approximants orders, making it difficult to compose two-point Padé tables. For the case of a few terms, this ambiguity makes the two-point Padé approximants practically very difficult to use. For example, it has been shown [49] that, for the same problem, one may construct different two-point Padé approximants, all having correct left and right-side limits, but differing from each other in the intermediate region by a factor of 40, which gives 1000% uncertainty. This demonstrates that in the case of short series the two-point Padé approximants do not allow for getting a reliable description. Fifth, the two-point Padé approximants can be used for interpolating between two different expansions not always, but only when these two expansions have compatible variables [5].

The question of compatibility of expansions in the small-variable and large-variable limits should not be ignored. The standard situation in many problems is when, in the small-variable limit, one has an expansion in integer powers, x^n, while the large-variable expansion exhibits the behaviour x^β, with a non-integer power β. Since the large-variable behaviour of a Padé approximant $P_{M,N}$ is x^{M-N}, this implies that the integer power $M - N$ is not compatible with the non-integer β. To overcome the problem of incompatibility, Baker and Gammel [4] suggested to use the fractional powers of Padé approximants $P^\gamma_{M,N}$, choosing the power γ so that $(M - N)\gamma = \beta$. The simplest case of the Baker–Gammel method is the polynomial approximant $P^\gamma_{M,0}$ in a fractional power $\gamma = \frac{\beta}{M}$. The Baker–Gammel method allows one to correctly represent the leading term of the large-variable behaviour, although the sub-leading terms not always can be uniquely defined [32, 50].

Another method that allows for the extrapolation of divergent series is the optimized perturbation theory, based on the introduction of control functions defined by an optimization condition and guaranteeing the transformation of divergent series into convergent series [55, 56]. There have appeared a number of variants of introducing control functions. It is worth mentioning the variational perturbation theory [36, 37], where control functions are introduced through a variable transformation and variational optimization conditions. This method provides good accuracy for the extrapolation of weak-coupling expansions to the strong-coupling limit, especially when a number of terms in the weak-coupling perturbation theory are available [33].

We address the problem of extrapolation by advancing another approach, method of self-similar approximants based on the self-similar approximation theory [57, 58,

59, 60, 68, 69, 70]. The main difference of this approach from the optimized perturbation theory is that such relatively new approximation methods can be used without
introducing control functions, which makes calculations essentially simpler.

There exists another principal problem, when one accomplishes an extrapolation
in the case for which the exact solution is not known and only a few terms of weak-
coupling perturbation theory are available. This is the problem of the reliability of
the obtained extrapolation. In such a case, it is important to be able to make the
extrapolation by several methods, with comparing their results. If these results yield
close values, this suggests that the extrapolation is reliable.

In line with this idea, different variants of self-similar approximations are applied to
the same problems. If the approximants for a problem, obtained by different methods,
are close to each other, this would suggest that the derived values are reliable.

Thus, it is prudent to apply various self-similar approximants for each problem and
show that they are really close to each other, hence they can successfully extrapolate
asymptotic expansions, valid as $x \to 0$, to their effective limits as $x \to \infty$. It makes
sense to concentrate on the more difficult strong-coupling limit, where approximate
methods usually are the least accurate, leading to the maximal errors. Even in this
least favorable situation, the self-similar extrapolation methods provide reasonable
accuracy, with just a few perturbative terms available. Generally speaking, the self-
similar methods allow to construct the approximants displaying good accuracy in the
whole domain $[0, \infty]$. For instance, effective equations of state can be derived, being
in good agreement with experimental data.

2. Algebraic renormalization and self-similar bootstrap

The main steps leading to the notion of self-similar approximants are presented following [20, 21, 64]. Without loss of generality one may assume that the critical point
is located at infinity, where it can be mapped, if necessary, by a suitable transformation. Suppose we are interested in the behaviour of a real function $\Phi(x)$ of a real
variable $x \in [0, \infty)$. Let the function be defined by a complicated problem that does
not allow for an explicit derivation of $\Phi(x)$. In order to proceed assume a perturbative
expansions representing function Φ for small $x > 0$ asymptotically as follows

$$\Phi(x) \simeq \Phi_k(x), \quad \text{as } x \to 0 \quad (k = 0, 1, \ldots), \tag{5.2.3}$$

with k being the expansion order. For example, expanding in integral powers of x
obtains

$$\Phi_k(x) = \sum_{n=0}^{k} a_n x^n. \tag{5.2.4}$$

Assume further that at infinity the function has asymptotic form

$$\Phi(x) \simeq Ax^\beta, \quad \text{as } x \to \infty, \tag{5.2.5}$$

where A is called the critical amplitude, β is the critical exponent (index). For definiteness, we consider the case of $\beta > 0$. With very little exception the approximants and methods of their application could be extended to the case of $\beta < 0$ without any modifications. The case of $\beta = 0$ should be treated with any other approximant when it is possible to express non-trivially the latter condition through the parameters of approximant.

Let us employ the self-similar renormalization for the reduced function (5.2.3) and obtain the self-similar approximant $\Phi^*(x)$ for the sought function. For the latter, consider the limit $x \to \infty$ and find the related approximation for the critical amplitude and critical index. When β is known from other arguments only A has to be found.

Now let us look for such a procedure that will allow for the reconstruction of the whole sequence of $\{\Phi_k(x)\}$ based on its few starting terms in such a way that its convergence properties will improve. To this end apply to an asymptotic series a transformation in order to construct an analytic approach which satisfy the following three conditions:

(1) Be general, i.e., applicable to any function without requiring knowledge of its specific properties, only assuming existence of the sought function;

(2) Be simple to permit analytic investigation;

(3) Be invertible, i.e., to possess a uniquely defined inverse transform in order to return from the transform to the function itself.

These requirements are satisfied by the algebraic transformation

$$S_c\{\Phi(x)\} = a(x, c) + b(x, c)\Phi(x), \tag{5.2.6}$$

which obviously possesses an inverse. Here c stands for a set of parameters such that the transformed system has better convergence properties when applied to $\{p_k(x)\}$. One of the most straightforward and simple variants as applied to a term of $\Phi_k(x)$ of the sequence is

$$S_c\{\Phi_k(x)\} = F_k(x, c) = x^c \Phi_k(x), \tag{5.2.7}$$

with an inverse

$$\Phi_k(x) = F_k(x, c)x^{-c}. \tag{5.2.8}$$

Such transformation was chosen because it has a transparent meaning when $\Phi_k(x)$ is a k-order truncated power series in powers of x. Then it effectively increases the approximation order from k to $k + c$.

Let us understand in what sense the properties of $\{F_k\}$-sequence should be better than that of $\{\Phi_k\}$. The greatest achievement would be if the transformed sequence is such that one could notice a relation between subsequent terms F_k and F_{k+1}. If so it would be feasible to map the low-order terms to those of arbitrary high-order. That is having just a few initial terms of $\{F_k\}$-sequence it becomes possible to extrapolate them to higher orders of k defining an effective limit F^* of this sequence. To this end, let us invoke some further transformations.

Define the expansion function $x(\phi, c)$ by the equations

$$F_0(x, c) \equiv \phi, \quad x = x(\phi, c), \tag{5.2.9}$$

where F_0 is the first available approximation, ϕ is the new variable. Substitute $x(\phi, c)$ back to F_k and get a sequence bijective to $\{F_k(x, c)\}$,

$$y_k(\phi, c) \equiv F_k(x(\phi, c), c), \tag{5.2.10}$$

(inverse transformation still exists) with an initial condition

$$y_0(\phi, c) = \phi. \tag{5.2.11}$$

Consider the family $\{y_k\}$ as a dynamical system in discrete time where the role of time is played by the approximation number, and call it an approximation cascade. Embed the cascade into a continuous sequence $\{y(\tau, \phi, c)\}$ with $\tau \in [0, \infty]$, implying that

$$y(k, \phi, c) = y_k(\phi, c), \quad k = 0, 1, 2.... \tag{5.2.12}$$

In other words, $\{y(\tau, \phi, c)\}$ composes by definition a dynamical system with continuous time, whose trajectory passes through all points of the approximation cascade. It is called an approximation flow.

Evolution of the flow is given by the **self-similarity relation**

$$y(\tau + \tau', \phi, c) = y(\tau, y(\tau', \phi, c), c), \tag{5.2.13}$$

common for any autonomous dynamical system and it reflects a group property of motion (i.e., the function conserves its form under a change of variables).

Since the property of self-similarity is the central concept of our approach, let us remind where this notion is originated from. Recall that this property is common for any autonomous dynamical system, reflecting the group property of motion [9]. Such relations appear in the renormalization group approach of quantum field theory [13]. The concept of self-similarity is widely employed in the context of fractals. In all the cases, self-similarity is the group property of a function conserving its form under the change of its variable. In particular cases, this variable can be time, as for dynamical systems, momentum, as for field theory, or space scale, as for fractals. In our case, such a variable is the approximation number, playing the role of time, and the motion occurs in the space of approximations, where self-similarity is also a necessary condition for fastest convergence.

The evolution equation (5.2.13) can be rewritten in the differential form

$$\frac{\partial y(t, \phi, c)}{\partial t} = v(y(t, \phi, c), c),$$ (5.2.14)

where the velocity field is given as follows,

$$v(y(t, \phi, c), c) = \lim_{\phi \to y(t,\phi,c)} \lim_{\tau \to 0} \frac{\partial y(\tau, \phi, c)}{\partial \tau}.$$ (5.2.15)

Equation (5.2.14) can be integrated over "time" between t and some $t + \tau$,

$$\int_{y(t,\phi,c)}^{y(t+\tau,\phi,c)} \frac{dy'}{v(y', c)} = \tau.$$ (5.2.16)

We are interested in the effective limit of the sequence $\{F_k(x, c)\}$ bijective to the trajectory of the cascade $\{y_k(\phi, c)\}$ and the limit of $\{F_k\}$ is one-to-one correspondence with a stable fixed point of the cascade. A fixed point is defined as a zero of velocity. The cascade velocity

$$v_k(\phi, c) = v(y(k, \phi), c), c)$$ (5.2.17)

can be written as an Euler discrete expression for the flow velocity

$$v_k(\phi, c) = y_{k+1}(\phi, c) - y_k(\phi, c).$$ (5.2.18)

Since v_k is non-zero for finite k one can not find an exact zero of velocity. One cannot also take the

$$\lim_{k \to \infty} y_k(\phi, c) = y^*(\phi, c)$$ (5.2.19)

explicitly, since an expression for $y_k(\phi, c)$ for arbitrary k is not available. All one can do is to find some approximate fixed point, or quasi-fixed point.

After explicit integration in (5.2.16) with velocity given by (5.2.18), after some yet undetermined number of steps τ, the quasi-fixed point $F_{k+1}^*(x, c, \tau)$ can be reached satisfying

$$\int_{F_k(x,c)}^{F_{k+1}^*} \frac{dy'}{v_k(y', c)} = \tau.$$ (5.2.20)

As long as the cascade velocity is not exactly zero, the trajectory of the cascade does not stop at F_k and continues up to the approximate fixed point. The goal thus is to integrate (5.2.20) explicitly and express F_{k+1}^* as a function of x, c and τ, where c comes from the algebraic transformation and τ is the time necessary to reach the quasi-fixed point. Both are unknown parameters to be determined at the last stage of the procedure from the asymptotic conditions. After expressing F_{k+1}^* an inverse transform should be taken to obtain the self-similar approximant

$$\Phi_k^*(x, c, \tau) = x^{-c} F_k^*(x, c, \tau). \tag{5.2.21}$$

Let us obtain explicit expressions for the case of an expansion in integer powers

$$\Phi_k(x) = \sum_{n=0}^{k} a_n x^n. \tag{5.2.22}$$

An algebraic transform leads to

$$F_k(x, c) = \sum_{n=0}^{k} a_n x^{n+c}, \tag{5.2.23}$$

while the expansion function

$$F_0(x.c) = a_0 x^c \equiv \phi, \tag{5.2.24}$$

with

$$x = \left(\frac{\phi}{a_0}\right)^{1/c}. \tag{5.2.25}$$

Mind that the name *algebraic self-similar renormalization* method has been used to refer to choice wherein the control functions (parameters) are introduced in the exponents of perturbative polynomials [20].

For the approximation cascade the following representation is valid,

$$y_k(\phi, c) = \sum_{n=0}^{k} a_n \left(\frac{\phi}{a_0}\right)^{1+n/c}. \tag{5.2.26}$$

The cascade velocity equals

$$v_k(\phi, c) = a_{k+1} \left(\frac{\phi}{a_0}\right)^{1+(k+1)/c}. \tag{5.2.27}$$

After integrating (5.2.20) and applying the inverse algebraic transform, an explicit generic formula for the quasi-fixed point is obtained as follows,

$$\Phi_k^*(x, c, \tau) = \left[(\Phi_{k-1}(x))^{-k/c} - \frac{k a_k \tau}{c a_0^{1+k/c}} x^k \right]^{-c/k}. \tag{5.2.28}$$

One should not stop at this stage and apply same formula (5.2.28), but to $\Phi_{k-1}(x)$, transforming it into Φ_{k-1}^*, so that

$$\Phi_k^* = [(\Phi_{k-1}^*)^{1/n_k} + B_k x^k]^{n_k}, \quad n_k = -\frac{c_k}{k}, \quad B_k = \frac{a_k \tau_k}{n_k a_0^{1-1/n_k}}. \tag{5.2.29}$$

In the same way

$$\Phi_{k-1}^* = \left[(\Phi_{k-2}^*)^{\frac{1}{n_{k-1}}} + B_{k-1} x^{k-1} \right]^{n_{k-1}},$$ (5.2.30)

and so on down to

$$\Phi_2^* = \left[(\Phi_1^*)^{1/n_2} + B_2 x^2 \right]^{n_2}, \quad \Phi_1^* = \left[(\Phi_0)^{1/n_1} + B_1 x \right]^{n_1}, \quad \Phi_0 \equiv a_0.$$ (5.2.31)

The structure is the sequence of nested roots or self-similar root approximants is clear from observing the third-order root approximant,

$$\Phi_3^* = \left(\left(\left((\Phi_0)^{1/n_1} + B_1 x \right)^{n_1/n_2} + B_2 x^2 \right)^{n_2/n_3} + B_3 x^3 \right)^{n_3}.$$ (5.2.32)

The whole procedure leading to a sequence of nested roots (or self-similar roots) amounts to the so-called self-similar bootstrap. The k-order self-similar root approximant Φ_k^* depends on a set of $2k$ parameters

$$c_k = \{c_1, c_2, ...c_k\}, \quad \tau_k = \{\tau_1, \tau_2, ...\tau_k\}$$ (5.2.33)

to be obtained from the asymptotic conditions (5.2.4) and (5.2.5).

For example, when the critical index β is known, in order to calculate the critical amplitude A one needs to find the limit

$$\lim_{x \to \infty} \Phi_k^*(x, c, \tau) x^{-\beta} = A_k, \quad k = 0, 1...,$$ (5.2.34)

providing k-approximations for A.

Even after all these efforts leading to the algorithm described above, there are plenty of practical problems arising, requiring the development of the particular self-similar approximants, methods of their application and transformations of variables for sake of adaptability to certain physical situations. In what follows several most useful realizations of self-similar approximants are presented.

3. Extrapolation problem and self-similar approximants

The nested or simply self-similar roots as well as other self-similar approximants discussed below, can be applied directly do the typical extrapolation problem.

Suppose we are interested in the behaviour of a real function $\Phi(x)$ of a real variable $x \in [0, \infty)$. And let this function be defined by a complicated problem that does not allow for an explicit derivation of the function form. But what can be done is only the use of some kind of perturbation theory yielding asymptotic expansions representing the function

$$\Phi(x) \simeq \Phi_k(x)$$ (5.3.35)

as $x \to 0$, with $k = 0, 1, \ldots$ being perturbation order. The perturbative series of kth order can be written as an expansion in powers of x as

$$\Phi_k(x) = \Phi_0(x)\left(1 + \sum_{n=1}^{k} a_n x^n\right), \qquad (5.3.36)$$

where $\Phi_0(x)$ is chosen so that the series in the brackets would start with the term one. It is convenient to define the reduced expression

$$\overline{\Phi}_k(x) \equiv \frac{\Phi_k(x)}{\Phi_0(x)} = 1 + \sum_{n=1}^{k} a_n x^n, \qquad (5.3.37)$$

which will be subject to self-similar renormalization.

Note that practically any perturbative series can be represented in form (5.3.36). For instance, if there is a Laurent-type series

$$\Phi_{m+k}(x) = \sum_{n=-m}^{k} c_n x^n,$$

it can be transformed to (2.2) by rewriting it as

$$\Phi_{m+k}(x) = \frac{c_{-m}}{x^m}\left(1 + \sum_{n=1}^{m+k} a_n x^n\right).$$

Here consider the series in integer powers, or those that can be reduced to such, since this is the most frequent type of perturbation-theory expansions. Thus, the Puiseaux expansions [47] of the type

$$\Phi_k(t) = \sum_{n=n_0}^{k} c_n t^{n/m},$$

where n_0 is an integer and m is a non-zero natural number, can be reduced to form (5.3.36) by the change of the variable $t = x^m$. It is possible to generalize the approach to the series of the type

$$\Phi_k(x) = \sum_{n}^{k} c_n x^{\alpha_n} \qquad (\alpha_n < \alpha_{n+1}) \qquad (5.3.38)$$

with arbitrary real powers α_n arranged in an ascending order. When α_n pertains to an ordered group, the latter expression corresponds to the Hahn series [35, 41].

The most difficult region for approximating is that of the large variable, where approximants are usually the least accurate. This is why our main interest is to find the large-variable behaviour of the function, where its asymptotic form is

$$\Phi(x) \simeq A x^\beta \qquad \text{as } x \to \infty. \qquad (5.3.39)$$

Mind that the constant A will be called the critical amplitude, while the power β is the critical exponent.

After employing the self-similar renormalization for the reduced function (5.3.37), a self-similar approximant $\bar{\Phi}_k^*(x)$ transpires, which gives a self-similar approximant

$$\Phi_k^*(x) = \Phi_0(x)\overline{\Phi}_k^*(x) \tag{5.3.40}$$

for the sought function $\Phi(x)$. Considering for the latter the limit as $x \to \infty$, one finds the related approximation for the critical amplitude and critical exponent. In many cases, the exponent is known from other arguments. Then only the critical amplitude remains unknown, and the most often considered extrapolation problem consists in finding A [7].

Being based on the self-similar approximation theory, several types of approximants have been derived. Their derivation that can be found with all details in original publications. For practical purposes, let us just present the corresponding expressions and explain how they will be used for the problem of extrapolation to infinity.

3.1. Self-similar factor approximants

The self-similar factor approximants have been introduced in Refs. [22, 65]. For the reduced expansion (5.3.37), the kth order self-similar factor approximant reads as

$$\overline{\Phi}_k^*(x) = \prod_{i=1}^{N_k}(1 + \mathcal{P}_i x)^{n_i}, \tag{5.3.41}$$

where

$$N_k = \begin{cases} \frac{k}{2}, & k = 2, 4, \dots \\ \frac{k+1}{2}, & k = 3, 5, \dots \end{cases} \tag{5.3.42}$$

and the parameters $\mathcal{P}_i$ and n_i are defined from the accuracy-through-order procedure, by expanding expression (5.3.41) in powers of x, comparing the latter expansion with the given sum (5.3.37), and equating the like terms in these expansions. When the approximation order $k = 2p$ is even, the above procedure uniquely defines all $2p$ parameters. When the approximation order $k = 2p + 1$ is odd, the number of equations in the accuracy-through-order procedure is $2p$ which is by one smaller than the number of parameters. But then, using the scale invariance arguments [71], one sets $\mathcal{P}_1 = 1$, thus, uniquely defining all parameters. Another way is to find one of the coefficients $\mathcal{P}_i$ from the variational optimization of the approximant [61]. Both these ways give close results, though the scaling procedure of setting $\mathcal{P}_1$ to one, or even else setting

$$\mathcal{P}_1 = |a_1|$$

as in [61], is simpler. Fixing the parameters is helpful in particular when the series are short.

With approximant (5.3.41), the self-similar approximant for the sought function (5.3.40) becomes

$$\Phi_k^*(x) = \Phi_0(x) \prod_{i=1}^{N_k} (1 + \mathcal{P}_i x)^{n_i}.$$ (5.3.43)

If the zero-order factor has the large-variable form

$$\Phi_0(x) \simeq A_0 x^\alpha \qquad (x \to \infty),$$ (5.3.44)

then approximant (5.3.43) behaves as

$$\Phi_k^*(x) \simeq A_k x^\beta \qquad (x \to \infty).$$ (5.3.45)

Under a given exponent β, the powers n_i must satisfy the equality

$$\beta = \alpha + \sum_{i=1}^{N_k} n_i,$$ (5.3.46)

while the critical amplitude A is approximated by

$$A_k = A_0 \prod_{i=1}^{N_k} \mathcal{P}_i^{n_i}.$$ (5.3.47)

It is worth stressing that the factor $\Phi_0(x)$ in Eq. (5.3.43) is explicitly defined by the perturbative expansion (5.3.36), so it is known. The factor approximants (5.3.43) may have singularities when some $\mathcal{P}_i$ and n_i are negative. This makes it possible to associate such singularities with critical points and phase transitions.

3.2. Self-similar root approximants

The self-similar root approximants were derived and studied in Refs. [21, 73] and their derivation is presented above. The self-similar renormalization (algebraic renormalization and bootstrap) of the reduced expansion (5.3.37) yields the following form

$$R_k(x) = \left(\left(\left(\cdots (1 + \mathcal{P}_1 x)^{n_1} + \mathcal{P}_2 x^2 \right)^{n_2} + \mathcal{P}_3 x^3 \right)^{n_3} + \cdots + \mathcal{P}_k x^k \right)^{n_k}.$$ (5.3.48)

So that the kth-order approximant for the sought function becomes

$$\Phi_k^*(x) = \Phi_0(x) R_k(x).$$ (5.3.49)

The parameters $\mathcal{P}_i$ and n_i are uniquely defined, provided that k terms of the large-variable expansion as $x \to \infty$ are known, and the condition $pn_p - p + 1 = const$ holds for $p = 1, 2, ..., k - 1$. Then expression (3.8) leads to

$$R_k(x) \simeq \mathcal{P}_k^{n_k} x^{kn_k} \qquad (x \to \infty).$$ (5.3.50)

With the given exponent β, the power n_k satisfies the relation

$$\beta = \alpha + k n_k \tag{5.3.51}$$

and the k-th order approximation for the critical amplitude is

$$A_k = A_0 \mathcal{P}_k^{n_k}. \tag{5.3.52}$$

3.3. Iterated root approximants

Self-similar root approximants are uniquely defined, when their parameters are prescribed by the large-variable behaviour of the sought function. But, if attempted to find these parameters from the small-variable expansion (2.2), then the problem of multiple solutions is met [66].

In order to remove the multiplicity of solutions, arising in the re-expansion procedure applied to the general form of the root approximants, it is necessary to impose some restrictions on the definition of the related parameters. Thus, we can keep the same parameters of the lower-order approximants in the higher-order approximants, leaving there unknown only the highest-order parameter that is to be found from the re-expansion procedure. And the highest-order power has to be such that to satisfy the limiting condition as $x \to \infty$. Since in this method, the lower-order root approximants are inserted into the higher-order approximants, the resulting expressions can be called, iterated root approximants.

To achieve the very same goal one can look for an additional conditions on the parameters. Such a condition would be the requirement that all k terms in root (5.3.48) would contribute to the large-variable amplitude [23]. For this, it is necessary and sufficient that the internal powers n_j be defined as

$$n_j = \frac{j+1}{j} \qquad (1 \le j \le k-1), \tag{5.3.53}$$

with the external power related to the exponent β as

$$n_k = \frac{\gamma}{k} \qquad (\gamma = \beta - \alpha). \tag{5.3.54}$$

Then expression (5.3.48) becomes the iterated root approximant

$$R_k(x) = \left(\left(\left(\cdots \left(1 + \mathcal{P}_1 x \right)^2 + \mathcal{P}_2 x^2 \right)^{3/2} + \mathcal{P}_3 x^3 \right)^{4/3} + \cdots + \mathcal{P}_k x^k \right)^{\gamma/k}, \tag{5.3.55}$$

where all parameters $\mathcal{P}_j$ are uniquely defined by the accuracy-through-order procedure.

In the large-variable limit, Eq. (5.3.55) yields

$$R_k \simeq \frac{A_k}{A_0} x^\gamma \qquad \text{as } x \to \infty, \tag{5.3.56}$$

with the critical amplitude

$$A_k = A_0 \left(\left(\ldots \left(\mathcal{P}_1^2 + \mathcal{P}_2 \right)^{3/2} + \mathcal{P}_3 \right)^{4/3} + \ldots + \mathcal{P}_k \right)^{\gamma/k}. \qquad (5.3.57)$$

The iterative nature of R_k makes it really easy to calculate the parameters $\mathcal{P}_i$. It may happen that the iterated root approximants are well defined up to an order k, after which they do not exist because some of the parameters A_p are negative. At the same time, the higher-order terms of perturbation-theory expansion can be available up to an order $k + p$. How then these additional terms can be utilized for constructing the higher-order approximants?

3.4. Corrected self-similar approximants

The general idea is rather simple. Let us ensure the correct critical index β already in the starting approximation, while all other additional parameters should be obtained by matching it asymptotically with the truncated series. In place of starting approximation one can assume iterated root approximants, or factor approximants. One can obtain an initial guess, or some value for the amplitude, as described above for such types. Instead of increasing the order of approximation as prescribed by accuracy-through-order approach [5], one can adopt different idea of Corrected approximants [23, 24] as described below.

Let an approximant $\Phi_k^*(x)$ be given, whose parameters are defined by the accuracy-through-order procedure, which implies that the approximant $\Phi_k^*(x)$ is re-expanded in powers of x and compared, term by term, with the initial expansion, so that

$$\Phi_k^*(x) \simeq \Phi_k(x), \quad \text{as } x \to 0. \qquad (5.3.58)$$

This is also called the re-expansion procedure. In constructing the approximant $\Phi_k^*(x)$, we keep in mind the limiting condition

$$\lim_{x \to \infty} \frac{\ln \Phi_k^*(x)}{\ln x} = \beta. \qquad (5.3.59)$$

Our intermediate aim is to find the strong-coupling amplitude

$$A_k \equiv \lim_{x \to \infty} \frac{\Phi_k^*(x)}{x^\beta}. \qquad (5.3.60)$$

Now, let us introduce a correcting function

$$C_{k+p}(x) \simeq \frac{\Phi_{k+p}(x)}{\Phi_k^*(x)} \quad \text{as } x \to 0, \qquad (5.3.61)$$

which is defined as an expansion in powers of x, such that

$$C_{k+p}(x) = \sum_{n=0}^{k+p} b_n x^n, \tag{5.3.62}$$

whose coefficients b_n are, evidently, functions of the initial coefficients a_n. Since the parameters of $\Phi_k^*(x)$ are defined through the re-expansion procedure, the correcting function contains not all powers of x but only those starting from the order $k + 1$, that is,

$$C_{k+p}(x) = 1 + \sum_{n=k+1}^{k+p} b_n x^n. \tag{5.3.63}$$

This fact essentially simplifies the construction of a self-similar approximant $C_{k+p}^*(x)$ from the correcting function. Constructing the approximant $C_{k+p}^*(x)$, we impose the limiting condition

$$\lim_{x \to \infty} C_{k+p}^*(x) = constant. \tag{5.3.64}$$

The corrected self-similar approximant is defined as

$$\widetilde{\Phi}_{k+p}(x) \equiv \Phi_k^*(x) C_{k+p}^*(x). \tag{5.3.65}$$

Because of the condition (5.3.64),

$$\lim_{x \to \infty} \frac{\ln C_{k+p}^*(x)}{\ln x} = 0. \tag{5.3.66}$$

Therefore the strong-coupling exponent does not change,

$$\lim_{x \to \infty} \frac{\ln \widetilde{\Phi}_{k+p}(x)}{\ln x} = \lim_{x \to \infty} \frac{\ln \Phi_k^*(x)}{\ln x} = \beta. \tag{5.3.67}$$

But the corrected strong-coupling amplitude

$$A_{k+p} \equiv \lim_{x \to \infty} \frac{\widetilde{\Phi}_{k+p}(x)}{x^\beta} \tag{5.3.68}$$

changes to

$$A_{k+p} = A_k \lim_{x \to \infty} C_{k+p}^*(x). \tag{5.3.69}$$

where A_k is given by (5.3.60).

For concreteness one can always consider iterated root as the starting approximation. Our goal then is to find the strong-coupling amplitude by correcting the value suggested by the original guess. Corrections to the iterated root approximant $R_k(x)$ (5.3.55), originating from the higher-order terms, can be constructed [23] by defining also for concreteness, the corrected root approximants

$$\widetilde{R}_{k+p}(x) = R_k(x)C^*_{k+p}(x),$$ (5.3.70)

with the correction function

$$C^*_{k+p}(x) = 1+$$

$$b_{k+1}x^{k+1}\left(\left(\left(\cdots(1+d_1x)^2 + d_2x^2\right)^{3/2} + d_3x^3\right)^{4/3} + \cdots + d_{p-1}x^{p-1}\right)^{-(k+1)/(p-1)},$$ (5.3.71)

where $p > 2$ and all parameters d_i are defined from the accuracy-through-order procedure, when the terms of the expansion of form (5.3.70) are equated with the corresponding terms of the perturbation theory expansion. Here, the critical exponent is defined by the iterated root approximant (5.3.56), so that the limit $x \to \infty$ of the correction function be finite:

$$C^*_{k+p}(\infty) = 1 + b_{k+1}((\cdots(d_1^2 + d_2)^{3/2} + d_3)^{4/3} + \cdots + d_{p-1})^{-(k+1)/(p-1)}.$$ (5.3.72)

The corresponding approximation for the sought function takes the form

$$\Phi^*_{k+p}(x) = \Phi_0(x)\widetilde{R}_{k+p}(x).$$ (5.3.73)

Its large-variable behaviour is

$$\Phi^*_{k+p}(x) \simeq A_{k+p}x^\beta, \qquad \text{as } x \to \infty,$$ (5.3.74)

with the corrected critical amplitude

$$A_{k+p} = A_k C^*_{k+p}(\infty).$$ (5.3.75)

The structure of corrected approximants is rather general. For example, in place of iterated roots $R_k(x)$, one can use factor approximants. Or, when more detailed information on large x behaviour is available one can employ self-similar roots, or additive self-similar approximants. On the other hand, the correction function maybe also expressed in terms of diagonal Padé approximants, allowing to calculate corrections to the leading amplitudes. Sometimes instead of the multiplicative corrections an additive corrections ansatz will be employed, in particular for calculation of the higher-order critical amplitude. The idea of "correction" can be re-formulated for the critical index and threshold.

3.5. Additive self-similar approximants

Let us discuss below yet different type of self-similar approximants, called *additive self-similar approximants*, to distinguish them from multiplicative factor approximants [22, 65]. They enjoy sufficiently fast numerical convergence and provide good accuracy of approximations. Additive approximants were originally motivated by the problems appearing in theory of regular composite materials but appeared useful for the most typical field-theoretical problems. Their general form was first presented in [25, 26]. We discuss below in detail how to construct the additive approximants of [26].

Let us as look for a solution that is a real function $\Phi(x)$ of a real variable x. In general, the function domain can be arbitrary. For concreteness, consider here the interval $0 \leq x < \infty$, since by a change of variables it is practically always possible to reduce a given interval to the ray $[0, \infty)$.

Suppose that the sought function is defined by complicated equations that allow us to find only its asymptotic expansion near one of the domain boundaries, say, for asymptotically small $x > 0$, where

$$\Phi(x) \simeq \Phi_k(x), \qquad \text{as } x \to 0, \tag{5.3.76}$$

with the kth-order finite series

$$\Phi_k(x) = \sum_{n=0}^{k} a_n x^n. \tag{5.3.77}$$

Or the large-variable expansion can be available, such that

$$\Phi(x) \simeq \Phi^{(p)}(x), \qquad \text{as } x \to \infty, \tag{5.3.78}$$

with the finite series

$$\Phi^{(p)}(x) = \sum_{n=1}^{p} b_n x^{\beta_n}. \tag{5.3.79}$$

The powers in the above series are arranged in the ascending order:

$$\beta_n > \beta_{n+1} \qquad (n = 1, 2, \ldots, p - 1). \tag{5.3.80}$$

The standard situation corresponds to the uniform power decrease with the constant difference

$$\Delta\beta \equiv \beta_n - \beta_{n+1} \qquad (n = 1, 2, \ldots). \tag{5.3.81}$$

For the problem of interpolation between the small-variable expansion (5.3.77) and large-variable expansion (5.3.79), the values of the coefficients b_n are needed. However, the most interesting and most complicated problem is that of the extrapolation of the small-variable expansion (5.3.77) into the large-variable limit, when the coefficients b_n are not known, although the powers β_n can be available. Consider the problem of extrapolation, with unknown coefficients b_n. In that procedure, when the small-variable expansion is obeyed by construction, but the large-variable coefficients are not known, the error of an approximation tends to zero, as $x \to 0$. Vice versa, the error increases when the variable tends to infinity, reaching a maximal value in the limit $x \to \infty$. Therefore, in the problem of extrapolation, the accuracy of the procedure as a whole is defined by the large-variable limit, that is, by the accuracy of the amplitude

$$A \equiv \lim_{x \to \infty} x^{-\beta_1} \Phi(x) = b_1 \qquad (5.3.82)$$

that has to be compared with the large-variable limits of the studied approximations. Only the main steps of the procedure are stressed below.

First, subject the variable x to the affine transformation $x \mapsto \mathcal{P}(1 + \lambda x)$, consisting of a scaling and shift. This transforms the terms of series (5.3.77) as $a_n x^n \mapsto \mathcal{P}_n(1 + \lambda x)^n$, where $\mathcal{P}_n = \mathcal{P} a_n$. Then the self-similar transformation of series (5.3.77) is just the affine transformation of its terms, which yields

$$\Phi_k^*(x) = \sum_i \mathcal{P}_i(1 + \lambda x)^{n_i}. \qquad (5.3.83)$$

The powers of the first k terms of this series correspond to the powers of series (5.3.79),

$$n_i = \beta_i \qquad (i = 1, 2, \ldots, k), \qquad (5.3.84)$$

while all coefficients $\mathcal{P}_i$ can be found by the accuracy-through-order procedure, expanding form (5.3.83) in powers of x and equating to expansion (5.3.77) . Expression (5.3.83) is the *additive approximant*, which is named for distinguishing it from the multiplicative *factor approximants* considered earlier in this chapter.

It is clear that in the large-variable limit, approximant (5.3.83) will reproduce the terms with the powers of series (5.3.79). However, except the terms with the correct powers β_i, there appear the terms with the powers $\beta_i - 1$. There can exist two situations. It may be that the powers $\beta_i - 1$ do not pertain to the set of the powers $\{\beta_i\}$. Then the terms with the incorrect powers should be canceled by including in approximant (5.3.83) correcting terms (counter-terms) with the powers

$$n_j = \gamma_j \equiv \beta_j - 1 \qquad (j = 1, 2, \ldots, q), \qquad (5.3.85)$$

where

$$\beta_{k+1} < \gamma_j < \beta_1, \qquad (5.3.86)$$

and the coefficients C_i are defined by the cancellation of the terms with incorrect powers in the large-variable expansion. In that way, the general form of the additive approximant is

$$\Phi_{k,q}^*(x) = \sum_{i=1}^{k} \mathcal{P}_i(1 + \lambda x)^{\beta_i} + \sum_{j=1}^{q} C_j(1 + \lambda x)^{\gamma_j}. \qquad (5.3.87)$$

The other possibility is when the powers $\beta_i - 1$ turn out to be the members of the set $\{\beta_i\}$, that is, the set $\{\beta_i\}$ is invariant under the transformation

$$\beta_i - 1 = \beta_j. \tag{5.3.88}$$

In that case, no correction terms are needed, and the additive approximant is

$$\Phi_k^*(x) \equiv \Phi_{k,0}^*(x) = \sum_{i=1}^{k} \mathcal{P}_i (1 + \lambda x)^{\beta_i}. \tag{5.3.89}$$

The coefficients $\mathcal{P}_i$ can be found by the accuracy-through-order procedure, comparing the expansion of the additive approximant with the small-variable expansion, or with the large-variable expansion, or using both of them.

As has been mentioned above, it is instructive to compare the exact amplitude (5.3.82) with the amplitude of the kth-order approximant

$$A_k = \lim_{x \to \infty} x^{-\beta_1} \Phi_k^*(x) = \mathcal{P}_1 \lambda^{\beta_1}. \tag{5.3.90}$$

Of course, not only the leading-order amplitude A_k, representing the coefficient b_1, can be found, but the sub-leading amplitudes, representing other coefficients b_k, can also be calculated.

Defining the coefficients of the additive approximant from the accuracy-through-order procedure, we confront with the non-uniqueness of solutions. Thus, when there are no counter-terms, we have k solutions in the kth order. In the case of q counter-terms, the kth-order approximant yields $k + q - 1$ solutions. Fortunately, the appearance of multiple solutions is not a serious obstacle, in this case.

Generally, among the solutions, there can happen real and also complex-valued solutions. The latter come in complex conjugate pairs, so that there sum is real. It turns out that all real solutions and the average sums of the complex conjugate pairs, in each order, are very close to each other. Then there can be two strategies. Either to consider only real solutions, or to take the average sums of all solutions of the given order.

Consider an example. The structure of the integral

$$Z(g) = \frac{1}{\sqrt{\pi}} \int_{-\infty}^{\infty} \exp\left(-z^2 - g z^4\right) dz \tag{5.3.91}$$

is typical for numerous problems in quantum chemistry, field theory, statistical mechanics, and condensed-matter physics dealing with the calculation of partition functions, where $g \in [0, \infty)$ plays the role of coupling parameter [36]. The integral expansion at small $g \to 0$ yields strongly divergent series, with the kth-order sums

$$Z_k(g) = \sum_{n=0}^{k} c_n g^n, \tag{5.3.92}$$

whose coefficients are $c_n = \frac{(-1)^n}{\sqrt{\pi}\, n!}\, \Gamma\left(2n + \frac{1}{2}\right)$. The coefficients c_n quickly grow with increasing n tending to infinity as n^n for $n \gg 1$, which makes the weak-coupling expansion strongly divergent. At strong coupling, we have

$$Z(g) \simeq b_1 g^{-1/4} + b_2 g^{-3/4} + b_3 g^{-5/4} + b_4 g^{-7/4} \qquad (g \to \infty), \tag{5.3.93}$$

where

$$b_1 = \frac{1}{2\sqrt{\pi}}\, \Gamma\left(\frac{1}{4}\right) = 1.022765, \qquad b_2 = \frac{1}{8\sqrt{\pi}}\, \Gamma\left(-\frac{1}{4}\right) = -0.345684,$$

$$b_3 = \frac{1}{16\sqrt{\pi}}\, \Gamma\left(\frac{1}{4}\right) = 0.127846, \qquad b_4 = \frac{1}{64\sqrt{\pi}}\, \Gamma\left(-\frac{1}{4}\right) = -0.043211.$$

The powers of the strong-coupling expansion,

$$\beta_n = -\frac{2n - 1}{4} \tag{5.3.94}$$

enjoy the uniform difference

$$\Delta\beta \equiv \beta_n - \beta_{n+1} = \frac{1}{2}.$$

The set $\{\beta_n\}$ is invariant with respect to transformation (5.3.88) because of the property

$$\beta_n - 1 = \beta_{n+2}.$$

Hence no correction terms are needed. All coefficients $\mathcal{P}_i$ of the additive approximant (5.3.89) are obtained from the accuracy-through-order procedure at weak coupling. The error of approximants grows as $g \to \infty$. Therefore, the accuracy of the method is defined by the accuracy of the strong-coupling amplitude

$$A_k = \lim_{g \to \infty} g^{1/4} Z_k^*(g) \tag{5.3.95}$$

that has to be compared with the exact value b_1.

Consider only real-valued solutions for $\mathcal{P}_i$. In each odd order, there is just one real solution. Then for the additive approximants (5.3.89), we have to third order

$$Z_3^*(g) = \mathcal{P}_1(1 + \lambda g)^{-1/4} + \mathcal{P}_2(1 + \lambda g)^{-3/4} + \mathcal{P}_3(1 + \lambda g)^{-5/4},$$

where

$$\mathcal{P}_1 = 1.510761, \qquad \mathcal{P}_2 = -0.717990, \qquad \mathcal{P}_3 = 0.207229, \qquad \lambda = 7.634834.$$

This gives the strong-coupling amplitude

$$A_3 = 0.908858 \qquad (Z_3^*).$$

To fifth order,

$$Z_5^*(g) = P_1(1 + \lambda g)^{-1/4} + P_2(1 + \lambda g)^{-3/4} + P_3(1 + \lambda g)^{-5/4} +$$

$$+ P_4(1 + \lambda g)^{-7/4} + P_5(1 + \lambda g)^{-9/4},$$

with the coefficients

$$P_1 = 1.808031, \qquad P_2 = -1.543729, \qquad P_3 = 1.134917, \qquad P_4 = -0.492745,$$

$$P_5 = 0.093526, \qquad \lambda = 12.297696.$$

The strong-coupling amplitude is

$$A_5 = 0.965495 \qquad (Z_5^*).$$

Continuing the procedure in higher orders it can be found that

$$A_7 = 0.992107 \ (Z_7^*), \qquad A_9 = 1.005760 \ (Z_9^*), \qquad A_{11} = 1.01312 \ (Z_{11}^*),$$

$$A_{13} = 1.01720 \ (Z_{13}^*), \qquad A_{15} = 1.01952 \ (Z_{15}^*), \qquad A_{17} = 1.02085 \ (Z_{17}^*),$$

$$A_{19} = 1.02072 \ (Z_{19}^*).$$

Comparing these amplitudes with the exact $A = 1.02277$, the corresponding errors are calculated

$$11\%, \quad 6\%, \quad 3\%, \quad 2\%, \quad 0.9\%, \quad 0.5\%, \quad 0.3\%, \quad 0.2\%.$$

As is seen, the accuracy improves with increasing order, which demonstrates good numerical convergence.

The accuracy of the additive approximants for the studied problem is much better than that of other approximants. Because of the incompatibility of the powers in the weak-coupling and strong-coupling limits, the standard Padé approximants are not applicable, but the modified Baker–Gammel approximants $P_{N/(N+1)}^{1/4}$ have to be used. The modified Padé approximant of 19th order (with $N = 9$) has an error of 10%, which is much worse than the error of 0.2% of the additive real approximant. Factor approximants are also less accurate. Thus, the factor approximant of ninth order yields an error of 11%, while the additive approximant in this order exhibits an error of 2%. Root approximants for this problem are not defined, resulting in complex solutions.

4. Corrected Padé approximants for indeterminate problem

Probably the simplest and direct way to extrapolate, is to apply the Padé approximants $P_{n,m}(x)$, which is nothing else but ratio of the two polynomials $P_n(x)$ and $P_m(x)$ of the order n and m, respectively. The coefficients are derived directly from the coefficients of the given power series [5, 52] from the requirement of asymptotic equivalence to the given series or function $\Phi(x)$. When there is a need to stress the last point, we simply write $PadeApproximant[\Phi[x], n, m]$, adopting a reduced notation from $Mathematica^{®}$.

Padé approximants locally are the best rational approximations of power series. Their poles determine singular points of the approximated functions [5, 52]. Calculations with Padé approximants are straightforward and can be performed with $Mathematica^{®}$.

One should select from the emerging sequences only approximants which are also holomorphic functions. Not all approximants generated by the procedure are holomorphic. The holomorphy of diagonal Padé approximants in a given domain implies their uniform convergence inside this domain [29].

4.1. Standard Padé scheme

Consider non-negative functions with asymptotic behaviour $\Phi(x) \simeq Ax^{\beta}$ at infinity, with known critical index β and given expansion at small $x > 0$,

$$\Phi(x) = a_0 + \sum_{n=1}^{N} a_n x^n + O(x^{N+1}). \tag{5.4.96}$$

Here N is integer and $N \geqslant 1$. Standard accepted suggestion for the solution of the problem of reconstructing the amplitude A, is based on the following conventional considerations [7].

Let us calculate the critical amplitude A. To this end, let us apply some transformation to the original series $\Phi(x)$ to obtain transformed series $T(x) = \Phi(x)^{-1/\beta}$, in order to get rid of the power law behaviour at infinity. Applying the technique of diagonal Padé approximants in terms of $xT(x)$, one can readily obtain the sequence of approximations A_n for the critical amplitude A,

$$A_n = \lim_{x \to \infty} (xPadeApproximant[T[x], n, n+1])^{-\beta}. \tag{5.4.97}$$

Uniform convergence of diagonal sequences of Padé approximants has been established, for example, for series of Stieltjes–Markov–Hamburger [5, 40]. The Stieltjes moment problem can possess a unique solution or multiple solutions (indeterminate problem), dependent on the behaviour of the moments, in contrast with the problem of moments for the finite interval [38], which is solved uniquely if the solution exists [53].

From the theory of S-fractions, one deduces a divergence by oscillation of the Padé approximants [53] in the indeterminate case. The phenomenon of spurious poles, when approximants can have poles which are not related to those of the underlying function, can also lead to divergence of the Padé sequence [5, 40, 53].

Generally, the form of the solution in the indeterminate case is covered by the Nevanlinna theorem, expressing a non-negative functions through a mixture of rational and non-rational contributions [1, 38, 43]. The former can be interpreted as the part covered by four entire functions related to construction of the Padé approximants, while the latter is only a non-rational function from the Nevanlinna class [10]. Thus in the indeterminate case Padé approximants are insufficient. Let us look for an approximation scheme which can express a non-rational part explicitly and also cover the rational part with Padé approximants.

4.2. Correction with Padé approximants

Let us ensure the correct critical index β already in the starting approximation $\mathcal{K}(x)$, while all other additional parameters should be obtained by matching it asymptotically with the truncated series for $\Phi(x)$. In place of $\mathcal{K}(x)$ one can assume iterated root approximants [23, 24], or factor approximants [22, 65]. One can extract corresponding value for the amplitude,

$$A_0 = \lim_{x \to \infty} (\mathcal{K}(x) x^{-\beta}). \tag{5.4.98}$$

Instead of increasing the order of approximation as prescribed by accuracy-through-order approach [5], one can adopt a different idea of Corrected approximants [23, 24]. Accordingly, to extract corrections to the critical amplitude, one has to divide the original series (5.4.96) for $\Phi(x)$ by $\mathcal{K}(x)$, and call the newly found series

$$G(x) = \frac{\Phi(x)}{\mathcal{K}(x)}.$$

Assuming that the non-rational part is already extracted through $\mathcal{K}(x)$, one can apply now the rational approximants. Thus, we finally build a sequence of the diagonal Padé approximants asymptotically equivalent to $G(x)$, so that the approximate amplitudes are expressed by the formula [27]

$$A_n = A_0 \lim_{x \to \infty} (PadeApproximant[G[x], n, n]). \tag{5.4.99}$$

The role of $\mathcal{K}(x)$ is crucial. It is supposed not only to approximate the non-rational part of the sought function, but also ensure convergence of the Padé approximants for the rational part .

The starting approximation should be viewed as a control function, selected to ensure the convergence of the sequence of Corrected Padé approximants, even if the Standard Padé scheme diverges. The suggestion is not only to construct controlled Padé sequences, but also in conjecturing that the control function should be chosen among the low-parametric subset of a irrational self-similar approximants.

When only single "critical index" β is known in advance, we are limited to the iterated roots and factor approximants, noting that self-similar approximants can be constructed uniquely. In all examples studied below, the control function is indeed found easily. More general self-similar roots and continued roots [21, 28, 67], can be applied to a more general situations, when in addition to β more detailed corrections to the leading scaling behaviour are available.

Note that Padé approximants by themselves cannot serve as a control function [2, 3]. Standard Padé scheme is uncontrolled and nothing can be improved withing the Standard scheme to modify it when it brings bad results. On the other hand, when Standard scheme is convergent, the Corrected scheme will also converge, and there is no significant difference in the behaviour of approximations for A, for the two schemes of applying Padé approximations.

Let us start with an illustration of this statement.

4.3. Convergent schemes. Mittag–Lefler function

Consider the Mittag–Leffler function from [42],

$$F(x) = \text{erfc}(x) \exp\left(x^2\right). \tag{5.4.100}$$

For small $x > 0$ in the lowest orders,

$$F(x) = 1 - \frac{2x}{\sqrt{\pi}} + x^2 - \frac{4x^3}{3\sqrt{\pi}} + \frac{x^4}{2} + O(x^5), \tag{5.4.101}$$

and $F(x) \simeq (\sqrt{\pi}x)^{-1}$, as $x \to \infty$.

Set the control function simply given by the second-order iterated root as follows

$$\mathcal{K}(x) = R_2^*(x) = \left[\left(\frac{2x}{\sqrt{\pi}} + 1\right)^2 - \frac{2(\pi - 4)x^2}{\pi}\right]^{-\frac{1}{2}}. \tag{5.4.102}$$

There is no significant difference in the behaviour of approximations for A, for the two schemes of applying Padé approximations. Good convergence is achieved as shown in Fig. 5.1.

4.4. Accelerating convergence. Quartic oscillator

Consider the dimensionless ground state energy $e(g)$ of the celebrated quantum one-dimensional quartic anharmonic oscillator [9]. Here g stands for the dimensionless coupling constant. The asymptotic expansion of $e(g)$ in the weak-coupling limit, when $g \to 0$, is

$$e(g) \simeq a_0 + a_1 g + a_2 g^2 + \cdots + a_{18} g^{18} + \cdots, \tag{5.4.103}$$

where a few starting coefficients are

$$a_0 = \frac{1}{2}, \quad a_1 = \frac{3}{4}, \quad a_2 = -\frac{21}{8}, \quad a_3 = \frac{333}{16}, \quad a_4 = -30885/128.$$

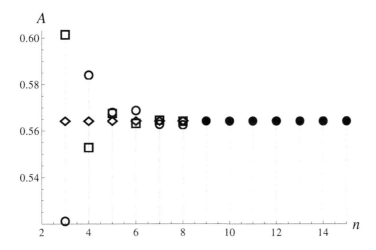

Figure 5.1 Mittag–Leffler function. Exact result is shown with diamonds. Amplitude with increasing approximation order for Corrected Padé Approximants is shown with empty circles and Standard scheme is shown with squares. Filled circles are shown when the results from two methods are virtually indistinguishable.

The ground-state energy diverges as $Ag^{\frac{1}{3}}$ with $A = 0.667986$, as $g \to \infty$.

Applying the standard Padé approximants, one obtains the following results,

$$A_2 = 0.759147, \quad A_3 = 0.734081, \quad A_4 = 0.720699, \quad A_5 = 0.712286,$$

$$A_6 = 0.706466, \quad A_7 = 0.702176, \quad A_8 = 0.698869, \quad A_9 = 0.696173.$$

The last point of the sequence brings an error of 4.21967%.

The control function for the method of Corrected Padé approximants is simply the second-order iterated root,

$$\mathcal{K}(g) = R_2^*(g) = \frac{1}{2}\sqrt[6]{\left(\frac{9g}{2} + 1\right)^2 - 18g^2}, \qquad (5.4.104)$$

and $A_0 = 0.572357$.

The results of calculations according to the Corrected Padé approximants methodology are given below, $A_1 = A_2 = A_0$,

$$A_3 = 0.587104, \quad A_4 = 0.63279, \quad A_5 = 0.655086, \quad A_6 = 0.660334,$$

$$A_7 = 0.661945, \quad A_8 = 0.663225, \quad A_9 = 0.665346.$$

The last point of the sequence brings an error just of 0.3952%, by order of magnitude

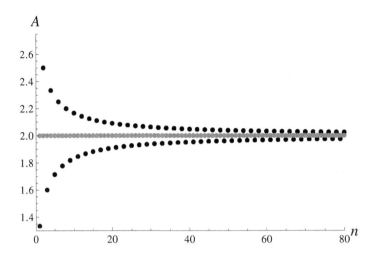

Figure 5.2 Debye–Hukel function. Amplitide with increasing approximation order obtained with Standard Padé approximants

.

better than standard approach.

4.5. Divergence by oscillation. Debye–Hukel function

Correlation function of the Gaussian polymer [30], is given in the closed form by Debye–Hukel function,

$$\Phi(x) = \frac{2}{x} - \frac{2(1 - \exp(-x))}{x^2}. \tag{5.4.105}$$

For small $x > 0$,

$$\Phi(x) = 1 - \frac{x}{3} + \frac{x^2}{12} - \frac{x^3}{60} + \frac{x^4}{360} + O(x^5), \tag{5.4.106}$$

and $\Phi(x) = 2x^{-1} + O(x^{-2})$, as $x \to +\infty$.

Standard Padé scheme gives oscillating solution with magnitude of oscillations slowly decreasing with approximation number, as shown in Fig. 5.2.

On the other hand, method of Corrected Padé approximants with rather simple control function,

$$\mathcal{K}(x) = R_2^*(x) = \frac{\sqrt{6}}{\sqrt{x^2 + 4x + 6}}, \tag{5.4.107}$$

demonstrates very good results shown in Fig. 5.3.

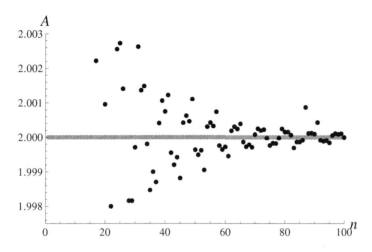

Figure 5.3 Debye- -Hukel function. Amplitide with increasing approximation order obtained with corrected approximants.

4.6. Divergence by oscillation. Hypergeometric function

The structure factor of three-dimensional branched polymers is given by the confluent hypergeometric function [39, 45]

$$S(x) = F_1\left(1; \frac{3}{2}; \frac{3}{2}x\right), \tag{5.4.108}$$

where x is a non-dimensional wave-vector modulus. At small $x > 0$ (long-wave) the coefficients in the expansion are given by general expression

$$a_n = \frac{\left(-\frac{3}{2}\right)^n \Gamma\left(\frac{3}{2}\right)}{\Gamma\left(n + \frac{3}{2}\right)}, \tag{5.4.109}$$

while for large x (short-wave) $S(x) \simeq \frac{1}{3}x^{-1}$.

Standard Padé scheme gives oscillating results for the amplitude, with increasing magnitude for larger orders, as demonstrated in Fig. 5.4.

The control function $\mathcal{K}(x)$ in this case is simplest factor approximant,

$$\Phi_4^*(x) = (1 + (0.14286 - 0.25555i)x)^{-0.5-1.677i}(1 + (0.14286 + 0.25555i)x)^{-0.5+1.677i}, \tag{5.4.110}$$

and the Corrected Padé approximants demonstrate very good convergence, see Fig. 5.5.

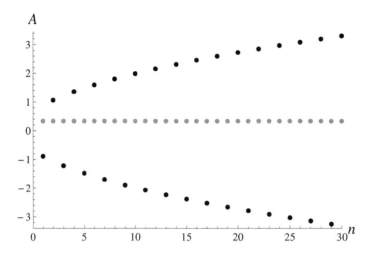

Figure 5.4 Branched polymer. Amplitide with increasing approximation order obtained with Padé approximants

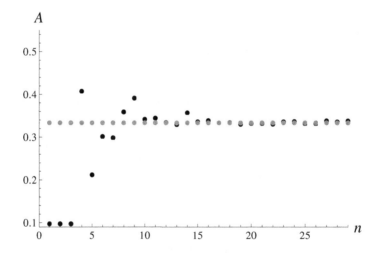

Figure 5.5 Branched polymer. Amplitude with increasing approximation order obtained with Corrected Padé approximants.

4.7. Marginal case

Formally the Standard Padé method converges very fast but to the wrong value. Also formally, this case can be viewed as divergent by oscillation, where the oscillations occur between this wrong value and ∞.

The following function [34]

$$\Phi(g) = \frac{\pi^2}{128}\left(1 + \frac{16}{\pi^4 g^2} + \sqrt{1 + \frac{64}{\pi^4 g^2}}\right) \tag{5.4.111}$$

arises in the calculation of the ground state energy of a quantum particle in a one-dimensional box. As $g \to 0$, this function possesses an expansion of the type,

$$\Phi(g) \simeq \frac{1}{8\pi^2 g^2} \sum_{n=0} a_n g^n \tag{5.4.112}$$

with

$$a_0 = 1, \quad a_1 = \frac{\pi^2}{4}, \quad a_2 = \frac{\pi^4}{32}, \quad a_3 = \frac{\pi^6}{512},$$

$$a_4 = 0, \quad a_5 = -\frac{\pi^{10}}{131072}, \quad a_6 = 0, \dots.$$

We shall be interested in finding the limiting value $\Phi(\infty)$ in the asymptotic expression

$$\Phi(g) = \Phi(\infty)\left(1 + O\left(g^{-2}\right)\right), \quad \text{as } g \to \infty, \tag{5.4.113}$$

whose exact limit is 0.077106.

The standard Padé scheme fails, converging very fast as $n = 2$, to the value of 0.0385531. Or else bringing rapid oscillations between the values of 0.0385531 and ∞ with changing approximation number.

For convenience, instead of the original function (5.4.111), use $8\pi^2 g^2\, \Phi(g)$. The control function for the method of Corrected Padé approximants appears to be simple third-order iterated root approximant

$$\mathcal{K}(g) = R_3^*(g) = \left(\frac{3\pi^6 g^3}{1024} + \left(\frac{\pi^4 g^2}{64} + \left(\frac{\pi^2 g}{8} + 1\right)^2\right)^{\frac{3}{2}}\right)^{\frac{2}{3}}. \tag{5.4.114}$$

Very good convergence is achieved in the method of Corrected Padé Approximants, as shown in Fig. 5.6.

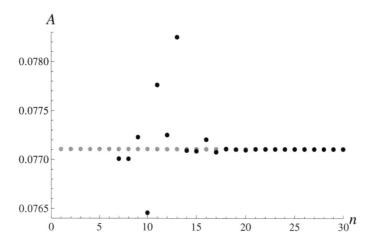

Figure 5.6 Function (5.4.111). Amplitide with increasing approximation order for Corrected Padé Approximants.

4.8. Divergent solution. Modified Bessel function

The $N = 4$ Super Yang–Mills Circular Wilson Loop [6] is given by the following expression

$$\Phi(y) = \frac{2 \exp\left(-\sqrt{y}\right) I_1\left(\sqrt{y}\right)}{\sqrt{y}}, \tag{5.4.115}$$

where I_1 is a modified Bessel function of the first kind. Let us set $\sqrt{y} = x$. For small $x > 0$,

$$\Phi(x) = 1 - x + \frac{5x^2}{8} - \frac{7x^3}{24} + \frac{7x^4}{64} + O(x^5), \tag{5.4.116}$$

and $\Phi(x) \simeq \sqrt{\frac{2}{\pi}} x^{-\frac{3}{2}}$, as $x \to \infty$.

Application of the Standard scheme brings a divergent results as is clearly seen in Fig. 5.7.

Let us introduce the control function

$$\mathcal{K}(x) = R_2^*(x) = \left(\frac{5x^2}{18} + \left(\frac{2x}{3} + 1\right)^2\right)^{-\frac{3}{4}}, \tag{5.4.117}$$

and construct Corrected Padé approximants with good numerical results for the amplitude, shown in Fig. 5.8.

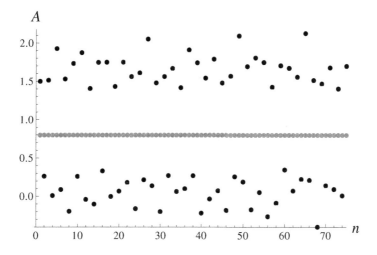

Figure 5.7 Function (5.4.115). Amplitide with increasing approximation order obtained with Standard Padé approximants

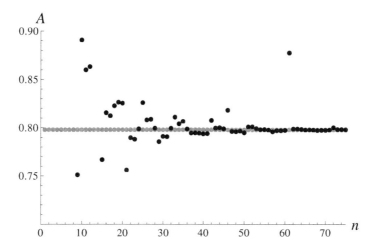

Figure 5.8 Function (5.4.115). Amplitide with increasing approximation order obtained with corrected approximants

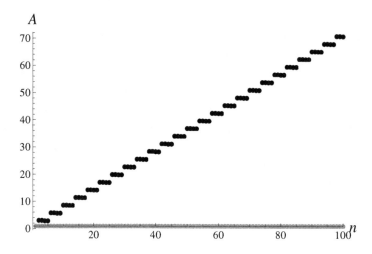

Figure 5.9 Error function. Amplitide with increasing approximation order obtained with Standard Padé approximants

4.9. Divergent case. Error function

Popular error function with many applications in statistics and physics is given as the following integral,

$$\Phi(x) = \int_0^x \exp\left(-u^2\right) du. \tag{5.4.118}$$

For small $x > 0$,

$$\Phi(x) = x - \frac{x^3}{3} + \frac{x^5}{10} + O(x^7), \tag{5.4.119}$$

also $\Phi(\infty) = \frac{\sqrt{\pi}}{2}$. Clearly divergent results are achieved in the framework of the Standard Padé scheme, see Fig. 5.9.

Numerically convergent results are achieved in the framework of the Corrected Padé scheme, with the following control function

$$\mathcal{K}(x) = R_3^*(x) = \frac{x}{\sqrt[6]{\frac{16x^6}{63} + \left(\frac{32x^4}{45} + \frac{4x^2}{3} + 1\right)^{3/2}}}, \tag{5.4.120}$$

as shown in Fig. 5.10.

With slightly different control function, corresponding to modified Pade approximant[4],

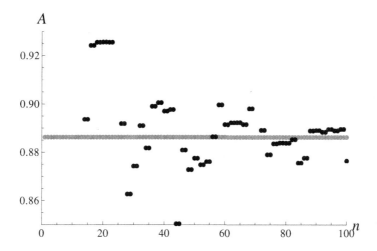

Figure 5.10 Error function. Amplitide with increasing approximation order obtained with corrected approximants

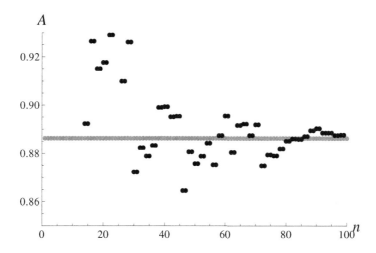

Figure 5.11 Error function. Amplitide with increasing approximation order obtained with corrected approximants corresponding to the control function 5.4.121.

equivalent to the simplest root approximant with all indices set to 1,

$$R_3^*(x) = \frac{x}{\sqrt[6]{\frac{772x^6}{945} + \frac{26x^4}{15} + 2x^2 + 1}}, \qquad (5.4.121)$$

the results appear to be even more stable, as shown below in Fig. 5.11.

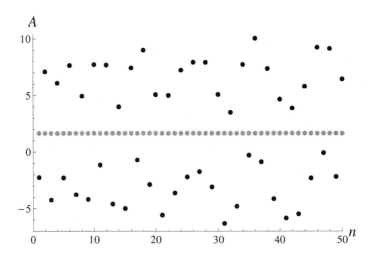

Figure 5.12 Debye function. Amplitide with increasing approximation order obtained with Standard Padé approximants

4.10. Divergent case. Debye function

Consider the following Debye function of the order 1,

$$\Phi_1(x) = \frac{1}{x} \int_0^x \frac{y}{\exp(y) - 1} \, dy. \qquad (5.4.122)$$

For large $x > 0$, $\Phi_1(x) \simeq Ax^{-1}$, with the amplitude at infinity, $A = \frac{\pi^2}{6} \approx 1.64493$.

Expansion for small $x > 0$,

$$\Phi_1(x) = 1 - \frac{x}{4} + \frac{x^2}{36} - \frac{x^4}{3600} + \frac{x^6}{211680} + O(x^8). \qquad (5.4.123)$$

Standard Padé approximation fails completely, following rather chaotic pattern, as shown in Fig. 5.12.

Starting approximation/control function is again simple,

$$\mathcal{K}(x) = R_2^*(x) = \frac{1}{\sqrt{\frac{5x^2}{72} + \left(\frac{x}{4} + 1\right)^2}}. \qquad (5.4.124)$$

Corresponding Corrected Padé approximants work well, bringing rather good accuracy, as shown in Fig. 5.13.

4.11. Divergent case. Connected moments

This example is borrowed from [16], and concerned with application of the method of connected moments to calculation of the ground state energy of the harmonic oscillator

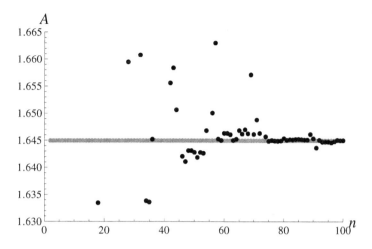

Figure 5.13 Debye function. Amplitide with increasing approximation order obtained with Corrected Padé approximants. The number n changes from $n = 1$ to 100.

as the limit of the moments generating function as "time" t goes to infinity.

Generating function for the harmonic oscillator,

$$E(t) = \frac{121u(t)^3 + 189199u(t)^2 + 8180919u(t) + 6561}{(81 - u(t))\left(121u(t)^2 + 20198u(t) + 81\right)}, \tag{5.4.125}$$

where $u(t) = \exp(-4t)$.

The Standard Padé scheme show no convergence, as shown in Fig. 5.14.

Corrected Padé approximants demonstrate good convergence, with simple control function corresponding to the simplest shifted root approximant

$$\mathcal{K}(t) = v_2 + v_3(v_4 t + 1)^{-c}, \tag{5.4.126}$$

where

$$v_2 = \frac{403171240048919}{85626857995920}, \quad v_3 = \frac{36337990380139}{85626857995920}, \quad v_4 = \frac{2331886111}{1340069829}, \quad c = \frac{9}{10}.$$

At first we considered a more general expression with all parameters including c, defined through accuracy-through-order conditions. Then, to make computations easier it appears expedient to replace c with the closest rational number, recalculating accordingly all other parameters.

The results of calculations according to the Corrected Padé scheme are shown in Fig. 5.15.

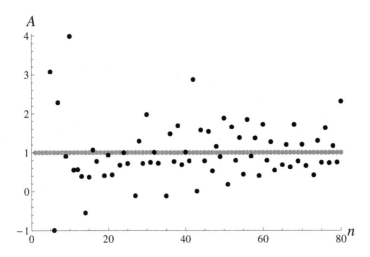

Figure 5.14 Function (5.4.125). Amplitide with increasing approximation order obtained with Padé approximants.

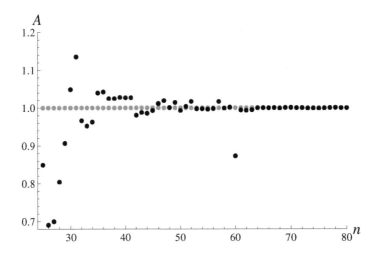

Figure 5.15 Function (5.4.125). Amplitide with increasing approximation order obtained with Corrected Padé approximants. The order n varies from 1 to 80.

5. Calculation of critical exponents

We have concentrated on the calculation of critical amplitudes, under the known critical exponents, by extrapolating the small-variable expansions to the large-variable limit, employing the techniques of self-similar approximants. Now let us show how the critical exponents can also be found by using the very same techniques with only minor restriction needed to separate this particularly important problem.

5.1. Scheme of general approach

When a function, at asymptotically large variable, behaves as

$$\Phi(x) \simeq A x^{\beta} \qquad \text{as } x \to \infty, \tag{5.5.127}$$

then the critical exponent can be represented by the limit

$$\beta = \lim_{x \to \infty} x \, \frac{d}{dx} \, \ln \Phi(x). \tag{5.5.128}$$

Assuming that the small-variable expansion for the function is given by the sum $\Phi_k(x)$, as in Eq. (2.2), the corresponding small-variable expression for the critical exponent can be constructed.

$$\beta_k(x) = x \, \frac{d}{dx} \, \ln \Phi_k(x), \tag{5.5.129}$$

which can be expanded in powers of x.

Applying to the obtained expansion the method of self-similar approximants, as has been treated above, we should select the self-similar approximant $\beta_k^*(x)$ whose limit, being by definition finite,

$$\beta_k^*(x) \to constant, \qquad \text{as } x \to \infty,$$

gives us the sought approximate expression for the critical exponent

$$\beta_k^* = \lim_{x \to \infty} \beta_k^*(x). \tag{5.5.130}$$

Note that the value of the critical amplitude A does not enter the consideration at all. Below, we illustrate this method of calculating the critical exponents by concrete example.

In practice, one can expand in powers of x only the derivative $\frac{d}{dx} \ln \Phi_k(x)$ and to construct the approximants behaving as x^{-1} as $x \to \infty$, see (5.5.135), (5.5.143). In such terminology the problem of critical index looks analogous to the problem of critical amplitude considered above. It implies that all methods developed for amplitude can be in principle applied to the index.

5.2. Critical index for the test function (5.5.131)

Consider an example with a true critical point, and the methodology described above van be tested against it. The following test function will be considered,

$$\Phi(x) = \frac{1}{8\left(\sqrt{1-x}+1\right) - 4\left(\sqrt{1-x}+2\right)x}. \tag{5.5.131}$$

Its low- and high-concentration expansions are similar to the corresponding expansions for conductivity. The coefficients a_i can be obtained in arbitrary order from Taylor expansion for small $x > 0$ is presented in the truncated form

$$\Phi(x) = \frac{1}{16} + \frac{x}{16} + \frac{15x^2}{256} + \frac{7x^3}{128} + \frac{105x^4}{2048} + \frac{99x^5}{2048} +$$
$$\frac{3003x^6}{65536} + \frac{715x^7}{16384} + \frac{21879x^8}{524288} + \frac{20995x^9}{524288} + \frac{323323x^{10}}{8388608} + O(x^{11}). \tag{5.5.132}$$

The threshold for the test function, $x_c = 1$. In addition to the "first" critical amplitude A one may be interested in the higher-order corrections in the expansion obtained in the vicinity of x_c

$$\Phi(x) = \frac{1}{4}(1-x)^{-1/2} - \frac{1}{2} + O((1-x)^{1/2}), \tag{5.5.133}$$

i.e., it diverges as a square root in the vicinity of the critical point $x \to x_c = 1$ with the constant correction term. Critical characteristics such as index $s = \frac{1}{2}$, amplitudes $A = \frac{1}{4}$ and $B = -\frac{1}{2}$, can be recovered from this expansion.

Is it possible to evaluate the character of singularity as $x \to x_c$ from the series (5.5.132), assuming only that it is a power law, $\Phi(x) \sim (1-x)^{-\beta}$?

Let us apply the following transformation,

$$z = \frac{x}{x_c - x} \Leftrightarrow x = \frac{zx_c}{z+1} \tag{5.5.134}$$

to the original series (5.5.132). In this new variable z the problem of critical index calculation is already familiar and analogous to the one considered above. Let us opt for the Padé approximants $P_{n,n+1}$ while intending to calculate $\beta_k(x)$.

To the transformed series $M_1(z)$ let us apply the *DLog* transformation (differentiate log of $M_1(z)$) and call the transformed series $M(z)$. In terms of $M(z)$. one can readily obtain the sequence of approximations β_n for the critical index β,

$$\beta_n = \lim_{z \to \infty}(zPadeApproximant[M[z], n, n+1]). \tag{5.5.135}$$

Namely,

$$\beta_{12} = 0.537089, \quad \beta_{13} = 0.534525,$$

and with increasing approximation order one observes monotonous convergence with the values of $\beta_{100} = 0.504926$ and $\beta_{150} = 0.5033$, as illustrated in Fig. 5.16.

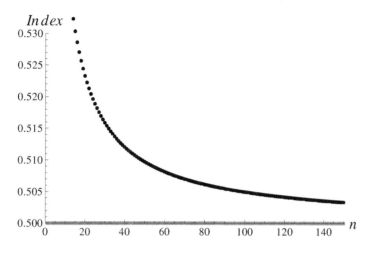

Figure 5.16 Index β_n calculated from (5.5.135), compared with the exact index.

5.3. Counterexample and constructive way out. Corrected critical index

Consider the different function,

$$\Phi(x) = \left(\sqrt{x^2 + 1} + x \right)^a,$$

(5.5.136)

with arbitrary a, it is used as a touch-stone generating function [54] for the sequence of numbers a_n corresponding to coefficients in the expansion of (5.5.136) at small $x > 0$,

$$a_n = \frac{2^n \left(\frac{a}{2} - \frac{n}{2} + 1 \right)^{\bar{n}}}{n! \left(\frac{n}{a} + 1 \right)},$$

(5.5.137)

while for large x, $\Phi(x) \simeq 2^a x^a$. Here $m^{\bar{k}}$ means $m(m + 1) \cdots (k + m - 1)$.

Consider the case of $a = \frac{1}{2}$ which is supposed to illustrate the failure of Padé approximants in evaluation of the critical index as well as constructive solution to the problem. Then

$$\Phi(x) = \sqrt{ \sqrt{x^2 + 1} + x },$$

(5.5.138)

and the following expansion is valid at small $x > 0$,

$$\Phi(x) = 1 + \frac{x}{2} + \frac{x^2}{8} - \frac{x^3}{16} - \frac{5x^4}{128} + \frac{7x^5}{256} + \frac{21x^6}{1024} - \frac{33x^7}{2048} - \frac{429x^8}{32768} + \frac{715x^9}{65536} + \frac{2431x^{10}}{262144} + O(x^{11}).$$

(5.5.139)

Critical characteristics of the (5.5.138), such as index $\beta = \frac{1}{2}$ and amplitudes $A = \sqrt{2}$, can be recovered from the expansion as $x \to \infty$,

$$\Phi(x) = \sqrt{2}\sqrt{x} + O(x^{\frac{3}{2}}). \tag{5.5.140}$$

Formally applying (5.5.135) brings meaningless results, i.e., the value of index alternates between 0 and ∞ with increasing approximation number. We are again confronted with the indeterminate problem when the Padé approximants are applied. So it necessary to go beyond the standard approach and develop some form of the Corrected Padé approximants.

Thus, in order to improve convergence of the Padé method, we suggest adapting the technique of corrected approximants [23], expressing the correction function in terms of the *DLog* Padé approximants [17, 18, 19, 23].

Let us first estimate the critical index by simplest non-trivial factor approximant [22, 61, 65, 71],

$$\Phi_3^*(x) = (B_1 x + 1)^{c_1} (1 + x)^{c_2}, \tag{5.5.141}$$

where

$$c_2 = -\frac{1}{2}, \quad c_1 = 2, \quad B_1 = \frac{1}{2}.$$

Let us divide then original series (5.5.139) by $\Phi_3^*(x)$, apply to the newly found series transformation the *DLog* transformation and call the transformed series $L(x)$. Finally one can obtain the following sequence of corrected approximations for the critical index,

$$\beta_n = \beta_0 + \lim_{x \to \infty}(xPadeApproximant[L[x], n, n + 1]), \tag{5.5.142}$$

where $\beta_0 = c_1 + c_2$, as usual for the factor approximants. The following corrected sequence of approximate values for the critical index can be calculated readily:

$$\beta_3 = 0.695364, \quad \beta_4 = 0.851802, \quad \beta_5 = 0.558687, \quad \beta_6 = 0.55874,$$

$$\beta_7 = 0.507656, \quad \beta_8 = 0.518442, \quad \beta_9 = 0.494053, \quad \beta_{10} = 0.511269,$$

$$\beta_{11} = 0.491185, \quad \beta_{12} = 0.509942, \quad \beta_{13} = 0.490645, \quad \beta_{14} = 0.509704.$$

The results point to the correct value of $\beta = \frac{1}{2}$ with only weak residual oscillations around it.

One can also attempt to apply iterated roots in place of Padé,

$$\beta_n = \lim_{x \to \infty}(x \, R_n(x)), \tag{5.5.143}$$

where $R_n(x)$ stands for the iterated root of nth order [23] and (5.3.55), constructed for the series $L(x)$ with such a power at infinity that defines a constant value standing for the index. Calculations with iterated roots are really easy since at each step we need to compute only one new coefficient, while keeping all preceding from previous steps. The power at infinity is selected in order to compensate for the factor x and extract constant correction to the initial approximation. Namely the few starting terms are shown below,

$$R_1^*(x) = 1, \quad R_2^*(x) = \frac{1}{2\sqrt{x^2+1}}, \quad R_3^*(x) \equiv R_2(x). \tag{5.5.144}$$

One can also combine calculations with Padé and roots, using low-order iterated roots as the control function. Already the first-order iterated root can be used a a control function. Define the new series

$$L_1(x) = \frac{L(x)}{R_2^*(x)},$$

and apply the technique of diagonal Padé to satisfy the new series asymptotically, order-by-order. The following expression for the critical index follows,

$$\beta_n = \lim_{x\to\infty}(x\,R_2^*(x)) * \lim_{x\to\infty}(PadeApproximant[L_1[x], n, n]). \tag{5.5.145}$$

Automatically we obtain the same correct answer $\beta = 1/2$ in all orders. On can also invert the $DLog$ transformation and find an approximation Φ^* for the function Φ,

$$\Phi^*(x) = \exp\left(\int_0^x R_2^*(X)\,dX\right) = e^{\frac{1}{2}\sinh^{-1}(x)}, \tag{5.5.146}$$

which is just an equivalent representation of the function (5.5.138)! Thus, both approaches work just fine in the dramatic situation when the standard approach fails completely. But the situation can be resolved constructively through control functions.

5.4. Corrected threshold. Test example (5.5.131)

Let us try to re-use the idea of a corrected approximants in order to estimate the threshold [17, 18, 19]. The value of critical index is assumed to be known. Although such assumption is not necessary, it makes sense when one is interested primarily in the value of threshold with the critical index available from different sources.

One can not start without formulating a initial guess, or starting approximation to be corrected. Let is look for the solution in the form of a simple additive approximant,

$$\Phi_2^*(x) = a(x_0 - x)^{-1/2} + b. \tag{5.5.147}$$

Return to the (5.5.131). All three unknowns may be found explicitly from the series (5.5.132)

$$a = \frac{1}{5\sqrt{5}}, \quad x_0 = \frac{4}{5}, \quad b = -\frac{3}{80}.$$

with an approximate threshold value of $x_0 = \frac{4}{5}$ to be corrected when the higher orders from the expansion will enter. Let us look for the solution in the same form but with an exact, yet unknown threshold X_c,

$$F_2^*(x) = a(X_c - x)^{-1/2} + b. \tag{5.5.148}$$

From here one can express formally,

$$X_c = \frac{a^2}{(b - F_2^*(x))^2} + x \tag{5.5.149}$$

since $F_2^*(x)$ is unknown. All we can do is to use for $F_2^*(x)$ the series (5.5.132), so that instead of a true threshold, we have an effective threshold presented in the form of expansion around x_0,

$$X_c(x) = \frac{4}{5} + \frac{13x^3}{128} + \frac{235x^4}{4096} + \frac{189x^5}{8192} + \frac{133x^6}{16384} + \frac{3055x^7}{1048576} + \frac{88173x^8}{67108864} + \frac{221285x^9}{268435456} + \frac{84967x^{10}}{134217728} + \cdots,$$

which should become a true threshold X_c as $x \to X_c$!

Moreover, let us apply resummation procedure to the expansion (5.5.150) using diagonal Padé approximants, and define the sought threshold X_c^* self-consistently,

$$X_c^* = P_{n,n}(X_c^*) \equiv PadeApproximant[X_c^*, n, n], \tag{5.5.150}$$

demanding that as we approach the threshold, the RHS should become the threshold. Since the Padé approximants are defined for arbitrary number of terms k, we will also have a sequence of $X_{c,k}^*$. Solving (5.5.150), we obtain

$$X_{c,3}^* = 0.979664, \quad X_{c,4}^* = 0.992787, \quad X_{c,5}^* = 0.994178, \quad X_{c,6}^* = 0.909891,$$

$$X_{c,7}^* = 0.997532, \quad X_{c,8}^* = 0.998231, \quad X_{c,9}^* = 0.998726, \quad X_{c,10}^* = 0.999166,$$

$$X_{c,11}^* = 0.999259, \qquad X_{c,13}^* = 0.999527.$$

The convergence of the series is good and the last value is closest to the exact result, with the relative percentage error of 0.0473%.

Ratio method [5], also works well. It evaluates the threshold through the value of index $\beta = 1/2$ and ratio of the series coefficients,

$$x_{c,n} = \frac{a_{n-1}}{a_n}\left(\frac{\beta - 1}{n} + 1\right). \tag{5.5.151}$$

The last point gives a good estimate, $x_{c,26} = 0.997904$. The percentage error achieved for the last point is equal to 0.20964%.

In place of the additive approximant (5.5.147), we can use different form and similar technique would apply as long as the effective threshold could be expressed explicitly and expanded into the series. In particular, factor approximants are amenable to such treatment in the context of random composites, see page 256 in Chapter 9.

5.5. Self-similar additive ansatz with critical point

An additive anzats can be applied to the test function most effectively since the parameter/shift is already defined by the position of critical point and the solution for the parameters of approximants is unique. Let us ensure the correct critical index and sub-critical indices already in the starting approximations, so that all parameters in (5.5.152) are obtained by matching it asymptotically with the truncated weak-coupling series (5.5.132). As above, we intend to reconstruct a few starting amplitudes in the strong-coupling expansion.

The additive anzats in low-orders can be readily written in terms of variable z (5.5.134)

$$
\begin{aligned}
\Phi_2^*(z) &= \alpha_{1,2}\sqrt{\alpha z + 1} + \alpha_{2,2}, \\
\Phi_3^*(z) &= \alpha_{1,3}\sqrt{\alpha z + 1} + \alpha_{2,3} + \alpha_{3,3}\frac{1}{\sqrt{\alpha z+1}}, \\
\Phi_4^*(z) &= \alpha_{1,4}\sqrt{\alpha z + 1} + \alpha_{2,4} + \alpha_{3,4}\frac{1}{\sqrt{\alpha z+1}} + \frac{\alpha_{4,4}}{\alpha z+1},
\end{aligned} \tag{5.5.152}
$$

and in terms of the original variable we have the following expressions,

$$
\begin{aligned}
\Phi_2^*(x) &= \frac{1}{8\sqrt{1-x}} - \frac{1}{16} \\
\Phi_3^*(x) &= \frac{3\sqrt{1-x}}{64} + \frac{11}{64\sqrt{1-x}} - \frac{5}{32}, \\
\Phi_4^*(x) &= \frac{x-1}{32} + \frac{9\sqrt{1-x}}{64} + \frac{13}{64\sqrt{1-x}} - \frac{1}{4}.
\end{aligned} \tag{5.5.153}
$$

We stop calculations at 16 terms from (5.5.132) being employed. Corresponding additive approximant is shown below,

$$
\begin{aligned}
\Phi_{16}^*(x) =&\; \frac{155955(1-x)^{3/2}}{131072} + \frac{348495(1-x)^{5/2}}{262144} + \frac{480429(1-x)^{7/2}}{524288} + \frac{85745(1-x)^{9/2}}{262144} + \frac{3289(1-x)^{11/2}}{65536} + \\
&\; \frac{645(1-x)^{13/2}}{262144} + \frac{17(1-x)^{15/2}}{1048576} - \frac{19(1-x)^7}{65536} - \frac{1729(1-x)^6}{131072} - \frac{4731(1-x)^5}{32768} - \frac{78755(1-x)^4}{131072} - \\
&\; \frac{77691(1-x)^3}{65536} - \frac{173223(1-x)^2}{131072} - \frac{16131(1-x)}{16384} + \frac{195867\sqrt{1-x}}{262144} + \frac{262125}{1048576\sqrt{1-x}} - \frac{65493}{131072}.
\end{aligned} \tag{5.5.154}
$$

Thus, we obtain a rapidly convergent sequence for the amplitudes A,

$$
A_2 = 0.125, \quad A_3 = 0.171875, \quad A_4 = 0.203125, \quad A_5 = 0.222656,
$$

$$
A_6 = 0.234375, \quad A_7 = 0.241211, \quad A_8 = 0.245117, \quad A_9 = 0.247314,
$$

$$
A_{10} = 0.248535, \quad A_{11} = 0.249207, \quad A_{12} = 0.249573, \quad A_{13} = 0.249771,
$$

$$
A_{14} = 0.249878, \quad A_{15} = 0.249966, \quad A_{16} = 0.249982.
$$

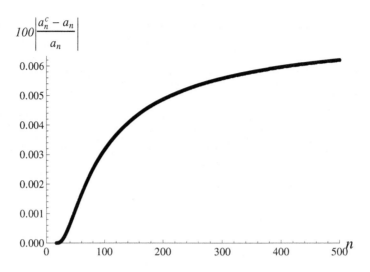

Figure 5.17 Relative percentage error for the coefficients a_n^c calculated from (5.5.154).

The percentage error achieved for the last point is just 0.0072%.

Simultaneously one can obtain rather good estimates for the second amplitude B,

$$B_2 = -0.0625, \quad B_3 = -0.15625, \quad B_4 = -0.25, \quad B_5 = -0.328125,$$

$$B_6 = -0.386719, \quad B_7 = -0.427734, \quad B_8 = -0.455078, \quad B_9 = -0.472656,$$

$$B_{10} = -0.48364, \quad B_{11} = -0.490356, \quad B_{12} = -0.494385, \quad B_{13} = -0.496765,$$

$$B_{14} = -0.498154, \quad B_{15} = -0.499413, \quad B_{16} = -0.499672.$$

The percentage error achieved for the last point is equal to 0.0656%.

From the formula (5.5.154), one can readily calculate the higher-order coefficients a_n^c, not employed in the derivation, and compare them with exact a_n, see Fig. 5.17 below. The quality of reconstruction is very good.

Additive approximants allow also for an interpolation between small-variable and large-variable expansions, when in addition to a_n the values of the coefficients b_n in the strong coupling are also known. There is no principal difference with the extrapolation considered above. One only should take care of the strong-coupling limit by re-expansion in terms of the variable $t = \frac{1}{x}$.

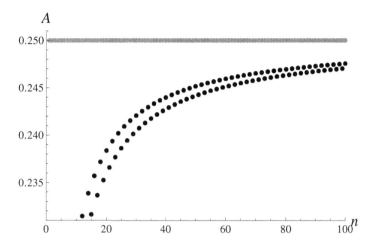

Figure 5.18 Critical amplitude A calculated using formula (5.5.155). A_n is compared with the exact result.

5.6. Correction to additive approximants. Critical amplitudes

Return now to the example (5.5.131). Assume that the values of critical index and threshold are known and "only" the critical amplitudes remain unknown.

Let us first calculate the critical amplitude A using standard approach with Padé approximants. To this end, let us again apply transformation (5.5.134) to the original series to obtain $M_1(z)$ as above. Then we apply to $M_1(z)$ another transformation to get

$$T(z) = M_1(z)^{-1/\beta},$$

in order to get rid of the square-root behaviour at infinity. In terms of $T(z)$, one can readily obtain the sequence of approximations A_n for the critical amplitude A,

$$A_n = x_c^\beta \lim_{z \to \infty} (zPadeApproximant[T[z], n, n+1])^{-\beta}. \tag{5.5.155}$$

The results are shown in Fig. 5.18. The results qualify as divergent by oscillation with the approximation number increasing. Several values are given below,

$$A_{12} = 0.231456, \quad A_{13} = 0.229056, \quad A_{99} = 0.247012, \quad A_{100} = 0.247537.$$

The percentage error achieved for the last point is equal to 0.985%.

Define first the starting approximation as additive approximant,

$$\Phi_2^*(x) = A_0(x_c - x)^{-1/2} + B_0, \tag{5.5.156}$$

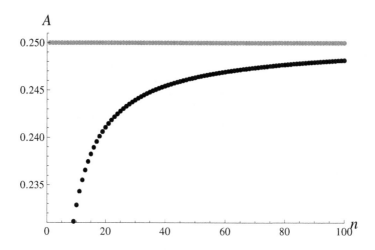

Figure 5.19 Critical amplitude A calculated using formula (5.5.158). A_n are compared with the exact result.

with the unknowns found from the two starting terms of (5.5.132),

$$\Phi_2(x)^* = \frac{1}{8\sqrt{1-x}} - \frac{1}{16}. \tag{5.5.157}$$

We intend to extract corrections to the approximate critical amplitude, $A_0 = \frac{1}{8}$.

First let us divide the original series (5.5.132) by (5.5.157)), then apply to the new series transformation (5.5.134); call the newly found series $G[z]$. Finally build a sequence of the diagonal Padé approximants, so that the corrected amplitudes are expressed by the formula,

$$A_n = A_0 \lim_{z \to \infty}(PadeApproximant[G[z], n, n]). \tag{5.5.158}$$

The results are shown in Fig. 5.19, and the sequence of amplitudes converges monotonously with the approximation number increasing.

Several values for the amplitude are given below,

$$A_{12} = 0.23551, \quad A_{13} = 0.236551, \quad A_{99} = 0.248124, \quad A_{100} = 0.248143.$$

The percentage error achieved for the last point is equal to 0.743%. The convergence is monotonous now.

Let us calculate the amplitude B assuming that amplitude A is already known. The way how we proceed is to look for additive corrections to some plausible "zero-order" approximate solution. Let us start with the choice of the simplest zero-approximation,

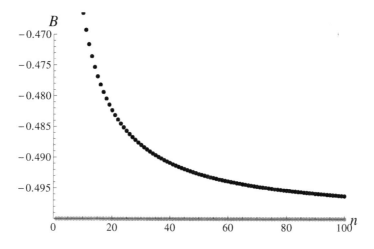

Figure 5.20 Sequence of approximations B_n calculated from (5.5.160).

when amplitude A is known

$$\Phi_2^*(x) = \frac{1}{4\sqrt{1-x}} - \frac{3}{16}.\tag{5.5.159}$$

Subtract (5.5.159) from (5.5.132) to get some new series $g(x)$. The diagonal Padé approximants to the series $g(x)$ are supposed to give a correction to the value of $B_0 = -\frac{3}{16}$, suggested by (5.5.159).

To calculate the correction one has to find the value of the corresponding approximant as $x \to x_c$. The following sequence of approximations for the amplitude B can be calculated now readily,

$$B_n = B_0 + PadeApproximant[g(x \to x_c, n, n].\tag{5.5.160}$$

The sequence of approximations is shown in Fig. 5.20 and it converges to the the known exact value.

6. Interpolation with self-similar root approximants

The self-similar approximants described above allow to interpolate between the small-variable and large-variable expansions. In particular, based on root approximants one can construct an analytical expression uniformly approximating the sought function in the whole domain $[0, \infty]$ and reproducing both the small-variable and large-variable expansions [25, 62]. Importantly, the uniqueness conditions are satisfied allowing for a unique definition of all parameters. The method of root approximants is more general than that of standard Padé approximants as well as the Baker–Gammel method of fractional Padé approximants. Therefore, by construction, the accuracy of the method is not worse, and often better, than that of the latter methods, with the advantage of being uniquely defined.

Let us be interested in a physical quantity represented by a real function $\Phi(x)$ of a real variable $x \in [0, \infty]$. However, the explicit form of this function is not known, since it is defined by complicated equations allowing only for deriving asymptotic expansions in the vicinity of two ends, where $x \to 0$ and $x \to \infty$.

For instance, in the small-variable limit, we have

$$\Phi(x) \simeq \Phi_k(x), \qquad \text{as } x \to 0, \tag{5.6.161}$$

with the series

$$\Phi_k(x) = \Phi_0(x)\left(1 + \sum_{n=1}^{k} a_n x^n\right), \tag{5.6.162}$$

where $\Phi_0(x)$ is a known function. In many cases, the latter enjoys the form

$$\Phi_0(x) = A_0 x^\alpha \qquad (A_0 \neq 0), \tag{5.6.163}$$

with α being any real number.

And in the large-variable limit, we can get

$$\Phi(x) \simeq \Phi^{(p)}(x), \qquad \text{as } x \to \infty, \tag{5.6.164}$$

with the series

$$\Phi^{(p)}(x) = \sum_{n=1}^{p} b_n x^{\beta_n}, \tag{5.6.165}$$

where the powers β_n are real numbers arranged in the descending order

$$\beta_{n+1} < \beta_n \qquad (n = 1, 2, \ldots, p - 1).$$

The values of the known powers β_n can be any, either integer or fractional. For interpolation problem all amplitudes b_n are known as well.

In what follows, it is convenient to deal with the reduced function $\frac{\Phi(x)}{\Phi_0(x)}$, normalized so that in the small-variable limit,

$$\frac{\Phi(x)}{\Phi_0(x)} \simeq \frac{\Phi_k(x)}{\Phi_0(x)}, \qquad \text{as } x \to 0,$$

we would have the simple asymptotic form

$$\frac{\Phi_k(x)}{\Phi_0(x)} = 1 + \sum_{n=1}^{k} a_n x^n .$$

The interpolation problem consists in constructing such a representation for the sought function $\Phi(x)$ that would reproduce the small-variable, as well as large-variable expansions (5.6.162) and (5.6.165), providing an accurate approximation for the whole domain $[0, \infty]$.

This formulation of interpolation problem is rather general. In the majority of cases, practically all realistic problems can be reduced to this representation employing a change of variables. Also, the small-variable and large-variable limits are conditional, since it is always possible to interchange them by introducing the variable $t = x^{-1}$, or more generally, $t = x^{-\mu}$, with a positive μ.

Consider the *self-similar root approximant*

$$\frac{\Phi_k^*(x)}{\Phi_0(x)} = \left(\left(\cdots (1 + \mathcal{P}_1 x)^{n_1} + \mathcal{P}_2 x^2 \right)^{n_2} + \cdots + \mathcal{P}_k x^k \right)^{n_k}. \tag{5.6.166}$$

Setting here all powers $n_j = \pm 1$, one can obtain different Padé approximants. And if the powers $n_j = \pm 1$, except the leading n_k that is found from the leading term of the large-variable expansion, then we get the modified Padé approximants of Baker and Gammel [4]. However, such a choice of the powers n_j is restrictive and arbitrary. The parameters of approximants, to be uniquely defined, have to follow from the available small-variable and large-variable expansions.

The root approximants (5.6.166) were used for extrapolation so that the powers n_i and parameters $\mathcal{P}_i$ were defined through the one-sided expansion, say, the large-variable expansion, while the other expansion, e.g., small-variable expansion, was not reproduced. The attempts to find all values of n_i and $\mathcal{P}_i$ from the small-variable expansion resulted in the equations with multiple solutions. Such a non-unique definition of the parameters is, of course, is desirable to avoid.

Nevertheless as far as the interpolation problem is concerned, it is possible to use the root approximants (5.6.166) in such a way that allows to uniquely define all powers n_i and parameters $\mathcal{P}_i$ and that both the small-variable as well as the large-variable expansions be reproduced.

The root approximant (5.6.166) can be identically rewritten as

$$\frac{\Phi_k^*(x)}{\Phi_0(x)} = \mathcal{P}_k^{n_k} x^{k n_k} \left(1 + \frac{B_{k-1}}{x^{m_{k-1}}} \left(1 + \frac{B_{k_2}}{x^{m_{k-2}}} \cdots \frac{B_1}{x^{m_1}} \left(1 + \frac{1}{\mathcal{P}_1 x} \right)^{n_1} \right)^{n_2} \cdots \right)^{n_k}, \tag{5.6.167}$$

with the parameters

$$B_j = \frac{\mathcal{P}_j^{n_j}}{\mathcal{P}_{j+1}} \qquad (j = 1, 2, \ldots, k - 1) \tag{5.6.168}$$

and powers

$$m_j = j + 1 - j n_j \qquad (j = 1, 2, \ldots, k - 1). \tag{5.6.169}$$

On the other hand, the large-variable expansion (5.6.165) can be represented in the form

$$\Phi^{(p)}(x) = b_1 x^{\beta_1} \left(1 + \frac{b_2}{b_1} x^{\beta_2 - \beta_1} \left(1 + \frac{b_3}{b_2} x^{\beta_3 - \beta_2} \cdots \right. \right.$$

$$\left. \left. \cdots \frac{b_{p-1}}{b_{p-2}} x^{\beta_{p-1} - \beta_{p-2}} \left(1 + \frac{b_p}{b_{p-1}} x^{\beta_p - \beta_{p-1}} \right) \right) \cdots \right). \tag{5.6.170}$$

After expanding the above expressions (5.3.82) and (5.6.170) in powers of $1/x$ and equating the similar terms, it turns out that these expansions are uniquely defined provided that Eq. (5.6.163) is valid,

$$\mathcal{P}_k^{n_k} = \frac{b_1}{A_0} \qquad (k = p), \tag{5.6.171}$$

the largest power n_k is given by the relation

$$k n_k = \beta_1 - \alpha \qquad (\alpha \neq \beta_1), \tag{5.6.172}$$

and the other powers satisfy the equations

$$m_j = \beta_{k-j} - \beta_{k-j+1} \qquad (j = 1, 2, \ldots, k - 1). \tag{5.6.173}$$

In this way, all powers n_j of the root approximant (5.6.166) can be uniquely defined through the *uniqueness conditions*

$$j n_j = j + 1 - \beta_{k-j} + \beta_{k-j+1} \qquad (j = 1, 2, \ldots, k - 1). \tag{5.6.174}$$

It may happen that the number of terms in the small-variable and large-variable asymptotic expansions are not the same, $k \neq p$. Sometimes, just a single term of the large-variable expansion is known ($p = 1$), while several terms of the small-variable expansion are available. How then the uniqueness conditions (5.6.174) will be changed?

Let us assume that just the leading term of the large-variable behaviour is known:

$$\Phi(x) \simeq A x^\beta, \qquad \text{as } x \to \infty. \tag{5.6.175}$$

It is easy to notice that the general expansion (5.6.165) is reducible to the asymptotic form (5.6.175) by setting $\beta_n = \beta$. Then the uniqueness condition (5.3.89) reduces to the equality

$$n_j = \frac{j + 1}{j} \qquad (k > p = 1), \tag{5.6.176}$$

where $j = 1, 2, \ldots, k - 1$, while the leading power reads as

$$n_k = \frac{\beta - \alpha}{k} \qquad (k = 1, 2, \ldots). \tag{5.6.177}$$

All parameters $\mathcal{P}_i$ are uniquely defined from the small-variable expansion.

In the general case, we have k terms of the small-variable expansion and p terms of the large-variable expansion. Therefore, to satisfy these expansions, the root approximant must be of order $k + p$, possessing $k + p$ parameters $\mathcal{P}_j$, among which k parameters $\mathcal{P}_j$ are defined by the accuracy-through-order procedure from the small-variable expansion and the remaining p parameters are defined from the large-variable expansion. But then we also need to have $k + p$ equations for determining $k + p$ powers n_j, while only p terms of the large-variable expansion are given. How all powers n_j could be found in such a case?

Fortunately, large-variable expansions practically always enjoy the following nice property. The difference

$$\Delta \beta_j \equiv \beta_j - \beta_{j+1} \qquad (j = 1, 2, \ldots, k - 1) \tag{5.6.178}$$

between the nearest-neighbor powers is invariant:

$$\Delta \beta_j = \Delta \beta = const \qquad (j = 1, 2, \ldots, k - 1). \tag{5.6.179}$$

In that case, with the leading power being always given by an equation of type (5.6.172) and with all remaining powers n_j being defined by the uniqueness condition (5.6.174), we now have

$$n_{k+p} = \frac{\beta_1 - \alpha}{k + p}, \qquad jn_j = j + 1 - \Delta \beta \qquad (j = 1, 2, \ldots, k + p - 1). \tag{5.6.180}$$

Then both the small-variable as well as the large-variable expansions can be satisfied, uniquely defining all parameters $\mathcal{P}_j$, with $j = 1, 2, \ldots, k + p$.

Consider an example of interpolation from the series presented in [14]. It is typical for finite quantum systems which often enjoy spherical symmetry. The example is important for applications. It is also a a difficult case for resummation since it corresponds to an incommensurate powers in different limits. Remarkably, the same technique can be applied to quantum and classical systems regardless.

An N-electron harmonium atom is described by the Hamiltonian

$$\hat{H} = \frac{1}{2} \sum_{i=1}^{N} \left(-\nabla_i^2 + v^2 r_i^2 \right) + \frac{1}{2} \sum_{i \neq j}^{N} \frac{1}{r_{ij}},$$

where dimensionless units are employed and

$$r_i \equiv |\mathbf{r}_i|, \qquad r_{ij} \equiv |\mathbf{r}_i - \mathbf{r}_j|.$$

This Hamiltonian provides a rather realistic modeling of trapped ions, quantum dots, and some other finite systems, such as atomic nuclei and metallic grains [11].

At a shallow harmonic potential, the energy can be expanded [14] in powers of v, so that

$$E(v) \simeq E_k(v), \qquad \text{as } v \to 0$$

with the truncated series

$$E_k(v) = \sum_{n=0}^{k} c_n v^{(2+n)/3}.$$

For instance, to third order,

$$E_3(v) = c_0 v^{2/3} + c_1 v + c_2 v^{4/3},$$

with the coefficients

$$c_0 = \frac{3}{2^{4/3}} = 1.19055, \qquad c_1 = \frac{1}{2}\left(3 + \sqrt{3}\right) = 2.36603, \qquad c_2 = \frac{7}{36} 2^{-2/3} = 0.122492.$$

And for a rigid potential, the energy is approximated [14] as

$$E(v) \simeq E^{(p)}(v) \qquad (v \to \infty),$$

where

$$E^{(p)}(v) = \sum_{n=0}^{p} b_n v^{(2-n)/2}.$$

To fourth order, one has

$$E^{(4)}(v) = b_0 v + b_1 v^{1/2} + b_2 + b_3 v^{-1/2},$$

where

$$b_0 = 3, \qquad b_1 = \sqrt{\frac{2}{\pi}} = 0.797885, \qquad b_2 = -\frac{2}{\pi}\left(1 - \frac{\pi}{2} + \ln 2\right) = -0.077891,$$

$$b_3 = \left(\frac{2}{\pi}\right)^{3/2}\left[2 - 2G - \frac{3}{2}\pi + (\pi + 3)\ln 2 + \frac{3}{2}\ln^2 2 - \frac{\pi^2}{24}\right] = 0.0112528,$$

with the Catalan constant

$$G \equiv \sum_{n=0}^{\infty} \frac{(-1)^n}{(2n+1)^2} = 0.91596559.$$

The root approximant, respecting all given small-v, as well as large-v expansions, is

$$E_6^*(v) = c_0 v^{2/3} \times$$

$$\left(\left(\left(\left(\left(\left(1 + \mathcal{P}_1 v^{1/3}\right)^{1/2} + \mathcal{P}_2 v^{2/3}\right)^{3/4} + \mathcal{P}_3 v\right)^{5/6} + \mathcal{P}_4 v^{4/3}\right)^{7/8} + +\mathcal{P}_5 v^{5/3}\right)^{9/10} + \mathcal{P}_6 v^2\right)^{1/6},$$

$$(5.6.181)$$

with the parameters

$$c_0 = 1.19055, \qquad \mathcal{P}_1 = 48.4532, \qquad \mathcal{P}_2 = 564.108,$$

$$\mathcal{P}_3 = 1088.39, \qquad \mathcal{P}_4 = 1221.08, \qquad \mathcal{P}_5 = 796.791, \qquad \mathcal{P}_6 = 256.$$

One can also estimate the accuracy of the root approximant comparing it with the numerical data from [44] and find that its maximal error is only 0.9%. Note that Padé approximants cannot be used in the case of harmonium, since the small-variable and large-variable asymptotic expansions are incompatible. The best additive approximant, respecting all given small-ν, as well as large-ν expansions, is

$$E^*_{ad}(\nu) = \nu^{2/3} \times$$

$$(1.99563(1.50329\sqrt[3]{\nu} + 1) + \frac{0.978275}{\sqrt{1.50329\sqrt[3]{\nu}+1}} - \frac{0.176023}{(1.50329\sqrt[3]{\nu}+1)^2} - \frac{0.352047}{(1.50329\sqrt[3]{\nu}+1)^3} -$$

$$\frac{0.52807}{(1.50329\sqrt[3]{\nu}+1)^4} + \frac{0.597767}{(1.50329\sqrt[3]{\nu}+1)^5} + \frac{0.489137}{(1.50329\sqrt[3]{\nu}+1)^{3/2}} + \frac{0.366853}{(1.50329\sqrt[3]{\nu}+1)^{5/2}} +$$

$$\frac{0.352582}{(1.50329\sqrt[3]{\nu}+1)^{7/2}} - \frac{0.537923}{(1.50329\sqrt[3]{\nu}+1)^{9/2}} - 1.99563).$$

$$(5.6.182)$$

To construct it, we employed all terms from both expansions, as well as counter-terms to suppress incorrect powers. There is no principal difference in the way how parameters are obtained, compared with the extrapolation problem for additive approximants and with the interpolation problem for roots. Formula (5.6.182) is slightly more accurate than root approximant. Its maximal error is about 0.33%. But the root approximant has more compact form than additives. Besides, it is defined uniquely. The advantage of additive ansatz is particularly clear though in the cases with fast growing coefficients such as zero-dimensional field theory or quartic oscillator described in preceding sections.

Both constructs involve no fitting parameters (although they could be introduced if needed). This is especially important in those complicated cases, where numerical data in the whole region of the variable are not available. The absence of fitting parameters makes the approach different from other interpolation methods, such as the Cioslowski method [14]. Generally, the suggested interpolation technique provide the accuracy not worse than the method of Padé approximants and in the majority of cases is more accurate than the latter.

In summary, we discussed in this chapter the problems of extrapolation and interpolation of the perturbation-theory expansions, obtained for asymptotically small variable as $x \to 0$, to the large-variable limit as $x \to \infty$. For this purpose, we concentrated on different variants of self-similar approximations, resulting in iterated root approximants, self-similar factor, self-similar root approximants, additive self-similar approximants and Corrected Padé approximants. Applying the Occam's razor, we would always start with the Padé approximants and their variations and then try more complex forms, usually in the same order as just presented above. The Padé approximants can be easily calculated in very high-orders using a variety of existing packages.

The accuracy and scope of the Padé approximants can be strongly improved by combining the method of self-similar approximants with that of Padé approximants, as is demonstrated in [27]. The Corrected Padé approximation scheme covers broader class of functions than Standard scheme, including indeterminate situations. It can still be rather easily constructed and its computational accuracy is essentially the same with the diagonal Padé approximants. Similar statement applies to the rate of convergence when Standard Padé scheme appears to converge. Most importantly the Corrected scheme works well when the Standard scheme fails completely. Computationally it is still is user friendly and allows to fully employ the existing packages for the Padé approximants.

Finally, only those problems, whose large-variable behaviour is of power law type were discussed. But there exists another class of problems possessing exponential behaviour and also demonstrating the Stokes phenomenon. For the problems of this class, it is necessary to use another variant of the self-similar approximation theory, involving the self-similar exponential approximants [63, 73]. The latter make it possible to derive accurate approximations for the functions of exponential behaviour as well as to treat the problems accompanied by the Stokes phenomenon.

REFERENCE

1. Adamyan VM, Tkachenko IM, Urrea M, Solution of the Stieltjes Truncated Moment Problem, Journal of Applied Analysis 2003; 9: 57–74.
2. Ambroladze A, Wallin H, Approximation by repeated Padé approximants, Journal of Computational and Applied Mathematics 1995; 62: 353–358.
3. Ambroladze A, Wallin H, Convergence Rates of Padé and Padé -Type Approximants, Journal of approximation theory 1996; 86: 310–319.
4. Baker GA, Gammel JL, The Padé approximant, J. Math. Anal. Appl. 1961; 2: 21–30.
5. Baker GA, Graves-Moris P, Padé Approximants. Cambridge, UK: Cambridge University; 1996.
6. Banks T, Torres TJ, Two Point Padé Approximants and Duality, arXiv:1307.3689v2 [hep-th] 13 Aug 2013.
7. Bender CM, Boettcher S, Determination of $f(\infty)$ from the asymptotic series for $f(x)$ about $x = 0$, J. Math.Phys 1994; 35, 1914–1921.
8. Bender CM, Orszag SA, Advanced Mathematical Methods for Scientists and Engineers. Asymptotic Methods and Perturmation Theory. New York: Springer; 1999.
9. Bender CM, Wu TT, Anharmonic oscillator, Phys. Rev. 1969; 184: 1231–1260.
10. Berg C, Indeterminate moment problems and the theory of entire functions, Journal of Computational and Applied Mathematics 1995; 65: 27–55.
11. Birman JL, Nazmitdinov RG, Yukalov VI, Effects of symmetry breaking in finite quantum systems, Phys. Rep. 2013; 526: 1–91.
12. Bogolubov NN, Mitropolsky YA, Asymptotic Methods in the Theory of Nonlinear Oscillations. New York: Gordon and Breach; 1961.
13. Bogolubov NN, Shirkov DV, Quantum Fields. London: Benjamin; 1983.
14. Cioslowski J, Robust interpolation between weak-and strong-correlation regimes of quantum systems, J. Chem. Phys 2012; 136: 044109.
15. Erdélyi A, Asymptotic Expansions. New York: Dover; 1955.
16. Fernandez FM, Prony's method and the connected-moments expansion, Physica Scripta 2011; 87:

025006.

17. Gluzman S, Mityushev V, Nawalaniec W, Cross-properties of the effective conductivity of the regular array of ideal conductors, Arch. Mech. 2014; 66: 287–301.

18. Gluzman S, Mityushev V, Series, index and threshold for random 2D composite, Arch. Mech. 2015; 67: 75–93.

19. Gluzman S, Mityushev V, Nawalaniec W, Starushenko G, Effective Conductivity and Critical Properties of a Hexagonal Array of Superconducting Cylinders, ed. P.M. Pardalos (USA) and T.M. Rassias (Greece), Contributions in Mathematics and Engineering. In Honor of Constantin Caratheodory. Springer: 255–297, 2016.

20. Gluzman S, Yukalov VI, Algebraic Self-Similar Renormalization in Theory of Critical Phenomena, Physical Review E 1997; 55: 3983–3999.

21. Gluzman S, Yukalov VI, Unified approach to crossover phenomena, Phys. Rev. E 1998; 58: 4197–4209.

22. Gluzman S, Yukalov VI, Sornette D, Self-similar factor approximants, Phys. Rev. E 2003; 67: 026109.

23. Gluzman S, Yukalov VI, Self-similar extrapolation from weak to strong coupling, J.Math.Chem. 2010; 48: 883–913.

24. Gluzman S, Yukalov VI, Extrapolation of perturbation theory expansions by self-similar approximants, Eur. J. Appl. Math. 2014; 25: 595–628.

25. Gluzman S, Yukalov VI, Effective summation and interpolation of series by self-similar approximants, Mathematics 2015; 3: 510–526.

26. Gluzman S, Yukalov VI, Additive self-similar approximants, Journal of Mathematical Chemistry 2016. DOI 10.1007/s10910-016-0698-4 2016.

27. Gluzman S, Yukalov VI, Self-Similarly corrected Padé approximants for indeterminate problem, European Physical Journal Plus 2016; 131: 340–361.

28. Gluzman S, Yukalov VI, Self-similar continued root approximants, Physics Letters A 2012; 377; 124–128.

29. Gonchar AA, Rational Approximation of Analytic Functions, Proceedings of the Steklov Institute of Mathematics 2011; 272, Suppl.2: S44–S57.

30. Grosberg AYu, Khokhlov AR, Statistical Physics of Macromolecules. NY: American Insitute of Physics; 1994.

31. Hardy GH, Divergent Series. Oxford: Oxford University; 1949.

32. Honda M, On perturbation theory improved by strong coupling expansion, J. High En. Phys. 2014; 14: 1–44.

33. Janke W, Kleinert H, Convergent strong-coupling expansions from divergent weak-coupling perturbation theory, Phys. Rev. Lett. 1995; 75: 2787–2791.

34. Kastening B, Fluctuation pressure of a fluid membrane between walls through six loops, Phys. Rev. E 2006; 73:, 011101.

35. Kedlava KS, The algebraic closure of the power series filed in positive characteristic, Proc. Amer. Math. Soc. 2001; 129: 3461–3470.

36. Kleinert H, Path Integrals in Quantum Mechanics, Statistics, Polymer Physics and Financial Markets . Singapore: World Scientific; 2006.

37. Kleinert H, Systematic corrections to the variational calculation of the effective classical potential, Phys. Lett. A 1993; 173: 332–342.

38. Krein MG, Nudel'man AA, The Markov moment problem and extremal problems. Translations of Mathematical Monographs Providence, RI.: American Mathematical Society; 1977.

39. Lam PM, The structure function of branched polymers in a good solvent: a lattice calculation, J. Chem. Phys. 1990; 92: 3136–3143.

40. Lubinsky D, Reflections on the Baker-Gammel-Wills (Padé) Conjecture, G. Milovanovic, M. Rassias, eds. Analytic Number Theory, Approximation Theory, and Special Functions. NY: Springer; 561–571, 2014.

41. MacLane S, The universality of formal power series fileds, Bull. Amer. Math. Soc. 1939; 45: 888–890.

42. Mainardi F, Goren R, On Mittag-Leffler-type functions in fractional evolution processes, Journal of Computational and Applied Mathematics 2000; 118: 283–299.
43. Malomuzh NP, Sushko MYa, On the character of narrowing of spectral lines near the phase transition isotropic liquid-nematic, Opt. and Spectr. (SSSR) 1987; 62: 386–391.
44. Matito E, Cioslowski J, Vyboishchikov SF, Properties of harmonium atoms from FCI calculations: Calibration and benchmarks for the ground state of the two-electron species, Phys. Chem. Chem. Phys. 2010; 12: 6712–6716.
45. Miller JD, Exact pair correlation function of a randomly branched polymer, Eur. Phys. Lett. 1991; 16: 623–628.
46. Nayfeh AH, Perturbation Methods. New York: John Wiley; 1981.
47. Puiseux VA, Recherches sur les fonctions algébriques, J. Math. Pures Appl. 1850; 15: 365–480.
48. Saff EB, Varga RS, On the sharpness of theorems concerning zero-free regions for certain sequences of polynomials, Numer. Math. 1976; 26: 245–354.
49. Selyugin OV, Smondyrev MA, Phase transition and Padé approximants for Fröhlich polarons, Phys. Stat. Sol. B 1989; 155: 155–167.
50. Sen A, S-duality improved superstring perturbation theory, J. High En. Phys. 2013; 11: 1–23.
51. Simon B, Fifty years of eigenvalue perturbation theory, Bull. Am. Math. Soc. (1991; 24: 303–319.
52. Suetin SP, Padé approximants and efficient analytic continuation of a power series, Russian Mathematical Surveys 2002; 57: 43–141.
53. Wall HS, Analytic Theory of Continued Fractions. Bronx N.Y.: Chelsea Publishing Company; 1948.
54. Wilf HS, Generatingfunctionology. 2nd ed. Boston, MA: Academic Press; 1994. ISBN 0-12-751956-4. Zbl 0831.05001.
55. Yukalov VI, Theory of perturbations with a strong interaction, Moscow Univ. Phys. Bull. 1976; 51: 10–15.
56. Yukalov VI, Model of a hybrid crystal, Theor. Math. Phys. 1976; 28: 652–660.
57. Yukalov VI, Statistical mechanics of strongly nonideal systems, Phys. Rev. A 1990; 42: 3324–3334.
58. Yukalov VI, Self-similar approximations for strongly interacting systems, Physica A 1990; 167: 833–860.
59. Yukalov VI, Method of self-similar approximations, J. Math. Phys. 1991; 32: 1235–1239.
60. Yukalov VI, Stability conditions for method of self-similar approximations, J. Math. Phys. 1992; 33: 3994–4001.
61. Yukalov VI, Gluzman S, Optimisation of self-similar factor approximants, Mol.Phys. 2009; 107: 2237–2244.
62. Yukalov VI, Gluzman S, Self-similar interpolation in high-energy physics, Phys. Rev. D 2015; 91: 125023.
63. Yukalov VI, Gluzman S, Self-similar exponential approximants, Phys. Rev. E 1998; 58: 1359–1382.
64. Yukalov VI, Gluzman S, Self-Similar Bootstrap of Divergent Series, Physical Review E 1997; 55: 6552-6570.
65. Yukalov VI, Gluzman S, Sornette D, Summation of Power Series by Self-Similar Factor Approximants, Physica A 2003; 328: 409–438.
66. Yukalov VI, Gluzman S, Extrapolation of power series by self-similar factor and root approximants, Int. J. Mod. Phys. B 2004; 18: 3027–3046.
67. Yukalov VI, Gluzman S, Self-similar crossover in statistical physics, Physica A 1999; 273: 401–415.
68. Yukalov VI, Yukalova EP, Self-similar approximations for thermodynamic potentials, Physica A 1993; 198: 573–592.
69. Yukalov VI, Yukalova EP, Higher orders of self-similar approximations for thermodynamic potentials, Physica A 1994; 206: 553–580.
70. Yukalov VI, Yukalova EP, Temporal dynamics in perturbation theory, Physica A 1996; 225: 336–362.

71. Yukalov, VI, Yukalova EP, Method of self-similar factor approximants, Phys. Lett. A 2007; 368: 341–347.
72. Yukalov VI, Yukalova EP, Self-similar structures and fractal transforms in approximation theory, Chaos Solit. Fract. 2002; 14: 839–861.
73. Yukalov VI, Yukalova EP, Gluzman S, Self-similar interpolation in quantum mechamics, Phys. Rev. A 1998; 58: 96–115.

CHAPTER 6

Conductivity of regular composite. Square lattice

At university,...., things are the opposite of the ways of the normal world:it isn't the sons who hate the fathers, but the fathers who hate the sons.

—Umberto Eco, Numero Zero

1. Introduction

In this chapter, we consider a two-dimensional composite corresponding to the regular square lattice (square array) arrangement of ideally conducting cylinders of radius r embedded into the matrix of a conducting material as shown in Fig. 6.1. The volume fraction (concentration) of cylinders is equal to $f = \pi r^2$ and corresponding periodicity cell is shown in Fig. 6.2.

The effective conductivity $\sigma_e = \sigma_e(f)$ of regular composites was discussed in the lowest orders in f by such giants of science as Maxwell [27] and Rayleigh [33]. Their work was continued in [32], resulting in rather good numerical solutions. The exact formulae (4.2.28), (4.2.29) on page 86 give the series representation of $\sigma_e(f)$ in the powers of f that can be considered as the final solution to the problem except the case $\rho = 1$ as $f \to f_c = \frac{\pi}{4}$. Considerable efforts have been dedicated to this particular array.

Already when the ratio of the conductivity of inclusions to the conductivity of matrix is greater than several hundreds, it can be taken to be infinite and inclusions could be considered as ideal conductors. Long series and rational approximations were proposed in [34] to find $\sigma_e(f)$. Two-point Padé approximants were applied in [2, 4, 39] to derive upper and lower bounds on the effective conductivity. Asymptotically equivalent functions were constructed in [6] and applied to the percolating random media [5].

In the high-concentration case, Keller [24] obtained close to the particle-phase threshold f_c, the following expression for the effective conductivity,

$$\sigma_e(f) \simeq \frac{\pi^{\frac{3}{2}}}{2\sqrt{\frac{\pi}{4} - f}},$$ (6.1.1)

Computational Analysis of Structured Media
http://dx.doi.org/10.1016/B978-0-12-811046-1.50006-X

Figure 6.1 Section perpendicular to the unidirectional cylinders is shown. Following [32, 33], we assume that the periodicity cell is the unit square and the radius r is given as a dimensionless value. The conductivity of matrix can be normalized to unity.

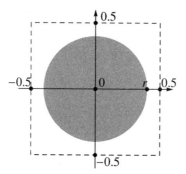

Figure 6.2 Periodicity cell with a disk of radius r.

and the value of amplitude is considered as exact

$$A = \frac{1}{2}\, \pi^{\frac{3}{2}}.$$

In the language of critical phenomena the divergence in (6.1.1) is controlled by the superconductivity critical index $\mathsf{s} = \frac{1}{2}$.

This formula can be supplemented by a constant term [29],

$$\sigma_e(f) \simeq \frac{\pi^{\frac{3}{2}}}{2\sqrt{\frac{\pi}{4} - f}} - \pi + 1. \tag{6.1.2}$$

Formula (6.1.2) follows from equation (49) of [29] for a dipole coefficient when the conductivity of inclusions tends to infinity. It also suggests the form of the higher-order terms. But the value of another amplitude $B = -\pi + 1$, is a subject of debate. In the case of a square lattice of inclusions, the popular approximate approach called lubrication theory gives the following asymptotic result [3],

$$\sigma_e(f) \simeq \frac{\pi^{\frac{3}{2}}}{2\sqrt{\frac{\pi}{4} - f}} - 1. \tag{6.1.3}$$

Formula (6.1.3) appears to be less accurate than (6.1.2).

In all the previous body of work such expressions for $\sigma_e(f)$ valid near f_c, were not matched with series expansions at $f = 0$, though tight bounds were obtained in [2, 39] for a sufficiently large f. We suppose that previous failure to construct the global approximation, is related to the principal impossibility of a uniform approximation by rational functions on the closed segment $[0, \frac{\pi}{4}]$, of the function $\sigma_e(f)$ having a fractional power singularity.

Our primary goal is to apply the approach of Chapter 5 to the series (4.2.28) in order to overcome such difficulty and derive an accurate, compact expression for the effective conductivity of the square array, valid for arbitrary concentrations. Moreover, we demonstrate that the series (4.2.28) derived rigorously near $f = 0$, capture implicitly the critical value $f_c = \frac{\pi}{4}$ and even the type of singularity.

From the phase interchange theorem [25], it follows that in two dimensions, the superconductivity index s is equal to the conductivity index t [40]. The latter case corresponds to the vanishing as a power law, effective conductivity of the conducting inclusions within a non-conducting matrix.

We also demonstrate how the most typical characteristics of the critical phenomena, the threshold value and the superconductivity critical index, can be calculated directly from the series (4.2.28) for $\rho = 1$. To this end, we primarily apply properly modified techniques of Padé approximants, widely used in the theory of critical phenomena.

The case is also relevant from the viewpoint of analogy with critical phenomena [37]. The critical behaviour of regular composites is practically never mentioned in such context. Relatively "simple" Laplace equation for the potential, when complemented with a non-trivial boundary conditions in the regular domain of inclusions (see, e.g., [34]), behaves critically even without explicit non-linearity or randomness, typical to the phase transitions and percolation phenomena [37].

The considered problem can be formulated as follows. Assume the polynomial approximation of the function $\sigma_e(f)$ is given. We have to estimate the convergence radius f_c of the Taylor series of $\sigma_e(f)$, and to determine parameters of the asymptotically equivalent approximation near $f = f_c$. For the sought quantities, we are going to obtain numerical sequences of approximations, and make conclusions based on convergence or semi-convergence of various sequences.

2. Critical point, square array

Underlying expressions for the series coefficients can be found from (4.2.28) for $\rho = 1$. Below, this expansion is presented in the truncated numerical form

$$
\begin{aligned}
\sigma_e(f) = \quad & 1 + 2f + 2f^2 + 2f^3 + 2f^4 + 2.6116556664543236 f^5 \\
+ \quad & 3.2233113329086476 f^6 + 3.8349669993629707 f^7 \\
+ \quad & 4.446622665817295 f^8 + 5.272062706160579 f^9 \\
+ \quad & 6.284564073656707 f^{10} + 7.484126768305675 f^{11} \\
+ \quad & 8.870750790107483 f^{12} + 11.376487693870311 f^{13} \\
+ \quad & 14.200048348474557 f^{14} + 17.39864131428396 f^{15} \\
+ \quad & 21.029475151662286 f^{16} + 26.277897666772887 f^{17} \\
+ \quad & 32.66592466836545 f^{18} + 40.31075089986001 f^{19} \\
+ \quad & 49.34706707473475 f^{20} + 62.10071128310876 f^{21} \\
+ \quad & 77.45700559231325 f^{22} + 95.88562104938396 f^{23} \\
+ \quad & 117.89818526393154 f^{24} + 149.04923618206843 f^{25} \\
+ \quad & 186.9598285755466 f^{26} + O(f^{27}).
\end{aligned}
\tag{6.2.4}
$$

Can we really extract a purely geometrical quantity, such as f_c, from the expansion (6.2.4) for the physical quantity $\sigma_e(f)$? The natural condition on f_c is to assume that it coincides with the point where σ diverges.

Since we are dealing with the limiting case of a perfectly conducting inclusions when the conductivity of inclusions tends to infinity, the effective conductivity is also expected to tend to infinity as a power law, as the concentration f tends to the maximal value f_c for the square array,

$$
\sigma_e(f) \simeq A(f_c - f)^{-s} + B.
\tag{6.2.5}
$$

The critical superconductivity index (exponent) s is believed to be $\frac{1}{2}$ for all lattices [13]. For sake of exploring how consistent are various resummation techniques, we will calculate the index. The critical amplitudes A and B will be treated as unknown non-universal parameters to be calculated below as well.

More generally we have to respect the high-concentration regime as much as the low-concentration regime. It turns out that detail knowledge of A and B is sufficient for derivation of very accurate formula. Based on acquired information on the amplitudes we are going to construct working expressions for the effective conductivity valid in the whole region of concentrations and equally respecting both regimes.

2.1. Padé approximants

Probably the simplest way to estimate the position of a critical point, is to apply the diagonal Padé approximants,

$$P_{1,1}(f) = \frac{A_1 f + 1}{B_1 f + 1}, \qquad P_{2,2}(f) = \frac{A_2 f^2 + A_1 f + 1}{B_2 f^2 + B_1 f + 1}, \qquad (6.2.6)$$

and so forth. Padé approximants locally are the best rational approximations of power series. Their poles determine singular points of the approximated functions. Calculations with Padé approximants are straightforward and can be performed with *Mathematica*®. They do not require any preliminary knowledge of the critical index and we have to find the position of a simple pole. There is a convergence within the approximations for the critical point generated by the sequence of Padé approximants, corresponding to their order increasing:

$$f_1 = 1, \quad f_2 = 1, \quad f_4 = 0.84447, \quad f_5 = 0.842471,$$

$$f_6 = 0.842781, \quad f_7 = 0.842471, \quad f_8 = 0.8446, \quad f_9 = 0.804535,$$

$$f_{10} = 0.804611, \quad f_{11} = 0.804535, \quad f_{12} = 0.80451, \quad f_{13} = 0.804536.$$

The percentage error given by the last approximant in the sequence equals 2.43649%.

2.2. Asymptotically equivalent functions with known critical index

We assume that incorporating explicitly the known value of critical index, will lead to improved accuracy in the threshold estimates. Let us use asymptotically equivalent functions in the form motivated by the diagonal Padé approximants, but raised into corresponding power [10],

$$P_1(f) = \sqrt{\frac{A_1 f + 1}{B_1 f + 1}}, \quad P_2(f) = \sqrt{\frac{A_2 f^2 + A_1 f + 1}{B_2 f^2 + B_1 f + 1}}, \qquad (6.2.7)$$

and so forth. The corresponding sequence of approximate values for the critical point is given as follows,

$$f_1 = 1/2, \quad f_2 = 1, \quad f_3 = 0.728069, \quad f_4 = 0.75074,$$

$$f_7 = 0.192259, \quad f_8 = 0.779385, \quad f_9 = 0.77997.$$

The percentage error given by the last approximant equals 0.691135%. In the next order there are two solutions $f_{10}^{(1)} = 0.7781474$ and $f_{10}^{(2)} = 0.643305$, and the computations were stopped.

  We suggest that further increase in accuracy is limited by "flatness" of the coefficients values in four starting orders of (6.2.4). We also consider another sequence of power-transformed Padé approximants, multiplied with the Clausius–Mossotti-type expression,

$$P_1^t(f) = \frac{1-f}{1+f} \sqrt{\frac{A_1 f + 1}{B_1 f + 1}}, \qquad P_2^t(f) = \frac{1-f}{1+f} \sqrt{\frac{A_2 f^2 + A_1 f + 1}{B_2 f^2 + B_1 f + 1}}, \qquad (6.2.8)$$

and so forth. The transformation which lifts the flatness, does improve convergence of the sequence of approximations for the threshold,

$$f_1 = 1/4, \quad f_3 = 0.568452, \quad f_5 = 0.826561, \quad f_6 = 0.803947,$$

$$f_7 = 0.827349, \quad f_8 = 0.750544, \quad f_9 = 0.762065, \quad f_{10} = 0.78493.$$

The percentage error achieved for the last point is equal to 0.0596084%. The calculations were stopped here because of the second solution emergence at $f = 0.887047$. Although it does not interfere with the correct result f_{10}, such branching gives a natural signal to stop.

3. Critical Index s

The series from [31] approximated by (6.2.4) diverge as $f \to f_c$. Is it possible to evaluate the character of singularity as $f \to f_c$ from the series (6.2.4), assuming only that it is a power law?

Let us apply the following transformation,

$$z = \frac{f}{f_c - f} \iff f = \frac{z f_c}{z + 1} \qquad (6.3.9)$$

to the original series (6.2.4). To such transformed series $M_1(z)$ let us apply the *DLog* transformation (differentiate log of $M_1(z)$) and call the transformed series $M(z)$. In terms of $M(z)$, one can readily obtain the sequence of approximations s_n for the critical index s,

$$s_n = \lim_{z \to \infty} (z PadeApproximant[M[z], n, n + 1]). \qquad (6.3.10)$$

Namely,

$$s_2 = 0.627161, \quad s_4 = 0.633025, \quad s_6 = 0.597334,$$

$$s_8 = 0.550467, \quad s_{10} = 0.551316, \quad s_{12} = 0.535407.$$

One may expect that adding more terms to the expansion (6.2.4) will also improve an estimate for s. In order to accelerate convergence of the method, we suggest adapting the technique of corrected approximants [19], also explained in the preceding chapter. Let us first estimate the critical index by simplest non-trivial factor approximant (see [20, 43] and page 113),

$$\Phi_4^*(f) = (B_1 f + 1)^{s_1} \left(1 - \frac{f}{f_c}\right)^{-s_0}, \tag{6.3.11}$$

where $s_0 = 0.59928, s_1 = 1.57496, B_1 = 0.785398$.

Let us divide original series (6.2.4) by $\Phi_4^*(f)$, apply to the newly found series transformation (6.3.9), then apply *DLog* transformation and call the transformed series $L(z)$. Finally one can obtain the following sequence of corrected approximations for the critical index,

$$s_n = s_0 + \lim_{z\to\infty}(zPadeApproximant[L[z], n, n + 1]), \tag{6.3.12}$$

The following corrected sequence of approximate values for the critical index can be calculated readily:

$$s_4 = 0.62284, \quad s_5 = 0.135551, \quad s_6 = 0.586051, \quad s_7 = 0.559834,$$

$$s_8 = 0.54658, \quad s_9 = 0.553502, \quad s_{10} = 0.549611, \quad s_{11} = 0.555069,$$

$$s_{12} = 0.503875, \quad s_{13} = 0.503789, \quad s_{14} = 0.503855.$$

The last three values are very close to 0.5 and are well within expectations with regard to approximate-numerical nature of such estimates.

4. Crossover formula for all concentrations

Our suggestion for the conductivity formula valid for all concentrations is based on the following standard considerations. Let us first calculate the critical amplitude A. To this end let us again apply transformation (6.3.9) to the original series to obtain $M_1(z)$ as above. Then apply to $M_1(z)$ another transformation to get $T(z) = M_1(z)^{-1/s}$, in order to get rid of the square-root behaviour at infinity. In terms of $T(z)$, one can readily obtain the sequence of approximations A_n for the critical amplitude A,

$$A_n = f_c^s \lim_{z\to\infty}(zPadeApproximant[T[z], n, n + 1])^{-s}; \tag{6.4.13}$$

$$A_1 = 2.28682, \quad A_2 = 2.24389, \quad A_3 = 2.4418, \quad A_4 = 2.57419,$$

$$A_5 = 2.35515, \quad A_6 = 2.34677, \quad A_7 = 2.52728, \quad A_8 = 2.63203,$$

$$A_9 = 2.69504, \quad A_{10} = 2.62364, \quad A_{11} = 2.55292, \quad A_{12} = 2.55224.$$

The ninth member of the sequence gives the best result for amplitude. Corresponding approximant to $\sigma_e(f)$, satisfying 19 starting terms from (6.2.4), can be readily written as follows,

$$\sigma^* = \sqrt{\frac{zQ_3(z)}{Q_1(z)}}, \tag{6.4.14}$$

where

$Q_1(z) = z(z(z(z(z\ Q_2(z) + 61.8908) + 27.588) + 7.03611) + 0.773878)$,
$Q_2(z) = z(z(z((0.10813z + 2.8373)z + 16.885) + 47.592) + 80.664) + 87.607$,
$Q_3(z) = z(z(z(z(z\ Q_4(z) + 325.32) + 160.471) + 51.0803) + 9.46732)$,
$Q_4(z) = z(z(z(z(z + 20.1693) + 97.8524) + 253.571) + 411.488) + 444.039$.

$$\tag{6.4.15}$$

Substitution of (6.3.9) in (6.4.14) yields

$$\sigma^* = 2.09382 \sqrt{\frac{fR_3(f)}{(\pi - 4f)R_1(f)}}, \tag{6.4.16}$$

where

$R_1(f) = f(f(f(f(f\ R_2(f) - 0.985604) + 0.83488) - 0.0555036) - 0.473743)$,
$R_2(f) = f(f(f(f(1.39918 - f) - 0.632586) + 0.648903) - 0.882948) + 1.45321$,
$R_3(f) = f(f(f(f(f\ R_4(f) + 0.53774) + 0.19359) - 0.49707) - 0.96545) - 0.33948$,
$R_4(f) = f(f(f(f(f + 1.44888) + 0.504626) + 0.0358506) + 0.24819) + 0.921202$.

$$\tag{6.4.17}$$

Maximal error of the formula (6.4.16) equals 0.306638% as $f = 0.76$.

For all practical purposes, this expression, which is a crossover between the low-concentration and high-concentration regimes, is indistinguishable from the modified Padé-based form,

$$\sigma^* = \sqrt{\frac{z\ U_3(z)}{U_1(z)}}, \tag{6.4.18}$$

where

$U_1(z) = z(z(z(z(z\ U_2(z) + 110.63) + 52.5492) + 14.5119) + 1.78014)$,
$U_2(z) = z(z(z((0.101321z + 3.76147)z + 24.5539) + 73.2536) + 130.155) + 148.405$,
$U_3(z) = z(z(z(z(z\ U_4(z) + 577.474) + 300.114) + 101.332) + 20.1044) + 1.78014$,
$U_4(z) = z(z(z(z(z + 27.4454) + 144.076) + 392.701) + 666.22) + 751.704$,

$$\tag{6.4.19}$$

or

$$\sigma^* = 1.91373 \sqrt{\frac{f\ T_3(f)}{(\pi - 4f)T_1(f)}}, \tag{6.4.20}$$

$$T_1(f) = f(f(f(f(f T_2(f) + 0.477564) - 0.489113) + 1.0752) - 0.995924),$$
$$T_2(f) = f(f(f(f(0.743958 - f) + 0.478232) - 0.398804) + 0.19758) + 0.308858,$$
$$T_3(f) = f(f(f(f(fT_4(f) + 1.21937) + 0.397015) - 0.388154) - 1.4072) - 0.8543,$$
$$T_4(f) = f(f(f(f(f + 2.24302) + 1.24708) + 0.216355) + 0.517846) + 1.25935.$$

$$(6.4.21)$$

The approximant (6.4.20) matches starting 18 terms from (6.2.4) and asymptotic form (6.1.1) (18 + 1), not unlike [7]. Here, the notation (18 + 1) means that 18 terms from the small concentration side and only single term from the high concentration limit are utilized. There is a clear convergence to the approximant (18+1), in the sub-sequence of approximants (2+1), (6+1), (8+1), (16+1) , (18+1); while all other approximants are discontinuous.

From the crossover formula (6.4.20) one can readily obtain the higher-order coefficients [45], not employed in the final formula,

$$a_{19} = 40.5543, \quad a_{20} = 49.768, \quad a_{21} = 62.9092, \quad a_{22} = 78.5999,$$
$$a_{23} = 98.0859, \quad a_{24} = 121.136, \quad a_{25} = 152.608, \quad a_{26} = 190.893.$$

The computed values in a fairly good agreement with the original series (6.2.4). Similar problem is important also in quantum field theory. We quote below from the paper [36]:

> It has long been a hope in perturbative quantum field theory (PQFT), first expressed by Richard Feynman, to be able to estimate, in a given order, the result for the coefficient, without the brute force evaluation of all the Feynman diagrams contributing in this order. As one goes to higher and higher order the number of diagrams, and the complexity of each, increases very rapidly. Feynman suggested that even a way of determining the sign of the contribution would be useful.

It is somewhat instructive that even by enforcing the correct amplitude A, one cannot further increase accuracy of the formulae of these type. In order to achieve actual improvement, one should incorporate more details of the conductivity behaviour at high concentrations, such as amplitude B.

Analytical expression suggested in [32], is the most appropriate for comparison with our suggestion,

$$\sigma_e(f) \approx 1 + \frac{2f}{1 - f - 0.013362f^8 - \frac{0.305827f^4}{1-1.1403f^8}}. \qquad (6.4.22)$$

The bounds for $\sigma_e(f)$ from [2] give similar numerical results for not critically high concentrations. Formula (6.4.20) deviates from the expression (6.4.22) in the region of concentrations $(0.7, f_c)$. Both formulae are positioned significantly higher than expansion (6.2.4). Most importantly, formula (6.4.20) smoothly interpolates between the two asymptotic expressions through the whole crossover region of $(0.7, f_c)$.

5. Expansion near the threshold

Throughout the chapter, we derived the expressions for the crossover properties from "left-to-right", i.e., extending the series from small f to large f. Alternatively, one can proceed from "right-to-left", i.e., extending the series from the large f (close to f_c) to small f [18]. The simplest way to proceed is to look for the solution in the whole region $[0, f_c)$. in the form

$$\sigma_{e1} = \alpha_0(f_c - f)^{-1/2} + \alpha_1, \qquad (6.5.23)$$

and obtain the unknowns from the two starting terms of (6.2.4), namely $\sigma \simeq 1 + 2x$. Then, $\alpha_0 = A$, $\alpha_1 = (1 - \pi)$, the same values as obtained in [28]. One can also see [28] for the discussion of different approximate asymptotic expressions. Generally speaking (6.5.23) implies a more general expansion

$$\sigma_{e,n}(f) = \sum_{k=0}^{n} \alpha_k (f_c - f)^{\frac{k-1}{2}} \qquad (6.5.24)$$

where $\alpha_0 \equiv A$, $\alpha_1 \equiv B$, and $n = 0, 1, \ldots, N$. The higher-order terms maybe accounted for as well when needed. For the elasticity problem, see Chapter 10, analogous expansions may turn to be useful as well.

In order to improve (6.5.23), let us divide series (6.2.4) by the approximant (6.5.23), apply to the new series transformation (6.3.9) and then use diagonal Padé (or two-point Padé) approximants. The following expression obtained from the simplest non-trivial two-point Padé approximant, satisfies five starting terms from (6.2.4) and Keller's asymptotic expression,

$$\sigma_e \approx \frac{2.05897(f(f(f + 0.912641) + 0.356584) + 0.349627)\left(1.3 - \sqrt{0.785398 - f}\right)}{(f(f(f + 0.847132) + 0.342827) + 0.336139)\sqrt{0.785398 - f}}. \qquad (6.5.25)$$

Another expression, obtained from the simplest non-trivial diagonal Padé approximant, satisfying the same terms, but with slightly higher value of amplitude $A = 2.80931$,

$$\sigma_e \approx \frac{2.17708(f(f + 0.414669) + 0.18083)\left(1.30004 - \sqrt{0.785398 - f}\right)}{(f(f + 0.42154) + 0.183827)\sqrt{0.785398 - f}}, \qquad (6.5.26)$$

works even better than (6.5.25). Formula (6.5.26) is almost as good as (6.4.16). Maximal error incurred by (6.5.26), equals 0.39%.

5.1. Discussion of the ansatz (6.5.23)

We looked for the solution in a simple form,

$$\sigma_1^{r-l} = \alpha_0(f_c - f)^{-1/2} + \alpha_1, \quad f_c = \frac{\pi}{4}, \tag{6.5.27}$$

and obtained the unknowns from the two starting terms of the corresponding series,

$$\sigma_4 = 1 + 2f + 2f^2 + 2f^3 + 2f^4 + \cdots. \tag{6.5.28}$$

Then, $\alpha_0 = \frac{\pi^{3/2}}{2} \approx 2.784$, $\alpha_1 = (1 - \pi)$, same form as obtained asymptotically in [28], with exactly the same value for the leading amplitude as obtained in [24].

Formula (6.5.23) despite its asymptotic nature, turned out to be rather accurate in the whole region of concentrations. We try to understand below why it is so. Even as we simply expand (6.5.27), the third-order coefficient can be easily calculated, $a_3 = 1.90986$. This value is in reasonable agreement with the best known value of 2. Thus, we can safely select σ_1^{r-l} as the reasonable starting approximation.

Let us then consider (6.5.27) as an initial approximation for the critical index calculation and calculate corrections by *DLog* Padé technique, as described above in Section 3. As expected the calculated values of the corrections are negligible (-0.0002), when all terms from the expansion are utilized.

Let us subtract the approximant (6.5.23) from the series (6.5.28), and apply to the new series transformation to the variable z (6.3.9). Then we apply to such transformed series another procedure, intended to find corrections to the values of amplitudes α_0 and α_1. Such task is non-trivial, especially when one is interested in analytical solutions. It can be solved using general form of root approximants derived in [18] and presented above (see (5.3.48) on page 114)

$$\sigma_{add} = b_0 z^2 \left((b_1 z + 1)^{c_1} + b_2 z^2 \right)^{c_2} \tag{6.5.29}$$

under asymptotic condition

$$\sigma_{add} \simeq p_1 \sqrt{z} + p_2, \quad \text{as } z \to \infty. \tag{6.5.30}$$

Elementary power-counting gives $c_1 = \frac{3}{2}$, $c_2 = -\frac{3}{4}$. All other unknowns can now be determined uniquely in a standard fashion from the condition of asymptotic equivalence as $z \to 0$. Final expression

$$\sigma_{add} = \frac{0.0556033 f^2}{(0.785398 - f)^2 \left(\frac{3.69302 f^2}{(0.785398-f)^2} + \left(\frac{1.98243 f}{0.785398-f} + 1 \right)^{\frac{3}{2}} \right)^{\frac{3}{4}}}, \tag{6.5.31}$$

can be re-expanded in the vicinity of f_c with the result

$$\sigma_{add} = \frac{0.0184973}{\sqrt{0.785398 - f}} - 0.0118315 + O(\sqrt{f_c - f}), \tag{6.5.32}$$

indicating only small corrections to the values of amplitudes. Such asymptotic stability of all amplitudes additionally justifies the ansatz, and final corrected expression $\sigma^{sq} = \sigma_1^{r-l} + \sigma_{add}$, appears to be just slightly larger than (6.5.23). Note that Padé approximants are only able to produce additive corrections in the form $\sigma_{add} \simeq p\sqrt{z} + O(\frac{1}{\sqrt{z}})$ as $z \to \infty$.

5.2. Corrected critical index from additive ansatz

Let us first estimate the critical index from the expression similar to (6.5.23), but not assuming now that we know its value. To this end, we have to employ one more coefficient from (6.2.4). Then the simple additive approximant can be written explicitly,

$$\sigma_1^{ad} = \frac{2.39749}{(0.785398 - f)^{0.570796}} - 1.75194. \tag{6.5.33}$$

To correct the value of index from (6.5.33), we apply again the procedure leading to the expression (6.3.12), starting with $s_0 = 0.570796$. Let us divide original series (6.2.4) by (6.5.33), apply to the newly found series transformation (6.3.9), then apply *DLog* transformation and call the transformed series $L(z)$. Let us also process the transformed series,

$$L_r(z) = -0.264754z^2 + 0.882515z^3 - 1.28539z^4 - 0.538199z^5 + 9.39947z^6 - \cdots, \tag{6.5.34}$$

with the diagonal Padé approximants.

The following sequence of approximate values for the critical index was found,

$$s_2 = 0.470413, \quad s_3 = 0.523229, \quad s_4 = 0.553386, \quad s_5 = 0.52638,$$

$$s_6 = 0.526343, \quad s_7 = 0.513666, \quad s_8 = 0.508803, \quad s_9 = 0.514609,$$

$$s_{10} = 0.511196, \quad s_{11} = 0.511804, \quad s_{12} = 0.464512, \quad s_{13} = 0.464579,$$

$$s_{14} = 0.511804, \quad s_{15} = 0.511196, \quad s_{16} = 0.514609, \quad s_{17} = 0.508798, \quad s_{18} = 0.513494.$$

One can also obtain the following sequence of corrected approximations to the critical index applying iterated roots (see formula (5.3.55) on page 115) in place of Padé approximants,

$$s_n = s_0 + \lim_{z \to \infty} (z\, R_n^*(z)), \tag{6.5.35}$$

where $R_n^*(z)$ stands for the iterated root of n-th order [19], constructed for the series $L_r(z)$ with such a power at infinity that defines constant correction to s_0.

Calculations with iterated roots are really easy since at each step we need to compute only one new coefficient, while keeping all preceding from previous steps. The power at infinity is selected in order to compensate for the factor z and extract constant correction to the initial approximation s_0. Namely the few starting terms are shown below,

$$R_1^*(z) = -\frac{0.264754z^2}{(1.11111z+1)^3}, \quad R_2^*(z) = -\frac{0.264754z^2}{\left(1.70159z^2+(1.11111z+1)^2\right)^{3/2}},$$

$$R_3^*(z) = -\frac{0.264754z^2}{\left(1.70159z^2+(1.11111z+1)^2\right)^{\frac{3}{2}}-1.5703z^3}. \tag{6.5.36}$$

A few good estimates are generated by the iterated roots,

$$s_1 = 0.37779, \quad s_2 = 0.518174, \quad s_3 = 0.494297, \quad s_4 = 0.53281,$$

$$s_5 = 0.50269, \quad s_6 = 0.539667, \quad s_7 = 0.448595, \quad s_8 = 0.544004,$$

and since $s_9 = 0.55058 + 0.0350158i$, the computations were stopped here. In order to improve results, let us combine calculations with Padé approximants and roots, using low-order iterated roots as the control function, as explained in the preceding chapter dedicated to various methods on page 125. Let us employ the first-order iterated root and find that stability of the results is improved, although the best values for the index are practically the same.

Thus, we define the new series

$$L_1(z) = \frac{L(z)}{R_1^*(z)},$$

and apply the technique of diagonal Padé to satisfy the new series asymptotically, order-by-order,

$$s_n = s_0 + \lim_{z\to\infty}(z\, R_1^*(z))\lim_{z\to\infty}(PadeApproximant[L_1[z],n,n]), \tag{6.5.37}$$

$$s_1 = 0.37779, \quad s_2 = 0.618109, \quad s_3 = 0.539355, \quad s_4 = 0.691756,$$

$$s_5 = 0.636668, \quad s_6 = 0.491483, \quad s_7 = 0.509117, \quad s_8 = 0.510457,$$

$$s_9 = 0.508025, \quad s_{10} = 0.512141, \quad s_{11} = 0.511125, \quad s_{12} = 0.511126, \quad s_{13} = 0.511125,$$

$$s_{14} = 0.512141, \quad s_{15} = 0.508026, \quad s_{16} = 0.51046, \quad s_{17} = 0.509131, \quad s_{18} = 0.492748.$$

The procedure brings more stable convergence to the value of 0.5.

5.3. Corrected threshold from additive ansatz

Let us in addition estimate the threshold from the expression (6.5.23), not assuming that we know its value in advance, as explained in the preceding chapter on page 144. To this end, we again have to use the three starting terms from (6.2.4), $\sigma \simeq 1 + 2f + 2f^2$. All three unknowns, f_c, α_0, α_1, may be found explicitly, leading to the following approximant,

$$\sigma = \frac{3\sqrt{3}}{2\sqrt{\frac{3}{4}-f}} - 2, \tag{6.5.38}$$

with approximate threshold value of $f_0 = 3/4$. The choice of zero approximation is always important and requires some intuitive knowledge about the sought solution. In our case we just operate with the high-concentration expansion and additive self-similar ansatz, which captures the most difficult part of the problem.

Let us look for the solution from the same family as (6.5.38), in the same form but with an exact, yet unknown threshold F_c,

$$\sigma' = \frac{3\sqrt{3}}{2\sqrt{F_c-f}} - 2. \tag{6.5.39}$$

From here one can express formally,

$$F_c = \frac{4f\sigma'(f)^2 + 16f\sigma'(f) + 16f + 27}{4(\sigma'(f)+2)^2}, \tag{6.5.40}$$

since $\sigma'(f)$ is also unknown. All we can do is to use for σ' the series (6.2.4), so that instead of a true threshold, we have an effective threshold as an expansion around f_0,

$$\begin{aligned}
F_c(f) = \quad & \frac{3}{4} + \frac{1}{9}f^3 + \frac{2}{27}f^4 - 0.268791f^5 + 0.0164609f^6 \\
+ \quad & 0.108801f^7 + 0.0933591f^8 - 0.0491905f^9 \\
+ \quad & 0.0306117f^{10} + 0.0816431f^{11} + \dots, \tag{6.5.41}
\end{aligned}$$

which should become a true threshold F_c as $f \to F_c$! Moreover, let us apply resummation procedure to the expansion (6.5.41) using factor approximants $\Phi^*(f)$, and define the sought threshold F_c^* self-consistently,

$$F_c^* = \frac{3}{4} + \frac{f^3}{9}\Phi^*(F_c^*), \tag{6.5.42}$$

as we approach the threshold the RHS should become the threshold. Since factor approximants are defined as Φ_k^* for arbitrary number of terms k, we will also have a sequence of $F_{c,k}^*$. For example, with

$$\begin{aligned}
\Phi_2^*(f) &= (7.92402f + 1)^{0.0841324}, \\
\Phi_4^*(f) &= (1 - 0.858374f)^{1.86127}(1.72735f + 1)^{1.31087}, \tag{6.5.43}
\end{aligned}$$

solving (6.5.42), we obtain $F_{c,2}^* = 0.823548$, $F_{c,4}^* = 0.770537$, $F_{c,6}^* = 0.778053$. There is no solution in the next even order, and the percentage error achieved for the last point is equal to 0.935%.

One can also define a sequence of "odd" factor approximants [44], starting from $\Phi_1^* = 1 + \frac{2}{3}f$, where $\frac{2}{3}$ comes as the ratio of fourth-and third-order coefficients in the series (6.5.41). The next-order odd approximant,

$$\Phi_3^*(f) = \frac{\left(\frac{2f}{3} + 1\right)^{31.0546}}{(0.425194f + 1)^{47.1229}},$$

(6.5.44)

brings the most accurate value of the threshold, $F_{c,3}^* = 0.782355$, while in the higher orders, $F_{c,5}^* = 0.774622$, $F_{c,7}^* = 0.779692$. The percentage error achieved for the last point is equal to 0.727%.

In the same manner, one can apply the diagonal Padé approximants, and obtain the following equation,

$$F_{c,n}^* = P_{n,n}(F_c^*).$$

(6.5.45)

Solving equation (6.5.45), we obtain

$$F_{c,4}^* = 0.779327, \quad F_{c,5}^* = 0.777817, \quad F_{c,7}^* = 0.781836, \quad F_{c,8}^* = 0.78705,$$

$$F_{c,9}^* = 0.784825, \quad F_{c,10} = 0.78499, \quad F_{c,11}^* = 0.784964, \quad F_{c,13}^* = 0.785012.$$

The convergence of the series is good and the last value is closest to the exact result, with the relative percentage error just of 0.049%.

Conventional ratio method [9], also works well. It evaluates the threshold through the value of index and ratio of the series coefficients,

$$f_{c,n} = \frac{\frac{s-1}{n} + 1}{\frac{a_n}{a_{n-1}}}.$$

(6.5.46)

The last point gives a good estimate, $f_{c,26} = 0.781895$, despite of the oscillations in the dependence on n, as seen in Fig. 6.3. The percentage error achieved for the last point is equal to 0.446%.

Let us now condition the effective threshold on exact threshold, i.e., require additionally that

$$F_c^*\left(\frac{\pi}{4}\right) = \frac{\pi}{4},$$

(6.5.47)

and obtain the following accurate expression for the effective conductivity valid in the correct interval of concentrations,

$$\sigma^* = \frac{3\sqrt{3}}{2\sqrt{F_c^*(f) - f}} - 2,$$

(6.5.48)

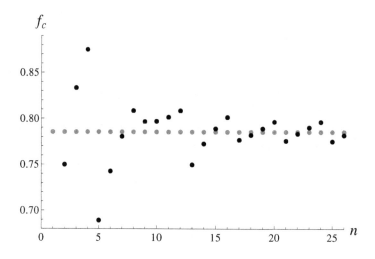

Figure 6.3 f_c calculated by ratio method, compared with the exact threshold.

where $F_c^*(f) = \frac{U_1(f)}{U_3(f)}$,

$$U_1(f) = \tfrac{3}{4} + 31.7213f - 65.0294f^2 + 39.5302f^3 +$$
$$3.80128f^4 - 32.6749f^5 + 56.1003f^6 + 1.91401f^7 - 8.87329f^8 - \qquad (6.5.49)$$
$$28.4318f^9 + 91.0295f^{10} - 57.8444f^{11} + 10.2101f^{12} + 14.3729f^{13},$$

and

$$U_3(f) = 1 + 42.2951f - 86.7059f^2 + 52.5587f^3 -$$
$$1.29633f^4 - 34.5401f^5 + 90.7136f^6 - 34.5946f^7 + 7.89328f^8 - \qquad (6.5.50)$$
$$42.1758f^9 + 111.09f^{10} - 55.4862f^{11} + 12.1403f^{12} + 4.84071f^{13}.$$

The maximum percentage error of the formula (6.5.48) is just 0.052%.

6. Additive ansatz. Critical amplitude and formula for all concentrations

We suggest below the particular resummation schemes based on further extensions of the additive ansatz already employed in this chapter, also explained on page 148. They are based on accurately calculated critical amplitude and lead to the very accurate analytical expressions for the effective conductivity.

6.1. Correction with Padé approximants

Let us ensure the correct critical index already in the starting approximation for σ^{ad}, so that all parameters in (6.6.51) are obtained by matching it asymptotically with the truncated series $\sigma_4 = 1 + 2f + 2f^2 + 2f^3 + 2f^4$.

We employ below an extended form of the additive self-similar ansatz, expressed in terms of the variable z,

$$\sigma_3^{ad}(z) = \alpha_1 \sqrt{\alpha z + 1} + \alpha_2 + \alpha_3 \frac{1}{\sqrt{\alpha z + 1}} + \frac{\alpha_4}{\alpha z + 1}, \tag{6.6.51}$$

or in the original variable,

$$\sigma_3^{ad}(f) = -1.90985 + 2.04589 \sqrt{\frac{2.24734 f}{\frac{\pi}{4} - f} + 1} + \frac{1.07995}{\sqrt{\frac{2.24734 f}{\frac{\pi}{4} - f} + 1}} - \frac{0.21599}{\frac{2.24734 f}{\frac{\pi}{4} - f} + 1}. \tag{6.6.52}$$

It has the following expansion as $f \to f_c$,

$$\sigma_3^{ad}(f) \simeq -1.90985 + \frac{2.71807}{\sqrt{-f + \frac{\pi}{4}}}, \tag{6.6.53}$$

which also brings minimum value for the amplitudes if the whole sequence of approximations (not shown) with their expansions shown below,

$$\sigma_1^{ad}(f) \simeq -2.65979 + \frac{3.00503}{\sqrt{-f + \frac{\pi}{4}}}, \quad \sigma_2^{ad}(f) \simeq -2.03187 + \frac{2.77628}{\sqrt{-f + \frac{\pi}{4}}},$$
$$\sigma_4^{ad}(f) \simeq -1.962 + \frac{2.72321}{\sqrt{-f + \frac{\pi}{4}}}, \tag{6.6.54}$$

is considered. To extract corrections to the critical amplitude, it appears justified that we pick $\sigma_3^{ad}(f)$ as a starting approximation. Thus, we divide the original series (6.2.4) by (6.6.52), apply to the new series transformation (6.3.9). Call the newly found series $G[z]$. Finally build a sequence of the diagonal Padé approximants, so that the amplitudes are expressed by the formula

$$A_n = A_0 \lim_{z \to \infty} (PadeApproximant[G[z], n, n]), \tag{6.6.55}$$

with $A_0 = 2.71807$.

We select from the emerging sequences only approximants which are also holomorphic functions. Not all approximants generated by the procedure are holomorphic. The holomorphy of diagonal Padé approximants in a given domain implies their uniform convergence inside this domain. Thus, we obtain several reasonable estimates for the amplitudes A,

$$A_4 = 2.71807, \quad A_5 = 2.71278, \quad A_6 = 2.79022, \quad A_7 = 2.7802,$$

$$A_8 = 2.78547, \quad A_9 = 2.78545, \quad A_{10} = 2.77409,$$

$$A_{11} = 2.7809, \quad A_{12} = 2.78625, \quad A_{13} = 2.79002,$$

and simultaneous estimates fot the second amplitude B,

$$B_4 = -1.90985, \quad B_5 = -1.90613, \quad B_6 = -1.96054, \quad B_7 = -1.9535,$$

$$B_8 = -1.9572, \quad B_9 = -1.95719, \quad B_{10} = -1.9492,$$

$$B_{11} = -1.95402, \quad B_{12} = -1.95775, \quad B_{13} = -1.9604.$$

Complete expression for the effective conductivity corresponding to A_{11} can be reconstructed readily,

$$\sigma_{12}^*(f) = \sigma_3^{ad}(f) \frac{c_1(f)}{c_2(f)}, \tag{6.6.56}$$

where

$$c_1(f) = 0.468548 + 0.59755 f - 1.02047 f^2 - 1.06259 f^3 + 0.377654 f^4$$
$$+ 0.938238 f^5 + 2.08709 f^6 + 2.06594 f^7 + 0.470793 f^8 - 0.443561 f^9 \tag{6.6.57}$$
$$+ 2.84916 f^{10} + 2.88063 f^{11} + 0.475575 f^{12},$$

and

$$c_2(f) = 0.468548 + 0.59755 f - 1.02047 f^2 - 1.06259 f^3 + 0.377654 f^4$$
$$+ 1.20036 f^5 + 1.13454 f^6 + 2.98399 f^7 + 0.067621 f^8 + 0.32355 f^9 \tag{6.6.58}$$
$$+ 2.11352 f^{10} + 3.66841 f^{11} - f^{12}.$$

Maximum error of the formula (6.6.56) is at the point $f = 0.77$ and equals 0.0596%.

6.2. Modified additive approximants. Correction with Padé

While formulating an additive ansatz (6.5.24), we can employ instead of a linear "unit" with a critical behaviour taken into different powers, a quadratic one. Then, we can write down another, modified additive approximant which turns to be good both for square and hexagonal arrays. Its simplest non-trivial form is given below,

$$\sigma^g = \alpha_0 + \alpha_1 \sqrt[4]{1 + \alpha_2 z + \alpha_3 z^2}. \tag{6.6.59}$$

Formula (6.6.59) can be also viewed as a shifted iterated root approximant. One can obtain the unknowns from the three starting terms of the corresponding series. Numerically, the initial approximation for the square lattice is computed as follows,

$$\sigma^g(z) = -1.79583 + 2.79583 \sqrt[4]{1.41167 z^2 + 2.24734 z + 1}. \tag{6.6.60}$$

The ansatz (6.6.59) is going to be multiplicatively corrected through application of the Padé approximants, as illustrated below. Emerging diagonal Padé sequences for critical amplitudes are convergent. Good results are achieved simultaneously for amplitudes A and B, employing all 26 terms from the corresponding expansion. For the amplitude A, we register the following sequence,

$$A_1 = 2.70078, \quad A_2 = 2.70078, \quad A_3 = 2.70078, \quad A_4 = 2.68555,$$

$$A_5 = 2.78007, \quad A_7 = 2.7655, \quad A_8 = 2.7722, \quad A_9 = 2.77165,$$

$$A_{10} = 2.78125, \quad A_{11} = 2.79261, \quad A_{12} = 2.77267, \quad A_{13} = 2.78037.$$

and for the amplitude B obtain simultaneously,

$$B_1 = -1.79583, \quad B_2 = -1.79583, \quad B_3 = -1.79583, \quad B_4 = -1.78571,$$

$$B_5 = -1.84856, \quad B_7 = -1.83887, \quad B_8 = -1.84332, \quad B_9 = -1.84296,$$

$$B_{10} = -1.84934, \quad B_{11} = -1.8569, \quad B_{12} = -1.84363, \quad B_{13} = -1.84875.$$

We select from the emerging sequences only approximants which are also holomorphic functions, and present below a very accurate expression for the effective conductivity of the square lattice of inclusions. For $n = 10$, employing 20 terms from the corresponding expansion for the square lattice we can construct correction term $Cor_n^{sq}(f)$, given by the diagonal Padé approximant,

$$\sigma_{sq}^*(f) = \sigma^g(f)\, Cor_{10}^{sq}(f), \tag{6.6.61}$$

where

$$\sigma^g(f) = 2.79583 \frac{\sqrt[4]{f(2.62925f + 3.10816) + 9.8696}}{\sqrt{\pi - 4f}} - 1.79583, \tag{6.6.62}$$

and

$$Cor_{10}^{sq}(f) = \frac{Cor_1^{sq}(f)}{Cor_2^{sq}(f)}, \tag{6.6.63}$$

with

$$\begin{aligned} Cor_1^{sq}(f) = {}& 0.735053 - 0.775163f + 0.740469f^2 + 0.0939897f^3 - \\ & 0.807184f^4 + 1.18166f^5 - 1.24128f^6 + 0.838147f^7 - 1.23568f^8 + \\ & 1.15514f^9 - 1.18863f^{10}, \end{aligned} \tag{6.6.64}$$

and

$$\begin{aligned} Cor_2^{sq}(f) = {}& 0.735053 - 0.775163f + 0.740469f^2 + 0.0939897f^3 - \\ & 0.648383f^4 + 0.645544f^5 - 0.468209f^6 - 0.0318346f^7 - 0.606582f^8 + \\ & 0.800566f^9 - f^{10}. \end{aligned} \tag{6.6.65}$$

It works rather well in the whole interval of concentrations, with maximum error of 0.0334%, see Fig. 6.4 demonstrating various regimes.

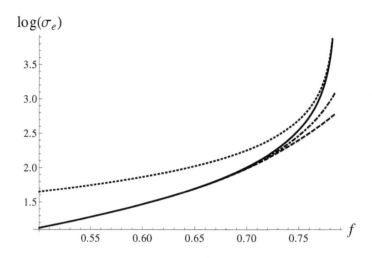

Figure 6.4 Our formula (6.6.61) (solid) is compared with (6.4.22) (dashed-dotted), asymptotic expression (6.1.2) (dotted) and expansion (6.2.4) (dashed).

The following expression for the conductivity follows readily

$$\sigma^*_{sq}(f) \simeq \frac{2.78125}{\sqrt{f - \frac{\pi}{4}}} - 1.84934, \quad \text{as } f \to f_c^{sq}. \tag{6.6.66}$$

The leading critical amplitude equals 2.781, and the next-order amplitude is equal to -1.849. Both amplitudes are in a good agreement with [28].

6.2.1. Compact formula

We also bring here an accurate and compact enough expression for the effective conductivity of the square lattice of inclusions, employing just 10 terms from the corresponding expansion for the square lattice,

$$\boxed{\sigma^*_{sq}(f) = \sigma^g(f)\, Cor_5^{sq}(f)}, \tag{6.6.67}$$

where $\sigma^g(f)$ is calculated by (6.6.62) and the correction $Cor_5^{sq}(f)$ is given as follows,

$$Cor_5^{sq}(f) = \frac{f(f(f(f(f(14.1698f - 2.6844) + 9.16247) - 12.0988) + 14.5921) + 32.5445}{f(f(f(f(f(f + 4.34651) + 9.16247) - 12.0988) + 14.5921) + 32.5445}. \tag{6.6.68}$$

It works rather good in the whole interval of concentrations, with maximum error of 0.088%.

The following expression for the conductivity follows readily

$$\sigma^*_{sq}(f) \simeq \frac{2.78007}{\sqrt{f - \frac{\pi}{4}}} - 1.84856, \quad \text{as } f \to f_c^{sq}. \tag{6.6.69}$$

The leading critical amplitude equals 2.78, and the next-order amplitude is equal to -1.849. Both amplitudes are still in a good agreement with [28].

Note also that from the formula (6.6.67), one can readily obtain the higher-order coefficients starting from a_{11}, not employed in the derivation, i.e., for small f

$$
\begin{aligned}
\sigma^*_{sq}(f) = {} & 1 + 2f + 2f^2 + 2f^3 + 2f^4 + 2.61166f^5 + \\
& 3.22331f^6 + 3.83497f^7 + 4.44662f^8 + 5.27206f^9 + 6.28456f^{10} + \\
& 7.64531f^{11} + 9.29831f^{12} + 11.4116f^{13} + 13.971f^{14} + 17.2054f^{15} + \\
& 21.169f^{16} + 26.1655f^{17} + 32.3365f^{18} + 40.094f^{19} + 49.7181f^{20} + \\
& 61.7964f^{21} + 76.8317f^{22} + 95.6911f^{23} + 119.229f^{24} + 148.753f^{25} + \\
& 185.673f^{26} + O(f^{27}).
\end{aligned}
\tag{6.6.70}
$$

The computed values of the coefficients are in a fairly good agreement with the original series (6.2.4).

7. Interpolation with high-concentration Padé approximants

When two expansions (6.2.4) and (6.2.5) are available, the problem of reconstruction greatly simplifies and can be solved in terms of Padé approximants in specially selected variable.

This approach requires as an input at least two parameters from weak and strong-coupling (high-concentration) regimes, including the value of amplitude $A = 2.78416$. Similar interpolation problem for random composites was considered in [5].

Assume that the next-order amplitude $B \approx -1.85$, is also known in advance. In terms of z-variable (6.3.9)) in the high-concentration limit it is simply

$$
\sigma = \frac{A}{\sqrt{f_c}} \sqrt{z} + B + O(z^{-1/2}).
\tag{6.7.71}
$$

The key point is also a natural choice for variable. If it chosen as a $\sqrt{z}$, then both weak and strong coupling limits can be satisfied in any required order. Note also that the Padé approximants should be conditioned to give a constant value as $z \to 0$. Several low-order approximant are given below,

$$
\begin{aligned}
P_{2,1}(z) &= \frac{\beta \sqrt{z}\left(1 + \beta_1 \frac{1}{\sqrt{z}} + \frac{\beta_2}{z}\right)}{1 + \beta_3 \frac{1}{\sqrt{z}}}, \\
P_{3,2}(z) &= \frac{\beta \sqrt{z}\left(1 + \beta_1 \frac{1}{\sqrt{z}} + \frac{\beta_2}{z} + \beta_3 z^{-3/2}\right)}{1 + \beta_5 \frac{1}{\sqrt{z}} + \frac{\beta_6}{z}}, \\
P_{4,3}(z) &= \frac{\beta \sqrt{z}\left(1 + \beta_1 \frac{1}{\sqrt{z}} + \frac{\beta_2}{z} + \beta_3 z^{-3/2} + \frac{\beta_4}{z^2}\right)}{1 + \beta_5 \frac{1}{\sqrt{z}} + \frac{\beta_6}{z} + \beta_7 z^{-3/2}}.
\end{aligned}
\tag{6.7.72}
$$

The unknowns in (6.7.72) will be obtained by the asymptotic conditioning to (6.2.5) and (6.2.4). In all orders $\beta = \frac{A}{\sqrt{f_c}}$. Explicitly, in original variables, the following expressions transpire,

$$P_{2,1}(f) = \frac{\sqrt{\frac{f}{0.785398-f}} + \frac{2.4674}{0.785398-f} - 2.23441}{\sqrt{\frac{f}{0.785398-f}} + 0.907183},$$

$$P_{3,2}(f) = \frac{\left(f\left(0.58116\sqrt{\frac{f}{-f+0.785398}} - 0.915318\right) - 0.243947\right)f + 0.545105}{f(f-1.33416) + 0.545105}, \qquad (6.7.73)$$

$$P_{4,3}(f) = \frac{8.31958z(f)^{3/2} + 3.14159z(f)^2 + 12.292z(f) + 4.65979\sqrt{z(f)} + 5.76453}{z(f)^{3/2} + 3.23708z(f) + 4.65979\sqrt{z(f)} + 5.76453}.$$

The approximants are strictly non-negative and respect the structure of (6.2.4), e.g.,

$$P_{4,3}(f) = 1 + 2f + 2f^2 + O(f^3), \quad \text{as } f \to 0, \qquad (6.7.74)$$

since all lower-order fractional powers generated by square-roots are suppressed by design. But in higher-orders, emerging fractional powers should be suppressed again and again, to make sure that only integer powers of f are present

$$P_{4,3}(f) = A(f_c - f)^{-1/2} + B + O((f_c - f)^{1/2}), \quad \text{as } f \to f_c, \qquad (6.7.75)$$

and only integer powers of a square-root appear in higher orders. Both $P_{3,2}(f)$ and $P_{4,3}(f)$ give reasonable estimates for the conductivity, from below and above respectively. The bounds hold from small concentrations till the very core of the high-concentration regime, as $f = 0.78$.

Particularly clear form is achieved for the resistivity, an inverse of conductivity, $r(z) = [\sigma_e(z)]^{-1}$, e.g.,

$$r_{3,4}(z) = \frac{0.31831z^{3/2} + 1.03039z + 1.48326\sqrt{z} + 1.83491}{2.64821z^{3/2} + z^2 + 3.91266z + 1.48326\sqrt{z} + 1.83491}. \qquad (6.7.76)$$

With the variable $X = \sqrt{z}$, the resistivity problem is reduced to studying the sequence of Padé approximants $R_n = r_{n,n+1}(X)$, $n = 1, 2, \ldots, \frac{1}{2}$, with $X \in [0, \infty)$, and analogy with the Stieltjes truncated moment problem [1, 16, 26], is complete as long as the resistivity expands at $X \to \infty$ in the Laurent polynomial with the sign-alternating coefficients, coinciding with the "Stieltjes moments" μ_k (see, e.g., [8, 42] where the original work of Stieltjes is explained very clearly).

The moments formally define corresponding Stieltjes integral

$$\int_0^\infty \frac{d\phi(u)}{u + X} \sim \sum_{k=0}^l (-1)^k \mu_k X^{-k-1} + O(X^{-l}), \quad \text{as } X \to \infty, \qquad (6.7.77)$$

l is even [16], and $\mu_k = \int_0^\infty u^k \, d\phi(u)$. Approximant $R_n(X)$ should match (6.7.77) asymptotically.

The Stieltjes moment problem can possess a unique solution or multiple solutions, dependent on the behaviour of the moments, in contrast with the problem of moments for the finite interval [11, 22, 38], which is solved uniquely if the solution exists. The role of variable is usually played by the contrast parameter, while in our case of a high-contrast composite, the variable is X.

In our setup there are just two moments available and resistivity is reconstructed using also a finite number of coefficients in the expansion at small X. That is the reduced (truncated) two point Padé approximation is considered, also tightly related to the moment problem [21, 23, 26]. In fact, even pure interpolation problem can be presented as a moment problem. We obtain here upper and lower bounds for resistivity (conductivity) in a good agreement with simulations [32].

It does seem interesting and non-trivial that the effective resistivity (conductivity) can be presented in the form of a Stieltjes integral [8, 42], when the variable (6.3.9) is used.

7.1. Independent estimation of the amplitude B. Additive corrections

We intend to calculate the amplitude B independent on previous estimates, as suggested on page 149. Start with the choice of the simplest high-concentration Padé approximant as zero-approximation,

$$P_{1,0}(z) = \beta \sqrt{z} + 1, \tag{6.7.78}$$

or in terms of the original variable,

$$P_{1,0}(f) = \pi \sqrt{\frac{f}{0.785398 - f} + 1}. \tag{6.7.79}$$

The way how we proceeded above was to look for multiplicative corrections to some plausible "zero-order" approximate solution. But we can also look for an additive corrections in a similar fashion.

To this end let us first subtract (6.7.79) from (6.5.28) to get some new series $g(f)$. Then let us change the variable $f = y^2$ to bring the series to a standard form. The diagonal Padé approximants to the series $g(y)$ are supposed to give a correction to the value of 1, suggested by (6.7.79). To calculate the correction one has to find the value of the corresponding approximant as $y \to \sqrt{f_c}$. The following sequence of approximations for the amplitude B can be calculated now readily,

$$B_n = 1 + PadeApproximant[g(y \to \sqrt{f_c}), n, n]. \tag{6.7.80}$$

The sequence of approximations is shown in Fig. 6.5.

There is clear saturation of the results for larger n, and $B_{28} = -1.84676$. One can

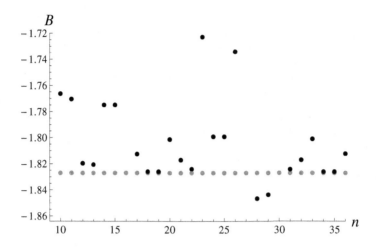

Figure 6.5 Sequence of approximations B_n calculated from (6.7.80).

reconstruct the expression for conductivity corresponding to B_{28} in additive form

$$\boxed{\sigma^*_{28}(f) = P_{1,0}(f) + \frac{F_2(f)}{F_8(f)}}, \qquad (6.7.81)$$

where $P_{1,0}$ is given by (6.7.79),

$$F_2(f) =$$
$$-3.54491\sqrt{f} + 9.50117f + 0.858369f^{3/2} + 11.7089f^2 -$$
$$511.27f^{5/2} + 1385.54f^3 + 517.845f^{7/2} - 2825.84f^4 +$$
$$248.757f^{9/2} + 1540.42f^5 - 82.9532f^{11/2} - 145.574f^6 +$$
$$475.127f^{13/2} - 1217.21f^7 - 516.99f^{15/2} + 2138.29f^8 -$$
$$89.0439f^{17/2} - 1010.14f^9 + 37.2112f^{19/2} + 75.2569f^{10} +$$
$$599.179f^{21/2} - 1704.33f^{11} - 619.738f^{23/2} + 3692.71f^{12} -$$
$$383.092f^{25/2} - 2130.91f^{13} + 107.833f^{27/2} + 170.982f^{14}, \qquad (6.7.82)$$

and

$$F_8(f) =$$
$$1 - 2.11604\sqrt{f} - 2.07261f - 2.56107f^{3/2} + 142.299f^2 -$$
$$308.258f^{5/2} - 412.612f^3 + 843.206f^{7/2} + 405.788f^4 -$$
$$704.774f^{9/2} - 174.033f^5 + 157.178f^{11/2} - 116.24f^6 +$$
$$289.94f^{13/2} + 342.72f^7 - 693.02f^{15/2} - 298.599f^8 +$$
$$500.384f^{17/2} + 124.396f^9 - 90.2296f^{19/2} - 176.742f^{10} +$$
$$371.909f^{21/2} + 523.732f^{11} - 1066.35f^{23/2} - 536.914f^{12} +$$
$$948.104f^{25/2} + 239.947f^{13} - 210.855f^{27/2} - 21.9942f^{14}. \qquad (6.7.83)$$

The maximum error for the formula (6.7.81) is very small and equals 0.0193%, slightly better compared with (6.6.61). The amplitude B is located firmly in the interval $(-1.95, -1.85)$, according to the two best formulae for conductivity.

8. Comment on contrast parameter

Effective conductivity of 2D conductors with arbitrary contrast parameters can be reconstructed from the expansion at small concentrations and from the critical behaviour at high concentrations. Consider the exact formulae (4.2.28), (4.2.29) with the contrast parameter $\rho = \frac{\sigma_i - \sigma_m}{\sigma_i + \sigma_m}$, entering the power series for conductivity explicitly. Usually for the matrix it is taken $\sigma_m = 1$, and for the inclusions $\sigma_i \equiv \sigma$, and $|\rho| \leq 1$.

There are two different limit cases for the effective conductivity. In the limiting case of a perfectly conducting inclusions, the effective conductivity is expected to tend to infinity as a power law, as the concentration f tends to the maximal value $f_c = \frac{\pi}{4}$ for the square array, and $\sigma_e(f) \simeq \frac{\pi^{3/2}}{2}(f_c - f)^{-1/2}$. From the phase interchange theorem [40] it follows that in two dimensions, the superconductivity index is equal to the conductivity index and another limit-case of the conducting inclusions embedded into the non-conducting matrix is simply obtained from the phase-interchange theorem by inverting the latter expression.

On the other hand, the following expansion in concentration f of the inclusions and contrast parameter ρ is valid for both regular and random cases

$$\sigma_e(f, \rho) = 1 + a_1(\rho)f + a_2(\rho)f^2 + \cdots . \tag{6.8.84}$$

The low-order coefficients depend only on ρ

$$a_1(\rho) = 2\rho, \quad a_2(\rho) = 2\rho^2.$$

Since there are two parameters f and ρ, we derive the final formula in two steps. First, let us guarantee the correct qualitative behaviour in concentration and contrast parameter. Mind that in 2D one has to respect celebrated Keller's phase-interchange relation [25, 40]. Since all dependence on the conductivity of inclusions and matrix is hidden within the contrast parameter, the phase interchange can be expressed even simpler as follows,

$$\frac{1}{\sigma_e(f, \rho)} = \sigma_e(f, -\rho),$$

and one should try to develop an expression which satisfy it by design. It is necessary to respect corresponding critical properties and then simply modify parameters of the solution to move away to a non-critical situations. Moving singularity to the non-physical values of f allows to preserve the form typical for critical regime for all values of ρ but suppressing singularity.

The simplest solution for the effective conductivity σ_e, to be derived along these lines, is going to be expressed as factor approximant. Such form is convenient when one has to include both types of critical behaviour, as $\rho \to 1$, and $\rho \to -1$, respectively,

$$\sigma_1^*(f,\rho) = (b_1 z_1(f) + 1)^{1/2} (b_2 z_2(f) + 1)^{-1/2}, \qquad (6.8.85)$$

where

$$z_1(f) = \frac{f}{\frac{f_c}{\rho} - f}, \quad z_2(f) = \frac{f}{\frac{-f_c}{\rho} - f}.$$

By introducing the two variables, we separate the two types of critical behaviour. In addition, we consider the simplest possible dependence of the effective thresholds leading to the correct value of f_c as $\rho \to 1$ and $\rho \to -1$, respectively. Such a dependence follows celebrated Clausius–Mossotti formula. Finally, let us require an asymptotic equivalence of (6.8.85) with the expansion (6.8.84) in order to find the parameters b_1 and b_2. In its final form

$$\sigma_1^*(f,\rho) = \frac{\sqrt{\frac{1.5708 f}{\frac{0.785398}{\rho} - f} + 1}}{\sqrt{\frac{1.5708 f}{-\frac{0.785398}{\rho} - f} + 1}}.$$

Even more general form of factor approximant could be suggested with each unit being a self-similar root approximant as described by formula (5.3.48) on page 114. Each root approximant respects not only the leading asymptotic term, but also the structure of the expansion at small and high concentrations as explained above in Subsection 5.1. Namely the following roots are involved,

$$R_1^*(f,\rho) = \left((b_1 z_1(f) + 1)^{3/2} + p_1 z_1(f)^2\right)^{1/4},$$
$$R_2^*(f,\rho) = \left((b_2 z_2(f) + 1)^{3/2} + p_2 z_2(f)^2\right)^{-1/4}, \qquad (6.8.86)$$

and they are organized into a factor approximant

$$\sigma_F(f,\rho) = R_1^*(f,\rho)\, R_2^*(f,\rho). \qquad (6.8.87)$$

Let us employ two asymptotic conditions on b_1 and b_2 and additional two conditions on the correct amplitude and inverse amplitude at the critical point as $\rho \to 1$ and $\rho \to -1$, respectively. In our case, there are two complex-valued solutions for b_1 and b_2. They come in complex conjugate pair, so that their sum is real. Consider the average sum of the complex conjugate pair,

$$\sigma_2^{**}(f,\rho) = \frac{1}{2}[\sigma_F(f,\rho) + \sigma_F^*(f,\rho)], \qquad (6.8.88)$$

where

$$
\sigma_F(f,\rho) = \frac{\sqrt[4]{\frac{(6.23057-5.25247i)f^2}{\left(\frac{0.785398}{\rho}-f\right)^2}+\left(1+\frac{(2.0944-3.19962i)f}{\frac{0.785398}{\rho}-f}\right)^{3/2}}}{\sqrt[4]{\frac{(6.23057+5.25247i)f^2}{\left(-\frac{0.785398}{\rho}-f\right)^2}+\left(1+\frac{(2.0944+3.19962i)f}{-\frac{0.785398}{\rho}-f}\right)^{3/2}}}
$$

$$
\sigma_F{}^*(f,\rho) = \frac{\sqrt[4]{\frac{(6.23057+5.25247i)f^2}{\left(\frac{0.785398}{\rho}-f\right)^2}+\left(1+\frac{(2.0944+3.19962i)f}{\frac{0.785398}{\rho}-f}\right)^{3/2}}}{\sqrt[4]{\frac{(6.23057-5.25247i)f^2}{\left(-\frac{0.785398}{\rho}-f\right)^2}+\left(1+\frac{(2.0944-3.19962i)f}{-\frac{0.785398}{\rho}-f}\right)^{3/2}}} .
$$

(6.8.89)

Formula (6.8.88) preserves phase-interchange relation with very high accuracy. To simplify consideration one can employ iterated roots in (6.8.87), with the final result

$$
\sigma_3^{**}(f,\rho) = \frac{\sqrt[4]{\frac{f\rho(0.239364\,f\rho+0.896605)+0.61685}{(f\rho-0.785398)^2}}}{\sqrt[4]{\frac{f\rho(0.239364\,f\rho-0.896605)+0.61685}{(f\rho+0.785398)^2}}} .
$$

(6.8.90)

Resulting formula (6.8.90) is much simpler and respects the phase-interchange relation.

Our formulae $\sigma_1^*(f,\rho)$ and $\sigma_3^{**}(f,\rho)$ are of a qualitative nature, and serve as an initial approximation. They are awaiting to be further corrected with higher-order terms to be obtained. The same approach is applicable to hexagonal arrays. In 3D, one can keep the basic idea but employ different critical indices s and t, or logarithmic dependence when needed.

In summary, we considered various problems for the regular square lattice arrangements of ideally conducting cylinders and concluded that series (6.2.4) is good enough to: calculate the position of a threshold for the effective conductivity, to calculate the value of a superconductivity critical index, and to obtain an accurate crossover expression valid for arbitrary concentrations. The latter task include factual reconstruction of the high-concentration expansion from the low-concentration expansion.

When two expansions around different points are available, the problem of reconstruction can be solved in terms of high-concentration Padé approximants, implying that the effective resistivity (conductivity) can be presented in the form of a Stieltjes integral, $\int_0^\infty \frac{d\phi(u)}{u+X}$, where $\phi(u)$ is a bounded non-decreasing function of u [8], in terms of the variable $X = \sqrt{\frac{f}{f_c-f}}$. Such Padé approximants give tight lower and upper bounds for the conductivity, valid up to the very high f. In our opinion, such approximants deserve further studies, especially in very high orders not yet reached.

Nonetheless, different non-standard techniques for resummation described in Chapter 5 work well and allow to reach very high orders. They give significantly better representation of conductivity than formula deduced by standard techniques. The obtained formula (6.7.81) is valid for all concentrations including touching cylinders, hence it completely solves the problem of the effective conductivity for the high contrast square array. For practical purposes one can turn to the more compact (6.6.67).

REFERENCE

1. Adamyan VM, Tkachenko IM, Urrea M, Solution of the Stieltjes Truncated Moment Problem, Journal of Applied Analysis 2003; 9: 57–74.
2. Andrianov IV, Two-point Padé approximants in the Mechanics of Solids, ZAMM 1994; 74: 121–122.
3. Andrianov IV, Awrejcewicz J, Starushenko GA, Application of an improved three-phase model to calculate effective characteristics for a composite with cylindrical inclusions, Latin American Journal of Solids and Structures 2013; 10: 197–222.
4. Andrianov IV, J. Awrejcewicz J, New trends in asymptotic approaches: summation and interpolation methods, Appl. Mech. Rev. 2001; 54: 69–92.
5. Andrianov IV, Danishevskyy VV, Kalamkarov AL, Analysis of the effective conductivity of composite materials in the entire range of volume fractions of inclusions up to the percolation threshold, Composites: Part B Engineering 2010; 41: 503–507.
6. Andrianov I, Danishevskyy V, Tokarzewski S, Quasifractional approximants in the theory of composite materials, Acta Applicandae Mathematicae 2000; 61: 29–35.
7. Andrianov IV, Starushenko GA, Danishevskyy VV, Tokarzewski S, Homogenization procedure and Padé approximants for effective heat conductivity of composite materials with cylindrical inclusions having square cross-section, Proc. R. Soc. Lond. A 1999; 455: 3401–3413.
8. W. Van Assche, The impact of Stieltjes work on continued fractions and orthogonal polynomials, Thomas Jan Stieltjes Oeuvres Complètes - Collected Papers. G. van Dijk, ed., Springer, 1993, Report number: OP-SF 5–37; 9 Jul 1993.
9. Baker GA, Graves-Moris P, Padé Approximants. Cambridge, UK: Cambridge University; 1996.
10. Bender CM, Boettcher S, Determination of $f(\infty)$ from the asymptotic series for $f(x)$ about $x = 0$, J. Math. Phys. 1994; 35: 1914–1921.
11. Bergman DJ, Dielectric-constant of a composite-material-problem in classical physics, Phys. Rep. 1978; 43: 378–407.
12. Berlyand L, Kolpakov A, Network approximation in the limit of small interparticle distance of the effective properties of a high contrast random dispersed composite, Arch. Ration. Mech. Anal. 2001; 159: 179–227.
13. Berlyand L, Novikov A, Error of the network approximation for densely packed composites with irregular geometry, SIAM J. Math. Anal. 2002; 34: 385–408.
14. Choy TC, Effective Medium Theory. Principles and Applications. Oxford. Clarendon Press. 1999.
15. Czapla R, Nawalaniec W, Mityushev V, Effective conductivity of random two-dimensional composites with circular non-overlapping inclusions, Comput. Mat. Sci. 2012; 63: 118–126.
16. Derkach V, Hassi S, de Snoo H, Truncated moment problems in the class of generalized Nevanlinna functions, Math. Nachr. 2012; 285: 1741–1769.
17. Gluzman S, Mityushev V, Nawalaniec W, Cross-properties of the effective conductivity of the regular array of ideal conductors, Arch. Mech. 2014; 66: 287–301.
18. Gluzman S, Yukalov VI, Unified approach to crossover phenomena, Phys. Rev. E 1998; 58: 4197–4209.
19. Gluzman S, Yukalov VI, Self-similar extrapolation from weak to strong coupling, J.Math.Chem. 2010; 48: 883–913.
20. Gluzman S, Yukalov VI, Sornette D, Self-similar factor approximants, Phys. Rev. E 2003; 67: 026109.
21. Gonzales-Vera P, Njastad O, Szego functions and multipoint Padé approximation, Journal of Computational and Applied Mathematics 1990; 32: 107–116.
22. Golden K, Papanicolaou G, Bounds for effective parameters of heterogeneous media by analytic

continuation, Commun. Math. Phys. 1983; 90: 473–491.

23. Hendriksen E, Moment methods in Two Point Padé Approximation, Journal of Approximation Theory. 1984; 40: 313–326.

24. Keller JB, Conductivity of a Medium Containing a Dense Array of Perfectly Conducting Spheres or Cylinders or Nonconducting Cylinders, Journal of Applied Physics 1963; 34: 991–993.

25. Keller JB, A Theorem on the Conductivity of a Composite Medium, J. Math. Phys. 1964; 5: 548–549.

26. Krein MG, Nudel'man AA. The Markov moment problem and extremal problems. Translations of Mathematical Monographs Providence, RI.: American Mathematical Society; 1977.

27. Maxwell JC, Electricity and magnetism. 1st edn. p.365. Clarendon Press. 1873.

28. McPhedran RC, Transport Properties of Cylinder Pairs and of the Square Array of Cylinders, Proc.R. Soc. Lond. 1986; A 408: 31–43.

29. McPhedran RC, Poladian L, Milton GW, Asymptotic studies of closely spaced, highly conducting cylinders, Proc. R. Soc. A 1988; 415: 185–196.

30. Mityushev V, Steady heat conduction of a material with an array of cylindrical holes in the nonlinear case, IMA Journal of Applied Mathematics 1998; 61: 91–102.

31. Mityushev V, Exact solution of the $\mathbb{R}$-linear problem for a disk in a class of doubly periodic functions, J. Appl.Functional Analysis 2007; 2: 115–127.

32. Perrins WT, McKenzie DR, McPhedran RC, Transport properties of regular array of cylinders, Proc. R. Soc.A 1979; 369, 207–225.

33. Rayleigh Lord, On the influence of obstacles arranged in rectangular order upon the properties of medium, Phil. Mag. 1892; 34, 481–502.

34. Rylko N, Transport properties of the regular array of highly conducting cylinders, J. Engrg. Math. 2000; 38: 1–12.

35. Rylko N, Structure of the scalar field around unidirectional circular cylinders, Proc. R. Soc. 2008; A 464: 391–407.

36. Samuel MA, Li G, Estimating perturbative coefficients in quantum field theory and the ortho-positronium decay rate discrepancy, Physics Letters B 1994; 331; 114–118.

37. Stanley HE, Scaling, universality, and renormalization:Three pillars of modern critical phenomena, Reviews of Modern Physics 1999; 71: 358–366.

38. Tokarzewski S, Andrianov I, Danishevsky V, Starushenko G, Analytical continuation of asymptotic expansion of effective transport coefficients by Padé approximants, Nonlinear Analysis 2001; 47: 2283–2292.

39. Tokarzewski S, Blawzdziewicz J, Andrianov I, Effective conductivity for densely parked highly condensed cylinders, Appl.Physics A 1994; 59: 601–604.

40. Torquato S, Random Heterogeneous Materials: Microstructure and Macroscopic Properties. New York. Springer-Verlag: (2002).

41. Torquato S, Stillinger FH, Jammed hard-particle packings: From Kepler to Bernal and beyond, Reviews of Modern Physics 2010; 82: 2634–2672.

42. Valent G, Van Assche W, The impact of Stieltjes' work on continued fractions and orthogonal polynomials: additional material, Journal of Computational and Applied Mathematics 1995; 65: 419–447.

43. Yukalov VI, Gluzman S, Sornette D, Summation of Power Series by Self-Similar Factor Approximants, Physica A 2003; 328: 409–438.

44. Yukalov VI, Gluzman S, Optimisation of self-similar factor approximants, Mol. Phys. 2009; 107: 2237–2244.

45. Yukalova EP, Yukalov VI, Gluzman S, Extrapolation and Interpolation of Asymptotic Series by Self-Similar Approximants, Journal of Mathematical Chemistry 2010; 47: 959–983.

CHAPTER 7

Conductivity of regular composite. Hexagonal array

1. Effective conductivity and critical properties of a hexagonal array of superconducting cylinders

Let us consider a two-dimensional composite corresponding to the regular hexagonal array of ideally conducting (superconducting) cylinders of concentration f embedded into the matrix of a conducting material and again use the exact formulae (4.2.28), (4.2.29) on page 86. Many features of the problem resemble the case of square array just considered. Naively one would expect the hexagonal array to present more challenges for resummation. Just because its threshold is higher one can think that longer expansion at small f are required to achieve similar accuracy for high concentrations.

In this chapter, we are interested primarily in the case of a high contrast regular composites, when the conductivity of the inclusions is much larger than the conductivity of the host. That is, the highly conducting inclusions are replaced by the ideally conducting inclusions with infinite conductivity. In this case, the contrast parameter is equal to unity. High contrast composites are extremely attractive for the design of new materials with physical properties better than those of their constituents. As soon as the ratio of the conductivity of inclusions to the conductivity of matrix is greater than several hundreds, it can be taken to be infinite.

Two-dimensional regular hexagonal-arrayed composites [3], much closer resemble the two-dimensional random composites, than their respective 3D counterparts do [29]. The tendency to order in the two-dimensional random system of disks, is a crucial feature in the theory of composites at high concentration. Most strikingly, it

Computational Analysis of Structured Media
http://dx.doi.org/10.1016/B978-0-12-811046-1.50007-1
Copyright © 2018 Elsevier Inc. All rights reserved.

appears that the maximum volume fraction of $\frac{\pi}{\sqrt{12}} \approx 0.9069$ is attained both for the regular hexagonal array of disks and for random (irregular) 2D composites [29]. Thus, we are able to evaluate in Chapter 9 the effect of randomness, or even use regular case as a starting point for the random case.

The higher threshold makes highly packed hexagonal array more attractive technologically than any other regular array. Their ability to achieve high concentration of the filling inclusions is particularly relevant to polymer–ceramic composites, because a polymer matrix compensates for the unsatisfying brittle nature of the highly conductive ceramics. Such materials are vitally important for an optimal design of electrical capacitors and thermal management in the electronics industry. In the latter case, an integrated circuit design requires highly efficient composites for heat removal.

A two-dimensional composite corresponding to the regular hexagonal array of ideally conducting cylinders of radius r embedded into the matrix of a conducting material as shown in Fig. 7.1. The volume fraction (concentration) of cylinders is equal to

$$f = \frac{2\pi}{\sqrt{3}} r^2$$

and corresponding periodicity cell is shown in Fig. 7.2.

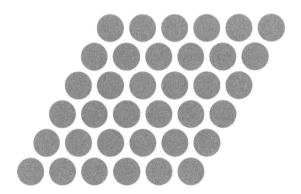

Figure 7.1 Hexagonal array. Section perpendicular to the unidirectional cylinders is shown. Disks of radius r is given as a dimensionless value.

A numerical study of the 2D hexagonal case can be found in [27]. It resulted in the following interpolation function

$$\sigma_{PMM}(f,\rho) = \frac{2f}{-\frac{0.071134\rho f^6}{0.943147 - \rho^2 f^{12}} - 0.000076\rho f^{12} + \frac{1}{\rho} - f} + 1. \qquad (7.1.1)$$

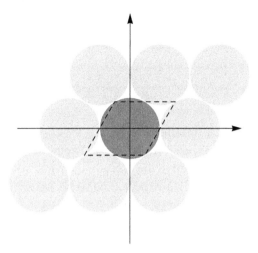

Figure 7.2 Periodicity cell bounded by the dashed line, rhombus generated by the vectors $\omega_1 = \sqrt[4]{\frac{4}{3}}$ and $\omega_2 = \sqrt[4]{\frac{4}{3}}e^{\frac{1}{3}\pi i}$ expressed in terms of the complex number. The area of the rhombus holds unity; the angle between the vectors ω_1 and ω_2 is equal to $\frac{\pi}{3}$.

In the case $\rho = 1$, we get

$$\sigma_e(f) = 1 - \frac{2f}{\frac{0.075422f^6}{1-1.06028f^{12}} + f - 1}. \tag{7.1.2}$$

Note that (7.1.2) diverges with critical exponent $s = 1$, as $f \to 0.922351$. This property on one hand makes the formula more accurate in the vicinity of a true critical point but, on the other hand, makes any comparison in the critical region meaningless. It remains rather accurate till $f = 0.85$, where the error is equal to 0.47%. For $f = 0.905$ the error equals 52%. Expression (7.1.2) was derived using only terms up to the 12th order in concentration. The expansion of (7.1.2) is characterized by a rather regular behaviour of the coefficients,

$$\begin{aligned}
\sigma^{reg}(f) &= 1 + 2f + 2f^2 + 2f^3 + 2f^4 + 2f^5 + 2f^6+ \\
&\quad 2.15084f^7 + 2.30169f^8 + 2.45253f^9 + 2.60338f^{10}+ \\
&\quad 2.75422f^{11} + 2.90506f^{12} + O(f^{13}).
\end{aligned} \tag{7.1.3}$$

One can, in principle, collect the higher order terms as well. However, such derivation of an additional terms cannot be considered as consistent since it relies on the agreement with numerical results. It turns out though, that (7.1.3) compares well with our exact result (7.2.6) shown below.

In a different limit of high concentration Keller [20] suggested a constructive asymptotic method for regular lattices, leading to very transparent, inverse square-root formula for the square array [20]. Berlyand and Novikov [8] extended Keller's method to the hexagonal array,

$$\sigma_e \simeq \frac{\sqrt[4]{3}\pi^{\frac{3}{2}}}{\sqrt{2}} \frac{1}{\sqrt{\frac{\pi}{\sqrt{12}} - f}}. \qquad (7.1.4)$$

Thus, the critical amplitude A (pre-factor), is equal to $A \approx 5.18$.

We will examine below this result for the critical amplitude from the perspective of re-summation techniques suggested in the preceding chapter for square regular arrays. By analogy with square lattice, we expect a constant correction in the asymptotic regime

$$\sigma_e \simeq \frac{\sqrt[4]{3}\pi^{\frac{3}{2}}}{\sqrt{2}} \frac{1}{\sqrt{\frac{\pi}{\sqrt{12}} - f}} + B, \qquad (7.1.5)$$

where the correction term B can not be found in the literature, to the best of our knowledge. It will be calculated below by different methods.

2. Series for hexagonal array of superconducting cylinders

We proceed to the case of a hexagonal array of inclusions, where rather long expansions in concentration will be presented an analysed systematically. The coefficients a_n in the expansion of

$$\sigma_e(f) = 1 + \sum_{n=1}^{\infty} a_n f^n,$$

are given by the exact series (4.2.28). Below, this expansion is presented in the truncated numerical form for $\rho = 1$

$$
\begin{aligned}
\sigma_e(f) = \quad & 1 + 2f + 2f^2 + 2f^3 + 2f^4 + 2f^5 + 2f^6 \\
+ \ & 2.1508443464271876f^7 + 2.301688692854377f^8 \\
+ \ & 2.452533039281566f^9 + 2.6033773857087543f^{10} \\
+ \ & 2.754221732135944f^{11} + 2.9050660785631326f^{12} \\
+ \ & 3.0674404324522926f^{13} + 3.2411917947659736`f^{14} \\
+ \ & 3.426320165504177f^{15} + 3.6228255446669055f^{16} \\
+ \ & 3.8307079322541555f^{17} + 4.049967328265928f^{18} \\
+ \ & 4.441422739726373f^{19} + 4.845994396051242f^{20} \\
+ \ & 5.264540375940583f^{21} + 5.69791875809444f^{22} \\
+ \ & 6.146987621212864f^{23} + 6.6126050439959f^{24} \\
+ \ & 7.135044602470776f^{25} + 7.700073986554016f^{26} \\
+ \ & O(f^{27}).
\end{aligned}
\tag{7.2.6}
$$

The first 12 coefficients of (7.2.6) and the Taylor expansions of (7.1.2) practically coincide. The next coefficients can be calculated by exact formula (4.2.28) first derived in [23, 24]. This requires use of the double precision and perhaps a computer more powerful than a standard laptop being employed.

Since we are dealing with the limiting case of a perfectly conducting inclusions when the conductivity of inclusions tends to infinity, the effective conductivity is also expected to tend to infinity as a power law, as the concentration f tends to the maximal value f_c for the hexagonal array,

$$
\sigma_e(f) \simeq A(f_c - f)^{-s} + B.
\tag{7.2.7}
$$

For sake of exploring how consistent are various resummation techniques, we will calculate the index. The critical amplitudes A and B will be considered below as unknown non-universal parameters to be calculated.

The problem of interest can be formulated similarly to the case of a square lattice as follows: given the polynomial approximation (7.2.6) of the function $\sigma_e(f)$, to estimate the convergence radius f_c of the Taylor series $\sigma_e(f)$; to determine critical index s and amplitudes A, B of the asymptotically equivalent approximation (7.2.7) near $f = f_c$.

When such extrapolation problem is solved, we proceed to solve an interpolation problem of matching the two asymptotic expressions for the conductivity and derive interpolation formula for all concentrations.

3. Critical Point

3.1. Padé approximants

The simplest and direct way to extrapolate power series, is to apply the Padé approximants $P_{n,m}(f)$ in general case leading to the whole table of estimates/predictions. The coefficients of the approximants are derived directly from the coefficients of the given power series from the requirement of asymptotic equivalence to the given series or function $\Phi(f)$. As above, when there is a need to stress the last point, we simply write *PadeApproximant*$[\Phi[f], n, m]$.

In order to estimate the position of a critical point, let us apply the diagonal Padé approximants,

$$P_{1,1}(f) = \frac{m_1 f + 1}{n_1 f + 1}, \quad P_{2,2}(f) = \frac{m_2 f^2 + m_1 f + 1}{n_2 f^2 + n_1 f + 1}, \tag{7.3.8}$$

and so forth. Padé approximants locally are the best rational approximations of power series. Their poles determine singular points of the approximated functions. Calculations with Padé approximants are straightforward and can be performed with *Mathematica*®. They do not require any preliminary knowledge of the critical index and we have to find the position of a simple pole. In the theory of periodic 2D composites [6, 18, 28], their application is justifiable rigorously away from the square-root singularity and from the high-contrast limit.

There is a convergence within the approximations for the critical point generated by the sequence of Padé approximants, corresponding to their order increasing:

$$f_1 = 1, \quad f_2 = 1, \quad f_3 = 1, \quad f_6 = 0.945958,$$

$$f_7 = 0.945929, \quad f_8 = 0.947703, \quad f_9 = 0.946772, \quad f_{10} = 0.942378,$$

$$f_{11} = 0.945929, \quad f_{12} = 0.945959, \quad f_{13} = 0.920878.$$

The main body of the approximations is well off the exact value. The percentage error given by the last/best approximant in the sequence equals 1.5413%. If only the first row of the Padé table is studied, then the best estimate is equal to 0.929867, close to the estimates with the diagonal sequence.

We suggest that further increase in accuracy is limited by triviality, or "flatness" of the coefficients values in six starting orders of (7.2.6). Consider another sequence of approximants, when diagonal Padé approximants are multiplied with Clausius–Mossotti-type expression,

$$P_1^t(f) = \frac{(1-f)}{(1+f)} \frac{(1+m_1 f)}{(1+n_1 f)}, \quad P_2^t(f) = \frac{(1-f)}{(1+f)} \frac{(1+m_1 f + m_2 f^2)}{(1+n_1 f + n_2 f^2)}, \tag{7.3.9}$$

and so forth. The transformation which lifts the flatness, does improve convergence of the sequence of approximations for the threshold,

$$f_7 = 0.94568, \quad f_9 = 0.948299, \quad f_{10} = 0.9287,$$

$$f_{11} = 0.945681, \quad f_{12} = 0.89793, \quad f_{13} = 0.903517.$$

The percentage error given by the last approximant in the sequence equals 0.373%. The estimate should be considered as rather good because the value of critical index was not employed.

In order to judge the quality of the latter estimate, let us try also highly recommended *DLog* Padé method, which also does not require a preliminary knowledge of the critical index value. One has to differentiate log of (7.2.6), apply the diagonal Padé approximants, and define the critical point as the position of the pole nearest to the origin. The best estimate obtained this way is $f_{12} = 0.919304$, with percentage error of 1.368%. One can also estimate the value of critical index as a residue [5], and obtain rather disappointing value of 0.73355.

3.2. Corrected threshold

An approach based on the Padé approximants produces the expressions for the cross-properties from "left-to-right", extending the series from the dilute regime of small f to the high-concentration regime of large f. Alternatively, one can proceed from "right-to-left", i.e., extending the series from the large f (close to f_c) to small f [13, 16, 31].

We will first derive an approximation to the high-concentration regime and then extrapolate to the lesser concentrations. There is an understanding that physics of a 2D high-concentration, regular and irregular composites is related to the so-called "necks", certain areas between closely spaced disks [7, 8, 20].

Assume also that the initial guess for the threshold value is available from previous Padé-estimates, and is equal to $f_6 = 0.945958$.

The simplest way to proceed is to look for the solution in the whole region $[0, f_c)$, in the form which extends asymptotic expression from [22], $\sigma_1^{r-l} = \alpha_0(f_c - f)^{-1/2} + \alpha_1$. This approximation works well for the square lattice of inclusions [13].

In the case of hexagonal array, we consider its further extension, with higher order term in the expansion

$$\sigma_2^{r-l} = \alpha_0(f_6 - f)^{-s} + \alpha_1 + \alpha_2(f_6 - f)^s, \tag{7.3.10}$$

where index s is considered as another unknown. All unknowns can be obtained from the three starting non-trivial terms of (7.2.6), namely $\sigma_3 \simeq 1 + 2f + 2f^2 + 2f^3$. Thus, the parameters equal

$$\alpha_0 = 2.24674, \quad \alpha_1 = -1.43401, \quad \alpha_2 = 0.0847261, \quad s = 0.83262.9.$$

Let us assume that the true solution σ may be found in the same form but with supposedly exact, yet unknown threshold F_c,

$$\Sigma = \alpha_0(F_c - f)^{-s} + \alpha_1 + \alpha_2(F_c - f)^s. \tag{7.3.11}$$

The expression (7.3.11) may be inverted and F_c expressed explicitly,

$$F_c = 2^{-\frac{1}{s}} \left(\frac{-\sqrt{(\alpha_1 - \Sigma)^2 - 4\alpha_0\alpha_2} - \alpha_1 + \Sigma}{\alpha_2} \right)^{\frac{1}{s}} + f. \tag{7.3.12}$$

Formula (7.3.12) is a formal expression for the threshold, since $\Sigma(f)$ is also unknown. We can use for Σ the series in f, so that instead of a true threshold, we have an effective threshold, $F_c(f)$, given in the form of a series in f. For the concrete series (7.2.6), the following expansion follows,

$$
\begin{aligned}
F_c(f) = \quad & f_6 + 0.0134664 f^4 + 0.00883052 f^5 \\
+ \quad & 0.00647801 f^6 - 0.0709217 f^7 + 0.0032732 f^8 \\
+ \quad & 0.00244442 f^9 + 0.00594779 f^{10} + 0.00482187 f^{11} \\
+ \quad & 0.00413887 f^{12} + \cdots , \tag{7.3.13}
\end{aligned}
$$

which should become a true threshold F_c as $f \to F_c$.

In order to take this limit, let us apply re-summation procedure to the expansion (7.3.13) using the diagonal Padé approximants. Finally let us define the sought threshold F_c^* self-consistently from the following equations dependent on the approximants order,

$$F_c^* = P_{n,n}(F_c^*), \tag{7.3.14}$$

meaning simply that as we approach the threshold, the RHS of (7.3.14) should become the threshold. Since the diagonal Padé approximants of the nth order are defined for an even number of terms $2n$, we will also have a sequence of $F_{c,n}^*$.

Solving equation (7.3.14), we obtain

$$F_{c,4}^* = 0.930222, \quad F_{c,5}^* = 0.855009, \quad F_{c,6}^* = 0.9483, \quad F_{c,7}^* = 0.932421,$$

$$F_{c,8}^* = 0.946773, \quad F_{c,9}^* = 0.941391, \quad F_{c,10}^* = 0.94682, \quad F_{c,11}^* = 0.932752,$$

$$F_{c,12}^* = 0.907423, \quad F_{c,13}^* = 0.903303.$$

The last two estimates for the threshold are good.

3.3. Threshold with known critical index

Also, one can pursue a slightly different strategy, assuming that critical index s is known and s = $\frac{1}{2}$, and is incorporated into initial approximation. Such situation is not far-fetched and can serve as an example, in general case when the critical index is known from different source and the value of threshold is of a primary interest.

Recalculated parameters of the approximant σ_2 equal

$$\alpha_0 = 5.12249, \quad \alpha_1 = -5.74972, \quad \alpha_2 = 1.52472.$$

For the series (7.2.6), in the vicinity of f_6, the following expansion follows,

$$
\begin{aligned}
F_c(f) = \quad & f_6 - 0.082561 f^3 + 0.0282108 f^4 - 0.000383173 f^5 \\
+ \quad & 0.0228241 f^6 - 0.0649593 f^7 + 0.01561635 f^8 \\
- \quad & 0.00911151 f^9 + 0.01874715 f^{10} + 0.00688507 f^{11} \\
+ \quad & 0.0169516 f^{12} + \cdots .
\end{aligned}
\tag{7.3.15}
$$

Let us apply re-summation procedure to the expansion (7.3.15) using super-exponential approximants $E^*(f)$ [32]. Namely we have the following sequence

$$E_1^*(f) = e^{t_1 f}, \quad E_2^*(f) = e^{t_1 e^{t_2 f} f}, \quad E_3^*(f) = e^{t_1 e^{t_2 e^{t_3 f} f} f}, \ldots,$$

where the parameters t_i will be found from the asymptotic equivalence and only one new value has to be calculated at each next step. These approximants are deeply connected with the self-similar approximation theory, algebraic renormalization and bootstrap expounded in Chapter 5.

Finally let us define the sought threshold F_c^* self-consistently,

$$F_c^* = 0.945958 - 0.082561 f^3 E^*(F_c^*).
\tag{7.3.16}$$

Since the super-exponential approximants are defined as E_k^* for arbitrary number of terms k, we will also have a sequence of $F_{c,k}^*$, e.g.,

$$
\begin{aligned}
E_1^* &= \exp(-0.341697 f), \quad E_2^* = \exp(-0.341697 e^{0.157266 f} f), \\
E_3^* &= \exp[-0.341697 \exp(0.157266 e^{5.28382 f} f) f], \ldots
\end{aligned}
\tag{7.3.17}
$$

and so on iteratively. Solving equation (7.3.16), we obtain

$$F_{c,1}^* = 0.901505, \quad F_{c,2}^* = 0.903321, \quad F_{c,3}^* = 0.945958,$$

$$F_{c,4}^* = 0.903404, \quad F_{c,5}^* = 0.916641, \quad F_{c,6}^* = 0.903412,$$

$$F_{c,7}^* = 0.903556, \quad F_{c,8}^* = 0.903412, \quad F_{c,9}^* = 0.903412.$$

There is a convergence in the sequence of approximations for the threshold. The percentage error achieved for the last point is equal to 0.3845%.

The method of corrected threshold produces good results based only on the starting 12 terms from the expansion (7.2.6), in contrast with the Padé-based approximations, requiring all available terms to gain similar accuracy. The task of extracting the threshold, a purely geometrical quantity, from the solution of the physical problem is

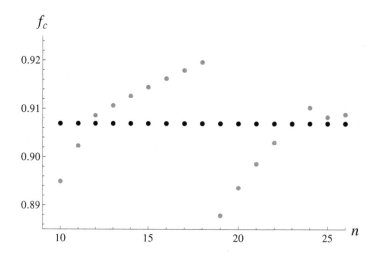

Figure 7.3 f_c calculated by ratio method, compared with the exact threshold.

not trivial and is relevant to similar attempts to find the threshold for random systems from the expressions for some physical quantities [29].

Instead of the super-exponential approximants, one can in the same manner apply the diagonal Padé approximants,

$$F_{c,n}^* = P_{n,n}(F_c^*).$$ (7.3.18)

Solving equation (7.3.18), we obtain

$$F_{c,3}^* = 0.908188, \quad F_{c,4}^* = 0.889169, \quad F_{c,5}^* = 0.889391, \quad F_{c,6}^* = 0.887983,$$

$$F_{c,7}^* = 0.899495, \quad F_{c,11}^* = 0.903011, \quad F_{c,12}^* = 0.90296, \quad F_{c,13}^* = 0.9057.$$

The last value is closest to the exact result.

The ratio method, also works well. It evaluates the threshold through the value of index and ratio of the series coefficients, $f_{c,n} = \frac{\frac{s-1}{n}+1}{\frac{a_n}{a_{n-1}}}$. The last point gives rather good estimate, $f_{c,26} = 0.908801$, despite of the oscillations in the dependence on n, as seen in Fig. 7.3.

4. Critical index and amplitude

Standard way to proceed with critical index calculations when the value of the threshold is known can be found in the Chapter 5. One would first apply the following transformation,

$$z = \frac{f}{f_c - f} \Leftrightarrow f = \frac{zf_c}{z + 1}, \qquad (7.4.19)$$

to the series (7.2.6) in order to make application of the different approximants more convenient.

Then, to such transformed series $M_1(z)$ apply the *DLog* transformation (differentiate log of $M_1(z)$) and call the transformed series $M(z)$. In terms of $M(z)$, one can readily obtain the sequence of approximations s_n for the critical index s,

$$\mathsf{s}_n = \lim_{z \to \infty}(zPadeApproximant[M[z], n, n+1]). \qquad (7.4.20)$$

Unfortunately, in the case of (7.2.6) this approach fails. There is no discernible convergence at all within the sequence of s_n. Also, even the best result $\mathsf{s}_{12} = 0.573035$, is far off the expected 0.5. Failure of the standard approach underscores the need for a new methods.

4.1. Additive ansatz. Critical index with *DLog* corrections

Let us look for a possibility of improving the estimate for the index along the same lines as were already employed in the case of a square lattice of inclusions [13], by starting with finding a suitable starting approximation for the conductivity and critical index. Mind that one can derive the expressions for conductivity from "left-to-right", i.e., extending the series from small f to large f. Alternatively, one can proceed from "right-to-left", i.e., extending the series from the large f (close to f_c) to small f [13, 16, 31].

Let us start with defining reasonable "right-to-left" zero-approximation, which extends the form used in [13, 22] for the square arrays. The simplest way to proceed is to look for the solution in the whole region $[0, f_c)$, as the formal extension of the expansion, which leads to an additive self-similar approximant

$$\sigma_3^{r-l} = \alpha_0(f_c - f)^{-\mathsf{s}} + \alpha_1 + \alpha_2(f_c - f)^{\mathsf{s}} + \alpha_3(f_c - f)^{2\mathsf{s}}. \qquad (7.4.21)$$

All parameters in (7.3.10) will be obtained by matching it asymptotically with the truncated series $\sigma_4 = 1 + 2x + 2x^2 + 2x^3 + 2x^4$, with the following result,

$$\sigma_3^{r-l}(f) = \frac{4.69346}{(0.9069 - f)^{0.520766}} - 5.86967 +$$
$$2.53246(0.9069 - f)^{0.520766} - 0.526588(0.9069 - f)^{1.04153}. \qquad (7.4.22)$$

We apply below a concrete scheme for calculating both critical index and amplitude, based on the idea of corrected approximants [17]. In this case we will attempt to correct the value of $\mathsf{s}_0 = 0.520766$ for the critical index by applying *DLog* Padé approximation to the remainder of series (7.2.6).

Let us divide the original series (7.2.6) by $\sigma_3^{r-l}(f)$ given by (7.4.22), apply to the

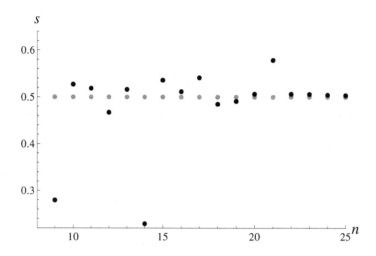

Figure 7.4 Critical index s is calculated by *DLog* corrections method, and compared with the exact value.

newly found series transformation (7.4.19), then apply *DLog* transformation and call the transformed series $L(z)$. Finally one can obtain the following sequence of the Padé approximations for the corrected critical index,

$$s_n = s_0 + \lim_{z \to \infty}(zPadeApproximant[L[z], n, n + 1]). \qquad (7.4.23)$$

The following "corrected" sequence of approximate values for the critical index can be calculated readily:

$$s_4 = 0.522573, \quad s_5 = 0.518608, \quad s_6 = 0.554342, \quad s_7 = 0.281015,$$

$$s_8 = -0.209639, \quad s_9 = 0.279669, \quad s_{10} = 0.527055, \quad s_{11} = 0.518543, \quad s_{12} = 0.488502.$$

The last two estimates frame the correct value.

Generally, one would expect that with adding more terms to the expansion (7.2.6), quality of estimates for s would improve. As was briefly discussed above, formula (7.1.2) can be expanded in arbitrary order in f, generating more terms in expansion (7.1.3). Such procedure, of course, is not a rigorous derivation of true expansion, but can be used for illustration of the convergent behaviour of s_n with larger n. Indeed, in the highest order corresponding to 50 terms in (7.1.3), we obtain $s_{25} = 0.503267$. Saturation of results is seen in Fig. 7.4 around this number of terms.

Let us put $\gamma_n(z) = PadeApproximant[L[z], n, n + 1]$ in the general formula

$$\sigma_n^*(f) = \sigma_3^{r-l}(f) \exp\left(\int_0^{\frac{f}{f_c-f}} \gamma_n(z)\, dz\right), \tag{7.4.24}$$

and compute numerically corresponding amplitude

$$A_n = \lim_{f \to f_c} (f_c - f)^{S_n} \sigma_n^*(f), \tag{7.4.25}$$

with $A_0 = 4.693$. Expressions of the type (7.4.24) have more general form than suggested before in [11, 12, 15], based on renormalization methods.

Convergence for the index above is expected to be supplemented by convergence in the sequence of approximate values for critical amplitude, but results are still a bit scattered to conclude about the amplitude value. For the last two approximations, we find

$$A_{11} = 4.80599, \quad A_{12} = 5.38288,$$

signaling possibility of a larger value than 4.82, originating from multiplication of the critical amplitude for the square lattice by $\sqrt{3}$, as suggested by O'Brien [27] (see also page 75).

To improve the estimates for amplitude A, assume that the value of critical index $s = \frac{1}{2}$ is given, and construct $\gamma_n(z)$ to satisfy the correct value at infinity. There is now a good convergence for the amplitude, i.e., in the highest orders,

$$A_{10} = 5.09584, \quad A_{11} = 5.1329, \quad A_{12} = 5.14063.$$

Corresponding expression for the approximant is given by the formula

$$\gamma_{12}(z) = \frac{b_1(z)}{b_2(z)}, \tag{7.4.26}$$

where

$$\begin{aligned}
b_1(z) = &-0.079533z^4 - 0.745717z^5 - 2.5712z^6 - 4.16091z^7 - \\
&2.88816z^8 + 0.36028z^9 + 1.74741z^{10} + 0.951728z^{11} - 0.0792987z^{12},
\end{aligned} \tag{7.4.27}$$

and

$$\begin{aligned}
b_2(z) = &1 + 14.3691z + 94.745z^2 + 380.2z^3 + 1037.51z^4 + \\
&2036.14z^5 + 2961.45z^6 + 3238.1z^7 + 2667.9z^8 + 1641.88z^9 + \\
&739.461z^{10} + 235.321z^{11} + 48.8016z^{12} + 3.81868z^{13}.
\end{aligned} \tag{7.4.28}$$

Corresponding effective conductivity can be obtained numerically,

$$\sigma_{12}^*(f) = \sigma_3^{r-l}(f) \exp\left(\int_0^{\frac{f}{f_c-f}} \gamma_{12}(z)\, dz\right). \tag{7.4.29}$$

The maximum error appears at $f = 0.905$ and equals 0.4637%. It turns out that formula (7.4.29) is rather good.

5. Critical amplitude and formula for all concentrations

For practical applications, we suggest below the particular re-summation schemes, leading to even more accurate analytical expressions for the effective conductivity.

5.1. Additive ansatz. Multiplicative correction with Padé approximants

Let us ensure the correct critical index and threshold already in the starting additive approximation for σ^{r-l}, so that all parameters in (7.5.30) are obtained by matching it asymptotically with the truncated series $\sigma_3 = 1 + 2x + 2x^2 + 2x^3$,

$$\sigma_3^{r-l}(f) = \frac{5.09924}{(0.9069 - f)^{1/2}} - 6.67022 + 3.04972(0.9069 - f)^{1/2} - 0.649078(0.9069 - f). \tag{7.5.30}$$

To extract corrections to the critical amplitude, we divide the original series (7.2.6) by (7.5.30), apply to the new series transformation (7.4.19), call the newly found series $G[z]$. Finally build a sequence of the diagonal Padé approximants, so that the amplitudes are expressed by the formula ($A_0 = 5.09924$),

$$A_n = A_0 \lim_{z \to \infty} (PadeApproximant[G[z], n, n]), \tag{7.5.31}$$

leading to a several reasonable estimates for amplitude

$$A_7 = 5.26575, \quad A_{11} = 5.23882, \quad A_{12} = 5.25781, \quad A_{13} = 5.25203.$$

Complete expression for the effective conductivity corresponding to A_{11} can be reconstructed readily,

$$\sigma_{11}^*(f) = \sigma_3^{r-l}(f) C_{11}(f), \tag{7.5.32}$$

where $C_{11}(f) = \frac{c_1(f)}{c_2(f)}$, and

$$c_1(f) = 1.15947 + 1.13125f + 1.12212f^2 + 1.1167f^3 + 3.8727f^4 + 0.824247f^5 - 2.62954f^6 + 1.19135f^7 + 1.21923f^8 + 1.42832f^9 + 1.0608f^{10} + 1.53443f^{11}; \tag{7.5.33}$$

and

$$c_2(f) = 1.15947 + 1.13125f + 1.12212f^2 + 1.1167f^3 + 3.86892f^4 + 0.849609f^5 - 2.58112f^6 + 1.11709f^7 + 1.18377f^8 + 1.36969f^9 + 1.06062f^{10} + f^{11}. \tag{7.5.34}$$

Formula (7.5.32) is practically as good as (7.4.29). Maximum error is at the point $f = 0.905$ and equals 0.563%.

5.2. Padé approximants. Standard scheme

Our second suggestion for the conductivity formula valid for all concentrations is based on the conventional considerations of the Standard approach to calculation of the amplitude as described in Chapter 5. Let us calculate the critical amplitude A. To this end, let us again apply transformation (7.4.19) to the original series (7.2.6) to obtain transformed series $M_1(z)$. Then apply to $M_1(z)$ another transformation to get yet another series, $T(z) = M_1(z)^{-1/s}$ with $s = 1/2$, in order to get rid of the square-root behaviour at infinity. In terms of $T(z)$, one can readily obtain the sequence of approximations A_n for the critical amplitude A,

$$A_n = f_c^s \lim_{z \to \infty} (zPadeApproximant[T[z], n, n + 1])^{-s}. \qquad (7.5.35)$$

There are only few reasonable estimates for the amplitude,

$$A_6 = 4.55252, \quad A_{11} = 4.49882, \quad A_{12} = 4.64665, \quad A_{13} = 4.68505.$$

The last value is the best if compared with the conjectured in [27], $A = 4.82231$. But it turns out to be way off the supposedly correct result of 5.18, underlying a deficiency of the Standard scheme. Thus, our eloquent proposition of the Corrected Approximants in Chapter 5 finds another supporting argument.

The effective conductivity can be easily reconstructed in terms of the Padé approximant (corresponding to A_{12}). The maximum error is at $f = 0.905$, and equals 5.6748%. On the other hand, if the conjectured value most accurate value of A is enforced at infinity, through the two-point Padé approximant, the results improve and the maximum error at the same concentration is equal to 3.1851%. Corresponding formula for all concentrations, which also respects 24 terms from the series $T[z]$ is given as follows,

$$\sigma_p^*(f) = \frac{1.02555}{\sqrt{0.9069 - f}} \sqrt{\frac{V_1(f)}{V_2(f)}}, \qquad (7.5.36)$$

where

$$V_1(f) = -0.927562 - 0.877939f + 0.0406992f^2 + 0.0440014f^3 + 0.0414973f^4 + 0.0436199f^5 + 0.319848f^6 + 0.0110109f^7 - 0.122646f^8 + 0.0351069f^9 + 0.0439523f^{10} + 0.0380654f^{11} + 1.01499f^{12} + f^{13}, \qquad (7.5.37)$$

and

$$V_2(f) = -1.07571 + 2.09854f - 2.17187f^2 + 2.23064f^3 - 2.3122f^4 + 2.374f^5 - 2.1397f^6 + 1.87791f^7 - 1.78516f^8 + 1.86446f^9 - 1.94838f^{10} + 2.03264f^{11} - f^{12}. \qquad (7.5.38)$$

In order to improve the quality of general formula further we will need to incorporate the second amplitude B, as will be seen shortly.

5.3. Accurate final formula

According to our calculations, based on various re-summation techniques applied to the series (7.2.6), we conclude that the critical amplitude is in the interval from 5.14 to 5.24, that is by $6 - 9\%$ higher than following naively to O'Brien's 4.82.

Below we present an exceptionally accurate and compact formula for the effective conductivity valid for all concentrations.

Let us start with modified expression (7.5.30) taking into account also the O'Brien suggestion already in the starting approximation for the amplitude in σ^{r-l}. All remaining parameters are obtained by matching it asymptotically with the truncated series $\sigma_2 = 1 + 2x + 2x^2$,

$$\sigma_3^{r-l}(f) = \frac{4.82231}{(0.9069-f)^{1/2}} - 5.79784+ \\ 2.13365(0.9069 - f)^{1/2} - 0.328432(0.9069 - f). \tag{7.5.39}$$

Repeating the procedure developed in the preceding subsection 5.1, we receive several reasonable estimates for the critical amplitude,

$$A_7 = 5.18112, \quad A_{11} = 5.15534, \quad A_{12} = 5.19509, \quad A_{13} = 5.18766.$$

Complete expression for the effective conductivity corresponding to the first estimate for the amplitude, is given as follows

$$\sigma_7^*(f) = \sigma_3^{r-l}(f)F_7(f), \tag{7.5.40}$$

and $F_7(f) = \frac{F_1(f)}{F_2(f)}$, where

$$F_1(f) = 52.0141 + 10.3198f - 38.8957f^2 + 4.70555f^3+ \\ 4.89777f^4 + 4.6887f^5 + 0.476241f^6 + 7.49464f^7, \tag{7.5.41}$$

and

$$F_2(f) = 52.0141 + 10.3198f - 38.8957f^2 + 2.17078f^3+ \\ 5.80088f^4 + 6.03946f^5 + 1.80866f^6 + f^7. \tag{7.5.42}$$

The formulae predict a sharp increase from $\sigma_7^*(0.906) = 166.708$, to $\sigma_7^*(0.9068) = 513.352$, in the immediate vicinity of the threshold, where other approaches [3, 7, 27], fail to to produce an estimate. At the largest concentration $f = 0.9068993$ mentioned in [27], the conductivity is very large, 8375.34.

The maximum error for the formula (7.5.40) is small, $< 0.1\%$. Asymptotic expression can be extracted from the approximant (7.5.40),

$$\sigma_7^* \simeq \frac{5.18112}{\sqrt{0.9069 - f}} - 6.229231. \tag{7.5.43}$$

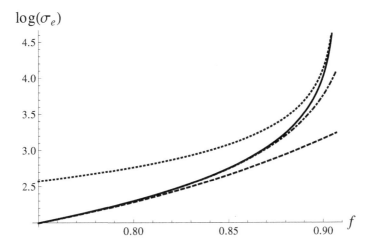

Figure 7.5 Our formula (7.5.44) (solid) is compared with the rational approximant (7.1.2) (dot-dashed) and with asymptotic formula (7.1.4) (dotted). The series (7.2.6) is shown with dashed line.

Even closer agreement with numerical results of [27] is achieved with approximant corresponding to A_{13}

$$\boxed{\sigma_{13}^*(f) = \sigma_3^{r-l}(f)\, \frac{F_1(f)}{F_3(f)}}, \qquad (7.5.44)$$

where $\sigma_3^{r-l}(f)$ is given by (7.5.39),

$$
\begin{aligned}
F_1(f) &= 1.49313 + 1.30576f + 0.383574f^2 + 0.467713f^3 + \\
&\quad 0.471121f^4 + 0.510435f^5 + 0.256682f^6 + 0.434917f^7 + \\
&\quad 0.813868f^8 + 0.961464f^9 + 0.317194f^{10} + 0.377055f^{11} - \\
&\quad 1.2022f^{12} - 0.931575f^{13}
\end{aligned}
\qquad (7.5.45)
$$

and

$$
\begin{aligned}
F_3(f) &= 1.49313 + 1.30576f + 0.383574f^2 + 0.394949f^3 + \\
&\quad 0.44785f^4 + 0.503394f^5 + 0.303285f^6 + 0.271498f^7 + 0.732764f^8 + \\
&\quad 0.827239f^9 + 0.25509f^{10} + 0.239752f^{11} - 1.26489f^{12} - f^{13}.
\end{aligned}
\qquad (7.5.46)
$$

It describes even more accurately than (7.5.40), the numerical data in the interval from $f = 0.85$ up to the critical point. The maximum error for the formula (7.5.44) is truly negligible and equals 0.042%.

Various expressions are shown in Fig. 7.5. Note, that significant deviations of the Corrected Padé formula (7.5.44) from the reference rational expression (7.1.2), start around $f = 0.85$. The formulae start to depart from the original series around $f = 0.8$.

Leading terms of the asymptotic expression can be extracted from for the approximant (7.5.44),

$$\sigma_{13}^* \simeq \frac{5.18766}{\sqrt{0.9069 - f}} - 6.2371. \tag{7.5.47}$$

6. Interpolation with high-concentration Padé approximants

When two expansions (7.2.6) and (7.5.43) are available, the problem of reconstruction greatly simplifies and can be solved upfront in terms of Padé approximants as demonstrated above for the case of square lattice.

The interpolation procedure requires as an input at least two parameters from weak and strong-coupling regimes, including the value of amplitude $A = 5.18112$ from (7.5.43).

Assume also that the next-order term, $B = -6.22923$ from (7.5.43), is known in advance. In terms of z-variable (7.4.19), the high-concentration limit can be expressed simply,

$$\sigma_e = \frac{A}{\sqrt{f_c}} \sqrt{z} + B + O(z^{-1/2}). \tag{7.6.48}$$

The Padé approximants all conditioned to give a constant value as $z \to 0$, are given below,

$$
\begin{aligned}
P_{2,1}(z) &= \frac{\beta \sqrt{z}\left(1+\beta_1 \frac{1}{\sqrt{z}}+\frac{\beta_2}{z}\right)}{1+\beta_3 \frac{1}{\sqrt{z}}}, \\
P_{3,2}(z) &= \frac{\beta \sqrt{z}\left(1+\beta_1 \frac{1}{\sqrt{z}}+\frac{\beta_2}{z}+\beta_3 z^{-3/2}\right)}{1+\beta_5 \frac{1}{\sqrt{z}}+\frac{\beta_6}{z}}, \\
P_{4,3}(z) &= \frac{\beta \sqrt{z}\left(1+\beta_1 \frac{1}{\sqrt{z}}+\frac{\beta_2}{z}+\beta_3 z^{-3/2}+\frac{\beta_4}{z^2}\right)}{1+\beta_5 \frac{1}{\sqrt{z}}+\frac{\beta_6}{z}+\beta_7 z^{-3/2}}.
\end{aligned}
\tag{7.6.49}
$$

The unknowns in (7.6.49) will be obtained by the asymptotic conditioning to (7.6.48) and (7.2.6). In all orders $\beta = \frac{A}{\sqrt{f_c}}$. Explicitly, in original variables, the following expressions transpire,

$$P_{2,1}(f) = \frac{\sqrt{\frac{f}{0.9069-f}} + \frac{4.9348}{0.9069-f} - 4.11284}{\sqrt{\frac{f}{0.9069-f}} + 1.32856},$$

$$P_{3,2}(f) = \frac{\left(0.608173\sqrt{\frac{f}{0.9069-f}} + 1.26563\right)f + 0.677749\sqrt{\frac{f}{0.9069-f}} + 1.13282}{-\left(0.747325\sqrt{\frac{f}{0.9069-f}} + 1\right)f + 0.677749\sqrt{\frac{f}{0.9069-f}} + 1.13282}, \tag{7.6.50}$$

$$P_{4,3}(f) = \frac{5.4414\sqrt{z(f)}\left(1 + 3.76414\frac{1}{\sqrt{z(f)}} + \frac{7.73681}{z(f)} + 1.97396z(f)^{-3/2} + \frac{3.76815}{z(f)^2}\right)}{1 + 4.90893\frac{1}{\sqrt{z(f)}} + \frac{10.7411}{z(f)} + 20.504z(f)^{-3/2}}.$$

These approximants respect the structure of (7.2.6), e.g., for small f,

$$P_{4,3}(f) = 1 + 2f + 2f^2 + O(f^3), \tag{7.6.51}$$

since all lower-order powers generated by square roots are suppressed by design. But in higher orders, emerging integer powers of roots should be suppressed again and again, to make sure that only integer powers of f are present. As $f \to f_c$,

$$P_{4,3}(f) = A(f_c - f)^{-1/2} + B + O((f_c - f)^{1/2}), \tag{7.6.52}$$

and only integer powers of a square-root appear in higher orders. Both $P_{3,2}(f)$ and $P_{4,3}(f)$ give good estimates for the conductivity, from below and above, respectively. Their simple arithmetic average works better than each of the approximants. The bounds hold till the very core of the high-concentration regime, till $f = 0.906$. In our opinion such approximants, their convergence, and accuracy deserve further studies in very high orders not yet reached.

6.1. Independent estimation of the amplitude B. Additive corrections

We intend to recalculate the amplitude B independent on previous estimates. Let us follow the same steps as in the analogous problem considered for the square lattice in Chapter 6. Start again with the choice of the simple high-concentration approximant as a zero approximation,

$$P_{1,0}(z) = 1 + \beta\sqrt{z}, \tag{7.6.53}$$

and equivalently

$$P_{1,0}(f) = 5.4414\sqrt{\frac{f}{0.9069 - f}} + 1. \tag{7.6.54}$$

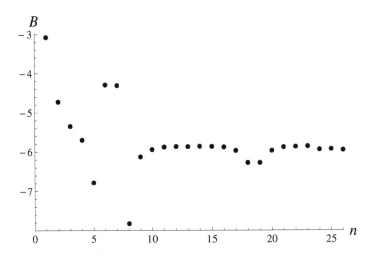

Figure 7.6 Sequence of approximations B_n calculated from (7.6.55).

The way how we often proceeded above was to look for multiplicative corrections to some plausible "zero-order" approximate solution. We can also look for an additive corrections in a similar fashion. To this end, subtract (7.6.54) from the series (7.2.6) to get some new series $g(f)$. The change of variable $f = y^2$ brings the series to a standard form. The diagonal Padé approximants to the series $g(y)$ are supposed to give a correction to the value of 1, suggested by (7.6.54). To calculate the correction one has to find the value of the corresponding approximant as $y \to \sqrt{f_c}$. The following sequence of approximations for the amplitude B can be calculated now readily,

$$B_n = 1 + PadeApproximant[g(y \to \sqrt{f_c}), n, n]. \qquad (7.6.55)$$

The sequence of approximations is shown in Fig. 7.6. There is clear saturation of the results for larger n, and $B_{26} = -5.94966$. One can reconstruct the expression for conductivity corresponding to B_{26} in additive form

$$\sigma^*_{26}(f) = P_{1,0}(f) + F_{26}(f), \qquad (7.6.56)$$

where $F_{26}(f) = \frac{F_2(f)}{F_6(f)}$,

$F_2(f) = -5.71388\sqrt{f}-$
$1.5564f - 0.358877f^{3/2} - 2.18519f^2 + 0.0918426f^{5/2} - 1.59468f^3-$
$0.149418f^{7/2} - 1.47691f^4 - 0.366848f^{9/2} - 1.49733f^5 - 0.56432f^{11/2}-$
$1.58738f^6 + 0.21344f^{13/2} - 1.31081f^7 - 0.366156f^{15/2} + 15.1037f^8-$
$15.1703f^{17/2} - 6.38227f^9 + 0.576004f^{19/2} - 2.5147f^{10} + 0.715526f^{21/2}-$
$1.53752f^{11} + 0.28655f^{23/2} - 1.19851f^{12} + 5.9511f^{25/2} + 0.558011f^{13},$

$$(7.6.57)$$

and

$F_6(f) = 1 + 0.622415\sqrt{f} - 0.27066f + 0.294568f^{3/2} - 0.00182913f^2+$
$0.0875453f^{5/2} + 0.0832152f^3 + 0.098912f^{7/2} + 0.114562f^4 + 0.116471f^{9/2}+$
$0.133685f^5 + 0.133737f^{11/2} - 0.0205003f^6 + 0.0451001f^{13/2}+$
$0.134146f^7 - 2.77482f^{15/2} + 1.6976f^8 + 3.12806f^{17/2} - 0.87267f^9+$
$0.16541f^{19/2} - 0.179645f^{10} + 0.0404152f^{21/2} + 0.000620289f^{11}+$
$0.0514796f^{23/2} - 0.986857f^{12} - 0.58415f^{25/2} + 0.388415f^{13}.$

$$(7.6.58)$$

The maximum error for the formula (7.6.56) is very small and equals 0.0824%, only slightly inferior compared with (7.5.44). The absolute value of amplitude B is firmly in the interval $(5.95, 6.22)$, according to our best two formulae.

7. Discussion of the ansatz (7.5.30)

For convenience we recapitulate the discussion from the preceding chapter on square lattice. In the case of a square lattice of inclusions, we looked for the solution in a simple form,

$$\sigma_1^{r-l} = \alpha_0(f_c - f)^{-1/2} + \alpha_1,$$

and obtained the unknowns from the two starting terms of the corresponding series. Despite its asymptotic nature, this formula works rather accurately in the whole region of concentrations and is quite stable. E.g., when subjected to a certain additive perturbation σ_{add} originating from the higher order terms, all amplitudes experience only slight change and final corrected expression $\sigma^{sq} = \sigma_1^{r-l} + \sigma_{add}$, appears to be just slightly larger than σ_1^{r-l}.

In the case of hexagonal array, such simple proposition as σ_1^{r-l} does not appear to be stable in the sense described above. We have to try lengthier expressions of the same type, such as σ_3^{r-l} (7.5.30).

If we simply expand (7.5.30), we find the fifth-order coefficient $a_5 = 1.99674$, in excellent agreement with exact value of 2. Thus, we can safely select σ_3^{r-l} as the reasonable starting approximation.

More support comes from the following argument. Let us consider (7.5.30) as an initial approximation for the critical index calculation and calculate corrections by *DLog* Padé technique. As expected, the calculated values of the corrections are small, and when all terms from the expansion are utilized is equal to just −0.0028.

Finally, just as in the case of square lattice, let us search for an additive correction to the initial approximation in the form of a root approximant

$$\sigma_{add} = b_0 z^4 \left((b_1 z + 1)^{c_1} + b_2 z^2 \right)^{c_2} \tag{7.7.59}$$

under asymptotic condition

$$\sigma_{add} \simeq p_1 \sqrt{z} + p_2, \quad \text{as } z \to \infty.$$

Elementary power-counting gives $c_1 = 3/2$, $c_2 = -7/4$. All other unknowns can now be determined uniquely in a standard fashion from the condition of asymptotic equivalence as $z \to 0$, so that

$$\sigma_{add} = b_0 z^4 \left(b_2 z^2 + (b_1 z + 1)^{3/2} \right)^{-7/4}.$$

It can can be explicitly expressed as follows,

$$\sigma_{add}(f) = \frac{0.00220821 f^4}{(0.9069 - f)^4 \left(\frac{21.8184 f^2}{(0.9069 - f)^2} + \left(\frac{3.48493 f}{0.9069 - f} + 1 \right)^{3/2} \right)^{7/4}},$$

and it leads to some very small, almost negligible asymptotic corrections to the ansatz (7.5.30). For example, the leading amplitude changes to the value of 5.09925. Such asymptotic stability of all amplitudes justifies the ansatz. Of course, it also appears to be reasonable when compared with the whole body of numerical results.

8. Square and hexagonal united

From the physical standpoint of there is no qualitative difference between the properties of hexagonal and square array of inclusions. Therefore, one might expect that a single expression exists for the effective conductivity of the two cases.

Mathematically one is confronted with the following problem: for the functions of two variables $\sigma_{sq}(f, f_c^{sq})$ and $\sigma_{hex}(f, f_c^{hex})$ to find the transformation or relation which connects the two functions. Here f_c^{hex}, f_c^{sq} stand for corresponding thresholds.

The problem is really simplified due to similarity of the leading asymptotic terms in the dilute and highly concentrated limits. On general grounds, one can expect that up to some simply behaving "correcting" function of a properly chosen non-dimensional concentration, the two functions are identical. Below we do not solve the problem from the first principles, but address it within the limits of some accurate approximate approach.

We intend to express σ_{sq} and σ_{hex} in terms of the corresponding non-dimensional variables,

$$Z_{sq} = \frac{f}{f_c^{sq} - f}, \quad Z_{hex} = \frac{f}{f_c^{hex} - f},$$

respectively. Each of the variables is in the range between 0 and ∞. Then, we formulate a modified additive ansatz which turns to be good both for square and hexagonal arrangements of disks, assuming also $Z_{sq} = Z_{nex} = Z$,

$$\sigma^g = \alpha_0 + \alpha_1 \sqrt[4]{1 + \alpha_2 Z + \alpha_3 Z^2}. \tag{7.8.60}$$

One can obtain the unknowns from the three starting terms of the corresponding series, which happen to be identical for both lattices under investigation. The form of (7.8.60) is motivated by a successful study of the square lattice in the corresponding Chapter 6. Ansatz (7.8.60) can be multiplicatively corrected through application of the Padé approximants. Emerging diagonal Padé sequences for corrected critical amplitudes are convergent for both lattices and good results are simultaneously achieved in the same order, employing 22 terms from the corresponding expansions.

We select from the emerging sequences only approximants which are also holomorphic functions. Not all approximants generated by the procedure are holomorphic and the holomorphy of diagonal Padé approximants in a given domain implies their uniform convergence inside this domain.

Corresponding corrective Padé approximants, Cor_{11}^{hex}, Cor_{11}^{sq}, are given below. For the hexagonal array,

$$\sigma_{hex}^*(Z) = \sigma^{g,hex}(Z) Cor_{11}^{hex}(Z), \tag{7.8.61}$$

and for the square lattice,

$$\sigma_{sq}^*(Z) = \sigma^{g,sq}(Z) Cor_{11}^{sq}(Z). \tag{7.8.62}$$

The initial approximation for the hexagonal array,

$$\sigma^{q,hex}(Z) = -6.44154 + 7.44154 \sqrt[4]{0.265686 Z^2 + 0.974959 Z + 1}, \tag{7.8.63}$$

and for the square lattice,

$$\sigma^{g,sq}(Z) = -1.79583 + 2.79583 \sqrt[4]{1.41167 Z^2 + 2.24734 Z + 1}. \tag{7.8.64}$$

Correction term for the hexagonal array has the following form,

$$Cor_{11}^{hex}(Z) = \frac{cor_1^{hex}(Z)}{cor_2^{hex}(Z)}, \tag{7.8.65}$$

and for the square lattice,

$$Cor_{11}^{sq}(Z) = \frac{cor_1^{sq}(Z)}{cor_2^{sq}(Z)}. \tag{7.8.66}$$

Numerators and denominators of these expressions are given by polynomials, in case of a hexagonal array,

$$
\begin{aligned}
cor_1^{hex}(Z) = {} & 1 + 11.8932Z + 64.7366Z^2 + 213.169Z^3 \\
& + 474.557Z^4 + 755.98Z^5 + 884.496Z^6 + 760.227Z^7 \\
& + 468.277Z^8 + 196.502Z^9 + 51.5454Z^{10} + 7.29645Z^{11},
\end{aligned} \tag{7.8.67}
$$

$$
\begin{aligned}
cor_2^{hex}(Z) = {} & 1 + 11.8932Z + 64.7366Z^2 + 213.169Z^3 \\
& + 474.565Z^4 + 756.051Z^5 + 884.793Z^6 + 760.882Z^7 \\
& + 469.097Z^8 + 197.06Z^9 + 51.7143Z^{10} + 7.12936Z^{11}.
\end{aligned} \tag{7.8.68}
$$

While for the square lattice the following expressions hold,

$$
\begin{aligned}
cor_1^{sq}(Z) = {} & 1 + 12.211Z + 66.2975Z^2 + 212.904Z^3 + 451.409Z^4 \\
& + 664.782Z^5 + 693.726Z^6 + 511.717Z^7 + 259.861Z^8 \\
& + 84.9746Z^9 + 15.2213Z^{10} + 0.73003Z^{11},
\end{aligned} \tag{7.8.69}
$$

$$
\begin{aligned}
cor_2^{sq}(Z) = {} & 1 + 12.211Z + 66.2975Z^2 + 212.904Z^3 + 451.492Z^4 \\
& + 665.308Z^5 + 694.974Z^6 + 513.153Z^7 + 260.53Z^8 \\
& + 84.8414Z^9 + 14.9845Z^{10} + 0.706023Z^{11}.
\end{aligned} \tag{7.8.70}
$$

The ratio of final expressions for the conductivity of corresponding lattices,

$$\frac{\sigma_{hex}^*(Z_{hex})}{\sigma_{sq}^*(Z_{sq})},$$

can be plotted (as $Z_{hex} = Z_{sq} = Z$), as shown in Fig. 7.7. It turns out that the ratio is bounded function of Z, and changes monotonously from 1 ($Z = 1$) to 1.7352 ($Z \to \infty$). The last number is very close to $\sqrt{3} \approx 1.73205$ [27]. This ratio can be simply expressed as

$$\frac{A^{hex}}{A^{sq}} \sqrt{\frac{f_c^{sq}}{f_c^{hex}}}.$$

Here $A^{hex} = 5.20709$ and $A^{sq} = 2.79261$ stand for critical amplitudes of their corresponding lattices.

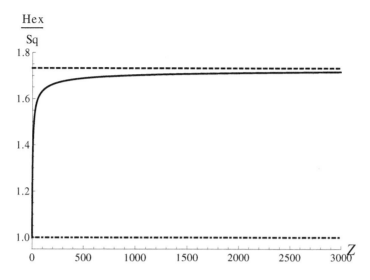

Figure 7.7 The ratio of final expressions for the conductivity of corresponding lattices, $\frac{\sigma_{hex}^*(Z_{hex})}{\sigma_{sq}^*(Z_{sq})}$, can be plotted with $Z_{hex} = Z_{sq} = Z$.

9. Dependence on contrast parameter

Consider dependence of the effective conductivity on the contrast parameter by using the exact formulae (4.2.28). There are two different limit cases for the effective conductivity. In the limiting case of a perfectly conducting inclusions $\rho \to 1$, the effective conductivity is expected to tend to infinity as a power law, as the concentration f tends to the maximal value $f_c = \frac{\pi}{\sqrt{12}}$ for the hexagonal array,

$$\sigma_e(f) \sim (f_c - f)^{-1/2}. \tag{7.9.71}$$

From the phase interchange theorem it follows that in 2D, the superconductivity index s is equal to the conductivity index t and as $\rho \to -1$

$$\sigma_e(f) \sim (f_c - f)^{1/2}. \tag{7.9.72}$$

The following expansion in concentration f of the inclusions also including contrast parameter can be deduced from (4.2.29) on page 87

$$\sigma_e(f, \rho) = \frac{1 + \rho f}{1 - \rho f} + 0.150844\rho^3 f^7 + 0.301688\rho^4 f^8 + 0.452532\rho^5 f^9 + \tag{7.9.73}$$

$$0.603376\rho^6 f^{10} + 0.75422\rho^7 f^{11} + \cdots .$$

Since there are two parameters f and ρ, we derive the final formula in two steps. Mind that in 2D one has to respect celebrated Keller's phase-interchange relation. Since all dependence on the conductivity of inclusions and matrix is hidden within the contrast parameter, the phase interchange can be expressed as $\frac{1}{\sigma_e(f,\rho)} = \sigma_e(f, -\rho)$. We will try to develop an expression which satisfy this relation by design. It is necessary to respect corresponding critical properties and then simply modify parameters of the solution to move away to a non-critical situations. The simplest solution along these lines is given by the factor approximant which includes both types of critical behaviour, as $\rho \to -1$, and $\rho \to 1$, respectively,

$$\sigma^*(f,\rho) = \left(1 + \frac{\rho f}{f_c}\right)^{\frac{1}{2}} \left(1 - \frac{\rho f}{f_c}\right)^{-\frac{1}{2}} \tag{7.9.74}$$

with the parameters defined from the asymptotic equivalence to the correct values of thresholds as $\rho \to -1$, and $\rho \to 1$. Such form is motivated by the celebrated Clausius–Mossotti (CM) formula valid for small concentrations. Moving singularity to the non-physical values of f allows to preserve the form typical for critical regime for all values of ρ but suppress the singularity except in the situations of extremely high contrast.

Let us further correct (7.9.74) with the Padé approximants by developing the following corrected factor approximant,

$$\sigma^{**}_{1c}(f,\rho) = \sigma^*(f,\rho)\frac{1+wf}{1-wf} = \frac{(0.448671\rho f+1)\sqrt{1.10266\rho f+1}}{\sqrt{1-1.10266\rho f}(1-0.448671\rho f)}, \tag{7.9.75}$$

obtained from asymptotic equivalence with series (7.9.73). The form of correction is motivated once again by CM formula which respects the phase interchange symmetry, formula (7.9.75) respects this symmetry as well. Assuming even more general form for the correcting Padé approximant, we obtain another more accurate formula.

$$\sigma^{**}_{2c}(f,\rho) = \sigma^*(f,\rho)\frac{w_1(\rho)f+w_2(\rho)f^3+1}{-w_1(\rho)f-w_2(\rho)f^3+1} = \frac{\sqrt{1.10266\rho f+1}(0.0797815\rho^3 f^3+0.448671\rho f+1)}{(-0.0797815\rho^3 f^3-0.448671\rho f+1)\sqrt{1-1.10266\rho f}}. \tag{7.9.76}$$

The phase interchange symmetry is again preserved. Finally in the highest-order compatible with (7.1.1), we derive the following corrected approximant

$$\boxed{\sigma^{**}_{11c}(f,\rho) = \sigma^*(f,\rho)\frac{U(f,\rho)}{W(f,\rho)},} \tag{7.9.77}$$

where

$$U(f,\rho) = -\rho^{11}f^{11} - 0.660339\rho^9 f^9 + \rho^7(1.44959f^4 - 0.232667)f^7 +$$
$$\rho^5(1.59231f^9 + 0.457218f^5) + \rho^3(1.99365f^7 + 2.10888f^3) + \tag{7.9.78}$$
$$11.8598\rho f + 26.4332,$$

and

$$W(f,\rho) = \rho^7 f^7 (0.232667 - 1.44959 f^4) + \rho^{11} f^{11} + 0.660339 \rho^9 f^9 +$$
$$\rho^5 (-1.59231 f^9 - 0.457218 f^5) + \rho^3 (-1.99365 f^7 - 2.10888 f^3) - \qquad (7.9.79)$$
$$11.8598 \rho f + 26.4332.$$

Our formula $\sigma_{11c}^{**}(f,\rho)$ is still of a semi-quantitative nature, awaiting to be further corrected with higher order terms to be obtained. The same approach is applicable to square and random arrays.

In summary, we derived an accurate and relatively compact formula for the effective conductivity in the high-contrast case (7.5.44) valid for all concentrations, including the most interesting regime of very high concentration. For the high-concentration limit, in addition to the amplitude value of 5.18112, we deduce also that the next-order (constant) term B, equals -6.22923. It is possible to extract more coefficients in the high-concentration expansion based on the exact formula (4.2.28).

Such properties as the superconductivity critical index and threshold for conductivity, can be calculated from the series (7.2.6). In the case of truncated series, the standard Padé approximants are not able to describe the correct asymptotic behaviour in the high-concentration limit, where in addition to the leading critical exponent also a non-trivial sub-leading exponent(s) plays the role [16, 31]. On the other hand, when such a non-trivial asymptotic behaviour is treated separately with different type of approximants, the Padé approximants are able to account for the correction. Such patchwork approximations appear to be more accurate and powerful than approximating conventionally with a single type of approximants.

Simple functional relation between the effective conductivity of the hexagonal and square arrays is suggested, expressed in terms of some bounded monotonous function of a non-dimensional concentration of inclusions. Getting an accurate formula in this case, means that correct asymptotic behaviour (7.5.43) is indeed can be extracted from the series (7.2.6), and together they determine the behaviour in the whole interval with good accuracy. Neglecting the high-concentration regime dominated by necks, is not admissible.

Here, the following remark concerning the paper [26] and others seems appropriate. As it is noted in item 1 on page 42, solution to a problem for a finite cluster on the plane can give only a formula for dilute clusters, i.e., the first-order approximation in f where f is the concentration of clusters on the plane. Consider a finite hexagonal type cluster as in [26]. The authors of [26] suppose that increasing the number of inclusions in the cluster allows to reach the exact formula (4.2.29) on page 87. However, as it is first demonstrated in [25] and in our book that this methodology cannot be applied to approximations exceeding $O(f^2)$. More less similar some numerical data for finite hexagonal clusters and for the hexagonal array can be explained simply observing the approximate formula (7.9.73). Consider the case $f = 0.85$ and $\rho = 0.999$ discussed in [26]. Then, the "dilute part" of (7.9.73) $\sigma_{dilute} = \frac{1+\rho f}{1-\rho f}$ gives the major contribution which is equal to 10.8726, into conductivity, while the rest contributes only 0.47779. The fact that dilute formulae work for sufficiently high f and ρ is the exceptional property of the hexagonal array. Besides (7.9.73) this property can be explained by the following reasons. Wall [30] observed that the numerical results are close to the lower bounds and noted that the hexagonal array can be approximated by a coated structure similar to lubrication approximation [4, 10]. Moreover, the coated structure is nothing else but the famous coated disks assemblage of Hashin [19], which has the effective conductivity coinciding with the lower bound calculated by the Clausius–Mossotti formula. Of course, if we consider the percolation regime, and not the special data of [26], the rest of the formula (7.9.73) dominates. In this case, one should take into account the singularity explicitly. As soon as the contrast parameter decreases the singularity starts moving to the positive non-physical region. For small-contrast composites its influence on the effective properties decreases. One should be concerned only with faithful description of the physical region which is now very weakly affected by the singularity. Formula (3.5.141) on page 75 qualitatively demonstrates this. Correspondingly, Padé approximations become more accurate, till to the point of a complete neglect of the singularity.

This remark stresses one of the principal lines of our book that analytical and asymptotic formulae show more details and trends than numerical results computed for special cases. For instance, in the case of hexagonal array, the exact analytical formula (4.2.29) on page 87 should be considered as the benchmark for further numerical and asymptotic study as it is done in the present section.

Computational approach and results of the present chapter are applicable everywhere. The results are weakly dependent on the initial approximations suggesting that the number of coefficients in the series is sufficient. Application of the various "corrected" Padé schemes allows to improve quality of approximations significantly.

REFERENCE

1. Andrianov IV, Awrejcewicz J, Starushenko GA, Application of an improved three-phase model to calculate effective characteristics for a composite with cylindrical inclusions, Latin American Journal of Solids and Structures 2013; 10: 197–222.
2. Andrianov IV, Danishevskyy VV, Kalamkarov AL, Analysis of the effective conductivity of composite materials in the entire range of volume fractions of inclusions up to the percolation threshold, Composites: Part B 2010; 41: 503–507.
3. Andrianov IV, Danishevskyy VV, Guillet A, Pareige P, Effective properties and micro-mechanical response of filamentary composite wires under longitudinal shear, Eur.J.Mech. A/Solids; 2005; 24: 195–206.
4. Andrianov IV, Starushenko GA, Danishevskyy VV, Tokarzewski S, Homogenization procedure and Padé approximants for effective heat conductivity of composite materials with cylindrical inclusions having square cross-section, Proc. R. Soc. Lond. A; 1999; 455: 3401–3413.
5. Baker GA, Graves-Moris P, Padé Approximants. Cambridge, UK: Cambridge University; 1996.
6. Bergman DJ, Dielectric-constant of a composite-material-problem in classical physics, Phys. Rep.

1978; 43: 378–407.

7. Berlyand L, Kolpakov A, Network approximation in the limit of small interparticle distance of the effective properties of a high contrast random dispersed composite, Arch. Ration. Mech. Anal. 2001; 159: 179–227.

8. Berlyand L, Novikov A, Error of the network approximation for densly packed composites with irregular geometry, SIAM J. Math. Anal. 2002; 34: 385–408.

9. Czapla R, Nawalaniec W, Mityushev V, Effective conductivity of random two-dimensional composites with circular non-overlapping inclusions, Comput. Mat. Sci. 2012; 63: 118–126.

10. Christensen RM, Mechanics of composite materials. Dover Publications. 1-384. 2005.

11. Dunjko V, Olshanii M, A Hermite-Padé perspective on the renormalization group, with an application to the correlation function of Lieb-Liniger gas, J. Phys. A: Math. Theor. 2011; 44: 055206.

12. Gluzman S, Karpeev DA, Berlyand LV, Effective viscosity of puller-like microswimmers: a renormalization approach, J. R. Soc. Interface 2013; 10: 20130720.

13. Gluzman S, Mityushev V, Nawalaniec W, Cross-properties of the effective conductivity of the regular array of ideal conductors, Arch. Mech. 2014; 66: 287–301.

14. Gluzman S, Mityushev V, Nawalaniec W, Starushenko G, Effective Conductivity and Critical Properties of a Hexagonal Array of Superconducting Cylinders, ed. P.M. Pardalos (USA) and T.M. Rassias (Greece), Contributions in Mathematics and Engineering. In Honor of Constantin Caratheodory. Springer: 255–297, 2016.

15. Gluzman S, Yukalov VI, Extrapolation of perturbation theory expansions by self-similar approximants, Eur. J. Appl. Math. 2014; 25: 595–628.

16. Gluzman S, Yukalov VI, Unified approach to crossover phenomena, Phys. Rev. E 1998; 58: 4197–4209.

17. Gluzman S, Yukalov VI, Self-similar extrapolation from weak to strong coupling, J.Math.Chem. 2010; 48: 883–913.

18. Golden K, Papanicolaou G, Bounds for effective parameters of heterogeneous media by analytic continuation, Commun. Math. Phys. 1983; 90: 473–491.

19. Hashin Z, The elastic moduli of heterogeneous materials, J. Appl. Mech. 1962; 29: 143–150.

20. Keller JB, Conductivity of a Medium Containing a Dense Array of Perfectly Conducting Spheres or Cylinders or Nonconducting Cylinders, Journal of Applied Physics 1963; 34: 991–993.

21. Keller JB, A Theorem on the Conductivity of a Composite Medium, J. Math. Phys. 1964; 5: 548–549.

22. McPhedran RC, Transport Properties of Cylinder Pairs and of the Square Array of Cylinders, Proc.R. Soc. Lond. 1986; A 408: 31–43.

23. Mityushev V, Steady heat conduction of a material with an array of cylindrical holes in the nonlinear case, IMA Journal of Applied Mathematics 1998; 61: 91–102.

24. Mityushev V, Exact solution of the $\mathbb{R}$-linear problem for a disk in a class of doubly periodic functions, J. Appl.Functional Analysis 2007; 2: 115–127.

25. Mityushev V, Rylko N, Maxwell's approach to effective conductivity and its limitations, The Quarterly Journal of Mechanics and Applied Mathematics 2013.

26. Mogilevskaya S, Nikolskiy D, Maxwell's equivalent inhomogeneity and remarkable properties of harmonic problems involving symmetric domains, Journal of Mechanics of Materials and Structures 2017; 12: 179–191.

27. Perrins WT, McKenzie DR, McPhedran RC, Transport properties of regular array of cylinders, Proc. R. Soc.A 1979; 369, 207–225.

28. Tokarzewski S, Andrianov I, Danishevsky V, Starushenko G, Analytical continuation of asymptotic expansion of effective transport coefficients by Padé approximants, Nonlinear Analysis 2001; 47: 2283–2292.

29. Torquato S, Stillinger FH, Jammed hard-particle packings: From Kepler to Bernal and beyond, Reviews of Modern Physics 2010; 82: 2634–2672.

30. Wall P, A Comparison of homogenization, Hashin-Shtrikman bounds and the Halpin-Tsai Equations, Applications of Mathematics 1997; 42: 245–257.

31. Yukalov VI, Gluzman S, Self-similar crossover in statistical physics, Physica A 1999; 273: 401–415.

32. Yukalov VI, Gluzman S, Self-similar exponential approximants, Phys. Rev. E 1998; 58: 1359–1382.

CHAPTER 8

Effective Conductivity of 3D regular composites

If the time comes when you are able to set up some device that may act to prevent the worst from happening, see if you can think of two devices, so that if one fails, the other will carry on.

— Isaac Asimov, Prelude to Foundation

In the present chapter, we discuss conductivity of 3D composites with spherical inclusions. First, the modified Dirichlet problem for a finite number of inclusions n in $\mathbb{R}^3$ is explicitly solved in terms of the 3D Poincaré series by the method of functional equations. In the next sections, we investigate the limiting form of functional equations as $n \to \infty$. At the end of the chapter, the known analytical formulae for periodic composites are outlined with the corresponding asymptotic treatment.

1. Modified Dirichlet problem. Finite number of balls

Consider mutually disjoint balls $\mathbb{D}_k = \{\mathbf{x} \in \mathbb{R}^3 : |\mathbf{x} - \mathbf{a}_k| < r_k\}$ $(k = 1, 2, \ldots, n)$ and the domain $\dot{\mathbb{D}} = \mathbb{R}^3 \setminus \cup_{k=1}^{n} (\mathbb{D}_k \cup \partial \mathbb{D}_k)$. It is convenient to add the infinite point and introduce the domain $\mathbb{D} = \dot{\mathbb{D}} \cup \{\infty\}$ lying in the one-point compactification of $\mathbb{R}^3$. We would like to find a $u_k(\mathbf{x})$ harmonic in $\dot{\mathbb{D}}$ and continuously differentiable in $\dot{\mathbb{D}} \cup_{k=1}^{n} \partial \mathbb{D}_k$ with the boundary conditions

$$u = c_k, \quad \mathbf{x} \in \partial \mathbb{D}_k \quad (k = 1, 2, \ldots, n). \tag{8.1.1}$$

Here, c_k are undetermined constants which should be found during solution to the boundary value problem. It is assumed that

$$u(\mathbf{x}) - x_1 \text{ tends to } 0, \text{ as } |\mathbf{x}| \to \infty \tag{8.1.2}$$

and

$$\int_{\partial \mathbb{D}_k} \frac{\partial u}{\partial \mathbf{n}} ds = 0 \iff \int_{\partial \mathbb{D}_k} \frac{\partial u}{\partial r} ds = 0 \quad (k = 1, 2, \ldots, n). \tag{8.1.3}$$

Computational Analysis of Structured Media
http://dx.doi.org/10.1016/B978-0-12-811046-1.50008-3

The inversion with respect to a sphere $\partial \mathbb{D}_k$ is introduced by formula

$$\mathbf{x}^*_{(k)} = \frac{r_k^2}{r^2}(\mathbf{x} - \mathbf{a}_k) + \mathbf{a}_k, \qquad (8.1.4)$$

where $r = |\mathbf{x} - \mathbf{a}_k|$ denotes the local spherical coordinate. The Kelvin transform with respect to $\partial \mathbb{D}_k$ has the form (see [2, p. 59]),

$$\mathcal{K}_k w(\mathbf{x}) = \frac{r_k}{r} w(\mathbf{x}^*_{(k)}). \qquad (8.1.5)$$

If a function $w(\mathbf{x})$ is harmonic in $|\mathbf{x} - \mathbf{a}_k| < r_k$, the function $\mathcal{K}_k w(\mathbf{x})$ is harmonic in $|\mathbf{x} - \mathbf{a}_k| > r_k$ and vanishes at infinity.

Let $v(\mathbf{x})$ be a given Hölder continuous function on a fixed sphere $|\mathbf{x} - \mathbf{a}_k| = r_k$. Consider an auxiliary Robin problem

$$\frac{1}{r_k} u_k + 2 \frac{\partial u_k}{\partial r} = v, \qquad r = r_k, \qquad (8.1.6)$$

for the function $u_k(\mathbf{x})$ harmonic in the ball $|\mathbf{x} - \mathbf{a}_k| < r_k$. This problem has a unique solution [12].

BOX A.8.8 Harmonic Continuation

Theorem 15 ([24, p. 200]). *Introduce a domain Ω divided by a simple Lyapunov curve L onto two domains Ω^+ and Ω^-. Let functions $V^+(\mathbf{x})$, $V^-(\mathbf{x})$ be harmonic in Ω^+, Ω^- and continuously differentiable in $\Omega^+ \cup L$, $\Omega^- \cup L$. Let*

$$V^+ = V^-, \qquad \frac{\partial V^+}{\partial \mathbf{n}} = \frac{\partial V^+}{\partial \mathbf{n}}, \qquad \mathbf{x} \in L. \qquad (8.1.7)$$

Then the function

$$V(\mathbf{x}) = V^+(\mathbf{x}), \ \mathbf{x} \in \Omega^+ \cup L, \qquad V^-(\mathbf{x}), \ \mathbf{x} \in \Omega^- \qquad (8.1.8)$$

is harmonic in the whole domain Ω.

The physical proof of the above theorem is evident. According to (8.1.7) the contact on the surface L between identical materials is perfect. Therefore, the material occupying the domain $\Omega^+ \cup L \cup \Omega^-$ conducts the heat flux as a monolith without the defect surface L. It is characterized then by $V(\mathbf{x})$.

Let $u_k(\mathbf{x})$ be a solution of the problem (8.1.6) with $v = \frac{\partial u}{\partial r}$, i.e.,

$$\frac{1}{r_k} u_k + 2 \frac{\partial u_k}{\partial r} = \frac{\partial u}{\partial r}, \qquad r = r_k. \qquad (8.1.9)$$

Introduce the piece-wise harmonic function in $\cup_{k=1}^{n} \mathbb{D}_k \cup \dot{\mathbb{D}}$

$$U(\mathbf{x}) = \begin{cases} u_k(\mathbf{x}) + \sum_{m \neq k} \frac{r_m}{|\mathbf{x} - \mathbf{a}_m|} u_m(\mathbf{x}^*_{(m)}) + c_k, & |\mathbf{x} - \mathbf{a}_m| \leq r_k, \; k = 1, 2, \ldots, n, \\ \\ u(\mathbf{x}) + \sum_{m=1}^{n} \frac{r_m}{|\mathbf{x} - \mathbf{a}_m|} u_m(\mathbf{x}^*_{(m)}), & \mathbf{x} \in \dot{\mathbb{D}}. \end{cases}$$

(8.1.10)

Let $U^+(\mathbf{x}) = \lim_{\cup_{k=1}^{n} \mathbb{D}_k \ni \mathbf{y} \to \mathbf{x}} U(\mathbf{y})$ and $U^-(\mathbf{x}) = \lim_{\mathbb{D} \ni \mathbf{y} \to \mathbf{x}} U(\mathbf{y})$. Calculate the jump $\Delta_k(\mathbf{x}) = U^+(\mathbf{x}) - U^-(\mathbf{x})$ across $\partial \mathbb{D}_k$ ($k = 1, 2, \ldots, n$). Using the definition of $U(\mathbf{x})$ and the boundary condition (8.1.1) we get

$$\Delta_k = u_k + c_k - u - \frac{r_k}{|\mathbf{x} - \mathbf{a}_k|} u_k = 0.$$

(8.1.11)

Now, we calculate $\Delta'_k = \frac{\partial U^+}{\partial r} - \frac{\partial U^-}{\partial r}$ across $\partial \mathbb{D}_k$. Using the definition of $U(\mathbf{x})$ we obtain for a fixed k

$$\Delta'_k(\mathbf{x}) = \frac{\partial u_k}{\partial r}(\mathbf{x}) - \frac{\partial u}{\partial r}(\mathbf{x}) - \frac{\partial}{\partial r}\left[\frac{r_k}{|\mathbf{x} - \mathbf{a}_k|} u_k(\mathbf{x}^*_{(k)})\right], \quad \mathbf{x} \in \partial \mathbb{D}_k.$$

(8.1.12)

In order to properly calculate the radial derivative in (8.1.12), we introduce the local spherical coordinates (r, θ, φ) near a fixed sphere $\partial \mathbb{D}_k$ and the functions

$$v(r) = u_k(\mathbf{x}) \equiv u_k(r, \theta, \varphi), \quad w(r) = u_k(\mathbf{x}^*_{(k)}) \equiv u_k\left(\frac{r_k^2}{r}, \theta, \varphi\right).$$

(8.1.13)

One can see that

$$\frac{\partial w}{\partial r} = -\frac{\partial v}{\partial r}, \quad r = r_k.$$

(8.1.14)

Therefore, (8.1.12) implies that

$$\Delta'_k = \left[\frac{\partial u_k}{\partial r} - \frac{\partial u}{\partial r} + \frac{r_k}{r}\frac{\partial u_k}{\partial r} - u_k \frac{\partial}{\partial r}\left(\frac{r_k}{r}\right)\right]\bigg|_{r=r_k} = 2\frac{\partial u_k}{\partial r} + \frac{1}{r_k} u_k - \frac{\partial u}{\partial r}.$$

(8.1.15)

Application of (8.1.9) yields $\Delta'_k = 0$. Then, Theorem 15 implies that the function $U(\mathbf{x})$ is harmonic in $\dot{\mathbb{D}}$. It follows from the definition of $U(\mathbf{x})$ near infinity and (8.1.2) that $U(\mathbf{x}) - x_1$ is harmonic in $\mathbb{R}^3 \cup \{\infty\}$. Therefore, $U(\mathbf{x}) - x_1$ is a constant by Liouville's theorem. Moreover, $U(\mathbf{x}) - x_1$ vanishes at infinity, hence, this constant is equal to zero and $U(\mathbf{x}) = x_1$ for all $\mathbf{x} \in \mathbb{R}^3 \cup \{\infty\}$. Equation (8.1.10) written in the considered domains gives the system of functional equations

$$u_k(\mathbf{x}) = -\sum_{m \neq k} \frac{r_m}{|\mathbf{x} - \mathbf{a}_m|} u_m(\mathbf{x}^*_{(m)}) + x_1 - c_k, \; |\mathbf{x} - \mathbf{a}_m| \leq r_k \; (k = 1, 2, \ldots, n) \quad (8.1.16)$$

and

$$u(\mathbf{x}) = -\sum_{m=1}^{n} \frac{r_m}{|\mathbf{x} - \mathbf{a}_m|} u_m(\mathbf{x}_{(m)}^*) + x_1, \quad \mathbf{x} \in \dot{\mathbb{D}}. \tag{8.1.17}$$

After solution to the system (8.1.16) the auxiliary functions $u_k(\mathbf{x})$ are substituted into (8.1.17). This gives the solution of the problem (8.1.2), the function $u(\mathbf{x})$.

Theorem 16 ([16, 17]). *Let $C = C\left(\cup_{k=1}^{n} \mathbb{D}_k\right)$ denote the space of functions harmonic in all $\mathbb{D}_k$ and continuous in their closures endowed with the norm $\|h\| = \max_k \max_{\mathbf{x} \in \partial \mathbb{D}_k} |h(\mathbf{x})|$. Let a given function $h(\mathbf{x})$ belong to C. The system of functional equations*

$$u_k(\mathbf{x}) = -\sum_{m \neq k} \frac{r_m}{|\mathbf{x} - \mathbf{a}_m|} u_m(\mathbf{x}_{(m)}^*) + h(\mathbf{x}), \ |\mathbf{x} - \mathbf{a}_m| \leq r_k \ (k = 1, 2, \ldots, n) \tag{8.1.18}$$

has a unique solution in C. This solution can be found by successive approximations converging in C, i.e., uniformly in $\cup_{k=1}^{n}(\mathbb{D}_k \cup \partial \mathbb{D}_k)$.

Let k_j run over $k = 1, 2, \ldots, n$. Consider the sequence of inversions with respect to the spheres $\partial \mathbb{D}_{k_1}, \partial \mathbb{D}_{k_2}, \ldots, \partial \mathbb{D}_{k_m}$ determined by the recurrence formula

$$x_{(k_m k_{m-1} \ldots k_1)}^* := \left(x_{(k_{m-1} k_{m-2} \ldots k_1)}^*\right)_{k_m}^*. \tag{8.1.19}$$

It is supposed there are no equal neighbor numbers in the sequence $k_1, k_2, \ldots, k_m$. The transformations (8.1.19) for $m = 1, 2, \ldots$ with the identity map form the so-called Schottky group S of maps acting in $\mathbb{R}^3$.

Straightforward application of the successive approximations described in Theorem 16 gives the exact formula

$$u_k(\mathbf{x}) = (P_k h)(\mathbf{x}) := (Ph)(\mathbf{x}) - \frac{r_k}{|\mathbf{x} - \mathbf{a}_k|}(Ph)(\mathbf{x}_{(k)}^*), \quad \mathbf{x} \in \mathbb{D}_k \cup \partial \mathbb{D}_k, \tag{8.1.20}$$

where the operator P acts in the space C and has the form

$$(Ph)(\mathbf{x}) = h(\mathbf{x}) - \tag{8.1.21}$$

$$\sum_{k=1}^{n} \frac{r_k}{|\mathbf{x} - \mathbf{a}_k|} h(\mathbf{x}_{(k)}^*) + \sum_{k=1}^{n} \sum_{k_1 \neq k} \frac{r_k}{|\mathbf{x} - \mathbf{a}_k|} \frac{r_{k_1}}{|\mathbf{x}_{(k)}^* - \mathbf{a}_{k_1}|} h(\mathbf{x}_{(k_1 k)}^*) -$$

$$\sum_{k=1}^{n} \sum_{\substack{k_1 \neq k \\ k_2 \neq k_1}} \frac{r_k}{|\mathbf{x} - \mathbf{a}_k|} \frac{r_{k_1}}{|\mathbf{x}_{(k)}^* - \mathbf{a}_{k_1}|} \frac{r_{k_2}}{|\mathbf{x}_{(k_1 k)}^* - \mathbf{a}_{k_2}|} h(\mathbf{x}_{(k_2 k_1 k)}^*) + \cdots,$$

where $\sum_{\substack{k_1 \neq k \\ k_2 \neq k_1}} := \sum_{k_1 \neq k} \sum_{k_2 \neq k_1}$. Every sum $\sum_{k_j \neq k_{j-1}}$ contains terms with $k_j = 1, 2, \ldots, n$ except $k_j = k_{j-1}$. Following [17, Chapter 4] one can prove compactness of P in C. The uniqueness of solution established in Theorem 16 for the linear system (8.1.18) implies that the successive approximations can be applied separately to $h_1(\mathbf{x})$ and to $h_2(\mathbf{x})$ when $h(\mathbf{x}) = h_1(\mathbf{x}) + h_2(\mathbf{x})$. The unique result will be the same. For instance, P can be applied to the piece-wise constant function $c(\mathbf{x}) = c_k$ where $x \in \mathbb{D}_k$ $(k = 1, 2, \ldots, n)$. Then, the solution of the system (8.1.16) can be written in the form

$$u_k(\mathbf{x}) = (P_k x_1)(\mathbf{x}) - (P_k c)(\mathbf{x}), \quad \mathbf{x} \in \mathbb{D}_k \cup \partial \mathbb{D}_k \ (k = 1, 2, \ldots, n). \tag{8.1.22}$$

Substitution of (8.1.22) into (8.1.17) yields

$$u(\mathbf{x}) = (P x_1)(\mathbf{x}) - (P_0 c)(\mathbf{x}), \tag{8.1.23}$$

where $(P_0 h)(\mathbf{x}) := (Ph)(\mathbf{x}) - h(\mathbf{x})$. Consider a class of functions $\mathcal{R}$ harmonic in $\mathbb{R}^3$ except a finite set of isolated points located in $\mathbb{D}$ where at most polynomial growth can take place. It follows from (8.1.21) that the operators $P_0 : C \to \mathcal{R}$ and $P : \mathcal{R} \to \mathcal{R}$ are properly defined.

The series $(Ph)(\mathbf{x})$ for $h \in \mathcal{R}$ including (8.1.23) converges uniformly in $\mathbf{x} \in \dot{\mathbb{D}} \cup \partial \mathbb{D}$. We call it the 3D Poincaré θ-series associated to the 3D Schottky group $\mathcal{S}$ for the function $h(\mathbf{x})$. The corresponding 2D theory of the Poincaré series is described in Chapter 2.

One can see from (8.1.23) that $u(\mathbf{x})$ contains the undetermined constants c_k only in the part

$$(P_0 c)(\mathbf{x}) = - \sum_{k=1}^{n} \frac{r_k}{|\mathbf{x} - \mathbf{a}_k|} c_k + \sum_{k=1}^{n} \sum_{k_1 \neq k} \frac{r_k}{|\mathbf{x} - \mathbf{a}_k|} \frac{r_{k_1}}{|\mathbf{x}_{(k)}^* - \mathbf{a}_{k_1}|} c_{k_1} - \tag{8.1.24}$$

$$\sum_{k=1}^{n} \sum_{\substack{k_1 \neq k \\ k_2 \neq k_1}} \frac{r_k}{|\mathbf{x} - \mathbf{a}_k|} \frac{r_{k_1}}{|\mathbf{x}_{(k)}^* - \mathbf{a}_{k_1}|} \frac{r_{k_2}}{|\mathbf{x}_{(k_1 k)}^* - \mathbf{a}_{k_2}|} c_{k_2} +$$

$$\sum_{k=1}^{n} \sum_{\substack{k_1 \neq k \\ k_2 \neq k_1 \\ k_3 \neq k_2}} \frac{r_k}{|\mathbf{x} - \mathbf{a}_k|} \frac{r_{k_1}}{|\mathbf{x}_{(k)}^* - \mathbf{a}_{k_1}|} \frac{r_{k_2}}{|\mathbf{x}_{(k_1 k)}^* - \mathbf{a}_{k_2}|} \frac{r_{k_3}}{|\mathbf{x}_{(k_2 k_1 k)}^* - \mathbf{a}_{k_3}|} c_{k_3} + \cdots .$$

$(P_0 c)(\mathbf{x})$ can be represented in the form $(P_0 c)(\mathbf{x}) = \sum_{m=1}^{n} c_m p_m(\mathbf{x})$. Let δ_{km} denote the Kronecker symbol. The functions $p_m(\mathbf{x})$ can be obtained from Lemma 4.3 from [17, p. 152], or by substituting into (8.1.24) the function $c(\mathbf{x}) = \delta_{km}, \mathbf{x} \in \mathbb{D}_k$ $(k = 1, 2, \ldots, n)$

$$p_m(\mathbf{x}) = - \frac{r_m}{|\mathbf{x} - \mathbf{a}_m|} + \sum_{k \neq m} \frac{r_k}{|\mathbf{x} - \mathbf{a}_k|} \frac{r_m}{|\mathbf{x}_{(k)}^* - \mathbf{a}_m|} - \tag{8.1.25}$$

$$\sum_{k=1}^{n} \sum_{k_1 \neq k, m} \frac{r_k}{|\mathbf{x} - \mathbf{a}_k|} \frac{r_{k_1}}{|\mathbf{x}_{(k)}^* - \mathbf{a}_{k_1}|} \frac{r_m}{|\mathbf{x}_{(k_1 k)}^* - \mathbf{a}_m|} +$$

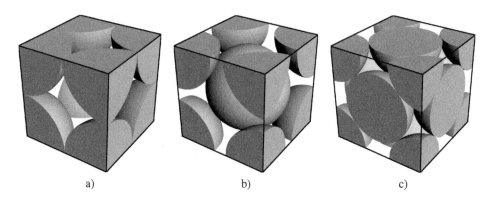

Figure 8.1 Unit cells for simple cubic (SC), body-centered cubic (BCC) and face-centered cubic (FCC) lattices are shown schematically

$$\sum_{k=1}^{n} \sum_{\substack{k_1 \neq k \\ k_2 \neq k_1, m}} \frac{r_k}{|\mathbf{x} - \mathbf{a}_k|} \frac{r_{k_1}}{|\mathbf{x}^*_{(k)} - \mathbf{a}_{k_1}|} \frac{r_{k_2}}{|\mathbf{x}^*_{(k_1 k)} - \mathbf{a}_{k_2}|} \frac{r_m}{|\mathbf{x}^*_{(k_2 k_1 k)} - \mathbf{a}_m|} + \dots.$$

Then, (8.1.3) yields the system of linear algebraic equations on the constants c_m ($m = 1, 2, \dots, n$).

$$\sum_{m=1}^{n} c_m \int_{\partial \mathbb{D}_k} p_m(\mathbf{x}) ds = \int_{\partial \mathbb{D}_k} (Px_1)(\mathbf{x}) ds, \quad k = 1, 2, \dots, n. \tag{8.1.26}$$

It follows from the general theory [15] that the system (8.1.26) has a unique solution.

2. 3D periodic problems

The present section is about 3D periodic structures represented by cubic cells. The main cubic cells are displayed in Fig. 8.1. On page 233, the theory is outlined for SC represented by the **0**-cell $O = \{\mathbf{x} \in \mathbb{R}^3 : -\frac{1}{2} < x_j < \frac{1}{2} \ (j = 1, 2, 3)\}$ formed by the vectors $\omega_1 = (1, 0, 0)$, $\omega_2 = (0, 1, 0)$ and $\omega_3 = (0, 0, 1)$. One can find the general theory for an arbitrary 3D cell in [5]. Formulae for the effective conductivity are discussed for arbitrary cubic cells at the end of this chapter.

2.1. Statement of the problem

Let mutually disjoint balls $\mathbb{D}_k = \{\mathbf{x} \in \mathbb{R}^3 : |\mathbf{x} - \mathbf{a}_k| < r_k\}$ ($k = 1, 2, \dots, N$) lie in O and $\mathbb{D}$ denote the complement of all the balls $|\mathbf{x} - \mathbf{a}_k| \leq r_k$ to O. Let $\mathbf{n} = (n_1, n_2, n_3)$ denote the unit outward normal vector to the sphere $\partial \mathbb{D}_k$ and $\frac{\partial}{\partial \mathbf{n}}$ the normal derivative. We have

$$\mathbf{n}(\mathbf{x}) = \frac{1}{r_k}(\mathbf{x} - \mathbf{a}_k), \quad \mathbf{x} \in \partial \mathbb{D}_k \quad (k = 1, 2, \dots, n). \tag{8.2.27}$$

We are looking for harmonic functions $u(\mathbf{x})$ in $\mathbb{D}$ and $u_k(\mathbf{x})$ in $\mathbb{D}_k$ $(k = 1, 2, \dots, N)$, respectively. These functions are also continuously differentiable in the closures of the considered domains with the conjugation conditions

$$u = u_k, \quad \frac{\partial u}{\partial \mathbf{n}} = \sigma \frac{\partial u_k}{\partial \mathbf{n}}, \quad |\mathbf{x} - \mathbf{a}_k| = r_k, \ k = 1, 2, \dots, N. \tag{8.2.28}$$

It is assumed that the function $u(\mathbf{x})$ is quasi-periodic, namely,

$$[u]_1 = 1, \quad [u]_2 = 0, \quad [u]_3 = 0, \tag{8.2.29}$$

where $[u]_j := u(\mathbf{x} + \omega_j) - u(\mathbf{x})$ stands for the jump of $u(\mathbf{x})$ per cell along the axis x_j.

Instead of the normal derivative in (8.2.28), we will also consider the derivative $\frac{\partial}{\partial r}$ where $r = |\mathbf{x} - \mathbf{a}_k|$ is the radial local coordinate near $\partial \mathbb{D}_k$. Let k be fixed and $\mathbf{a}_k = (a_{k1}, a_{k2}, a_{k3})$. We have

$$\frac{\partial}{\partial \mathbf{n}} = \frac{x_1 - a_{k1}}{r_k} \frac{\partial}{\partial x_1} + \frac{x_2 - a_{k2}}{r_k} \frac{\partial}{\partial x_2} + \frac{x_3 - a_{k3}}{r_k} \frac{\partial}{\partial x_3} = \frac{\partial}{\partial r}, \quad r = r_k. \tag{8.2.30}$$

Then, (8.2.28) becomes

$$u = u_k, \quad \frac{\partial u}{\partial r} = \sigma \frac{\partial u_k}{\partial r}, \quad r = r_k, \ k = 1, 2, \dots, N. \tag{8.2.31}$$

The averaged flux $\langle q_1 \rangle$ along the axis x_1 can be calculated by the Ostrogradsky–Gauss formula

$$\langle q_1 \rangle = \int_{\mathbb{D}} \frac{\partial u}{\partial x_1} d\mathbf{x} + \sigma \sum_{k=1}^{N} \int_{\mathbb{D}_k} \frac{\partial u_k}{\partial x_1} d\mathbf{x} = \tag{8.2.32}$$

$$\int_{\partial O} u \, n_1 ds + \sum_{k=1}^{N} \int_{\partial \mathbb{D}_k} (\sigma u_k - u) \, n_1 ds,$$

where $d\mathbf{x} = dx_1 dx_2 dx_3$ and ds is the surface differential. The integral over ∂O in (8.2.32) is equal to unity because of (8.2.29). Using the first relation (8.2.28) and again the Ostrogradsky–Gauss formula, we obtain

$$\langle q_1 \rangle = 1 + (\sigma - 1) \sum_{k=1}^{N} \int_{\mathbb{D}_k} \frac{\partial u_k}{\partial x_1} d\mathbf{x}. \tag{8.2.33}$$

The mean value theorem for harmonic functions yields

$$\langle q_1 \rangle = 1 + (\sigma - 1) \sum_{k=1}^{N} \frac{4}{3} \pi r_k^3 \frac{\partial u_k}{\partial x_1}(\mathbf{a}_k). \tag{8.2.34}$$

Because of the jump conditions (8.2.28), the latter formula gives the $(1,1)$-component σ_{11} of the effective conductivity tensor. In the case of equal radii we arrive at the formula

$$\sigma_{11} = 1 + (\sigma - 1)f \frac{1}{N} \sum_{k=1}^{N} \frac{\partial u_k}{\partial x_1}(\mathbf{a}_k), \tag{8.2.35}$$

where f denote the concentration of balls.

2.2. Highly conducting inclusions

Below, we study the case when $\sigma \gg 1$ and use the ansatz

$$u = u^{(0)} + \frac{1}{\sigma} u^{(1)} + \frac{1}{\sigma^2} u^{(2)} + \cdots, \quad u_k = u_k^{(0)} + \frac{1}{\sigma} u_k^{(1)} + \frac{1}{\sigma^2} u_k^{(2)} + \cdots. \tag{8.2.36}$$

Substitute (8.2.36) into (8.2.28) and keep the terms up to $O(\sigma^{-1})$. The zeroth coefficient yields the problem

$$u^{(0)} = u_k^{(0)}, \tag{8.2.37}$$

$$\frac{\partial u_k^{(0)}}{\partial \mathbf{n}} = 0, \quad |\mathbf{x} - \mathbf{a}_k| = r_k \ (k = 1, 2, \ldots, N). \tag{8.2.38}$$

It follows from equation (8.2.38) that $u_k^{(0)}(\mathbf{x})$ is constant for each k. Then, equation (8.2.37) can be considered as the modified Dirichlet problem discussed in the next section. The coefficient in σ^{-1}-term yields the problem

$$u^{(1)} = u_k^{(1)}, \tag{8.2.39}$$

$$\frac{\partial u^{(0)}}{\partial \mathbf{n}} = \frac{\partial u_k^{(1)}}{\partial \mathbf{n}}, \quad |\mathbf{x} - \mathbf{a}_k| = r_k \ (k = 1, 2, \ldots, N). \tag{8.2.40}$$

The relation (8.2.40) can be considered as N independent Neumann problems on $u_k^{(1)}(\mathbf{x})$ for the balls $\mathbb{D}_k \ (k = 1, 2, \ldots, N)$ where $u^{(0)}(\mathbf{x})$ is taken after solution to the modified Dirichlet problem (8.2.37). Solving the latter problem we arrive at the Dirichlet problem (8.2.39).

Using (8.2.35), we get

$$\sigma_{11} = 1 + (\sigma - 1)f \frac{1}{N} \sum_{k=1}^{N} \left[\frac{\partial u_k^{(0)}}{\partial x_1}(\mathbf{a}_k) + \frac{1}{\sigma} \frac{\partial u_k^{(1)}}{\partial x_1}(\mathbf{a}_k) \right] + O(\sigma^{-1}). \tag{8.2.41}$$

Remembering that $u_k^{(0)}(\mathbf{x})$ is a constant, we obtain

$$\sigma_{11} = 1 + \frac{\sigma - 1}{\sigma} f \frac{1}{N} \sum_{k=1}^{N} \frac{\partial u_k^{(1)}}{\partial x_1}(\mathbf{a}_k) + O(\sigma^{-1}). \qquad (8.2.42)$$

Formula (8.2.42) can be written in the form

$$\sigma_{11} = 1 + f \frac{1}{N} \sum_{k=1}^{N} \frac{\partial u_k^{(1)}}{\partial x_1}(\mathbf{a}_k) + O(\sigma^{-1}). \qquad (8.2.43)$$

In equations (8.2.42) and (8.2.43), $u_k^{(1)}(\mathbf{x})$ is a solution of the Neumann problem (8.2.40) where $u_k^{(0)}(\mathbf{x})$ is a solution of the modified Dirichlet problem (8.2.37). In order to avoid solving the two problems, we transform (8.2.43) as follows. Let $\mathbf{a}_k = (a_{k1}, a_{k2}, a_{k3})$. First, application of the Green identity

$$\int_{\mathbb{D}_k} (g\Delta f + \nabla f \cdot \nabla g) d\mathbf{x} = \int_{\partial \mathbb{D}_k} g \frac{\partial f}{\partial \mathbf{n}} ds \qquad (8.2.44)$$

for $f = u_k^{(1)}$, $g = x_1 - a_{k1}$ and the mean value theorem yield

$$\frac{4}{3}\pi r_k^3 \frac{\partial u_k^{(1)}}{\partial x_1}(\mathbf{a}_k) = \int_{\mathbb{D}_k} \frac{\partial u_k^{(1)}}{\partial x_1} d\mathbf{x} = \int_{\partial \mathbb{D}_k} (x_1 - a_{k1}) \frac{\partial u_k^{(1)}}{\partial \mathbf{n}} ds. \qquad (8.2.45)$$

Using (8.2.40) we can write (8.2.43) in the form

$$\sigma_{11} = 1 + \sum_{k=1}^{N} \int_{\partial \mathbb{D}_k} (x_1 - a_{k1}) \frac{\partial u^{(0)}}{\partial \mathbf{n}} ds + O(\sigma^{-1}), \qquad (8.2.46)$$

where $u^{(0)}(\mathbf{x})$ is a solution of the modified Dirichlet problem (8.2.37). Using the radial derivative (8.2.30), we rewrite (8.2.46) in the equivalent form

$$\sigma_{11} = 1 + \sum_{k=1}^{N} \int_{\partial \mathbb{D}_k} (x_1 - a_{k1}) \frac{\partial u^{(0)}}{\partial r} ds + O(\sigma^{-1}). \qquad (8.2.47)$$

2.3. Effective conductivity

The countable set of centers can be ordered in the following way $\mathcal{A} = \{\mathbf{a}_k + \sum_{j=1,2,3} m_j \omega_j\}$. Here, $(m_1, m_2, m_3) \in \mathbb{Z}^3$ is the coordinate of the cell; the points $\mathbf{a}_k$ $(k = 1, 2, \ldots, N)$ belong to the **0**-cell.

First, we consider the finite number of inclusions $n = NM$ in the space $\mathbb{R}^3$ using the linear-order numeration $\mathbf{a}_k$ $(k = 1, 2, \ldots, n)$. Here M is the number of cells and the fixed number N is the number of inclusions per cell. This allows us to use exact formulae deduced on page 221 for the local fields around a finite number of balls in

$\mathbb{R}^3$. Introduce the value

$$\sigma_{11}^{(n)} = 1 + \frac{1}{M} \sum_{k=1}^{n} \int_{\partial \mathbb{D}_k} (x_1 - a_{k1}) \frac{\partial u}{\partial \mathbf{n}} ds \qquad (8.2.48)$$

which corresponds to σ_{11} from (8.2.46) when

$$\sigma_{11} = \lim_{n \to \infty} \sigma_{11}^{(n)} + O(\sigma^{-1}). \qquad (8.2.49)$$

It follows from (8.1.17) that for every fixed k

$$\frac{\partial u}{\partial \mathbf{n}}(\mathbf{x}) = \frac{x_1 - a_{k1}}{r_k} - \frac{\partial}{\partial r}\left(\frac{r_k}{r} u_k(\mathbf{x}_{(k)}^*)\right) - \frac{\partial F_k}{\partial r}(\mathbf{x}), \quad \mathbf{x} \in \partial \mathbb{D}_k, \qquad (8.2.50)$$

where

$$F_k(\mathbf{x}) := \sum_{m \neq k} \frac{r_m}{|\mathbf{x} - \mathbf{a}_m|} u_m(\mathbf{x}_{(m)}^*). \qquad (8.2.51)$$

Using (8.1.14), we calculate

$$-\frac{\partial}{\partial r}\left(\frac{r_k}{r} u_k(\mathbf{x}_{(k)}^*)\right) = \frac{1}{r_k} u_k(\mathbf{x}) + \frac{\partial u_k}{\partial r}(\mathbf{x}), \quad \mathbf{x} \in \partial \mathbb{D}_k. \qquad (8.2.52)$$

Substitute (8.2.52) into (8.2.50), then multiply the obtained result by $(x_1 - a_{k1})$ and integrate over $\partial \mathbb{D}_k$

$$\int_{\partial \mathbb{D}_k} (x_1 - a_{k1}) \frac{\partial u}{\partial \mathbf{n}} ds = \int_{\partial \mathbb{D}_k} \frac{(x_1 - a_{k1})^2}{r_k} ds + \int_{\partial \mathbb{D}_k} \frac{x_1 - a_{k1}}{r_k} u_k ds + \qquad (8.2.53)$$
$$\int_{\partial \mathbb{D}_k} (x_1 - a_{k1}) \frac{\partial (u_k - F_k)}{\partial \mathbf{n}} ds.$$

It follows from the Ostrogradsky–Gauss formula that for a function U

$$\int_{\mathbb{D}_k} \frac{\partial U}{\partial x_1} d\mathbf{x} = \int_{\partial \mathbb{D}_k} \frac{x_1 - a_{k1}}{r_k} U ds, \qquad (8.2.54)$$

since the first component of the unit outward normal vector to the sphere $\partial \mathbb{D}_k$ is equal to $\frac{x_1 - a_{k1}}{r_k}$. Let $|\mathbb{D}_k|$ denote the volume of the ball $\mathbb{D}_k$. Application of (8.2.54) to the first integral from the right-hand side of (8.2.53) yields

$$\int_{\partial \mathbb{D}_k} \frac{(x_1 - a_{k1})^2}{r_k} ds = |\mathbb{D}_k|. \qquad (8.2.55)$$

The same arguments yield

$$\int_{\partial \mathbb{D}_k} \frac{x_1 - a_{k1}}{r_k} u_k ds = \int_{\mathbb{D}_k} \frac{\partial u_k}{\partial x_1} d\mathbf{x} = |\mathbb{D}_k| \frac{\partial u_k}{\partial x_1}(\mathbf{a}_k), \qquad (8.2.56)$$

where the mean value theorem is used. Green's identity (8.2.44) for $f = u_k - F_k$,

$g = x_1 - a_{k1}$ and the mean value theorem implies that

$$\int_{\partial \mathbb{D}_k} (x_1 - a_{k1}) \frac{\partial(u_k - F_k)}{\partial \mathbf{n}} ds = |\mathbb{D}_k| \left[\frac{\partial u_k}{\partial x_1}(\mathbf{a}_k) - \frac{\partial F_k}{\partial x_1}(\mathbf{a}_k) \right], \qquad (8.2.57)$$

since $F_k(\mathbf{x})$ is harmonic in the ball $\mathbb{D}_k$ and continuously differentiable in its closure. Substitution of (8.2.53)–(8.2.57) into (8.2.48) yields

$$\sigma_{11}^{(n)} = 1 + \frac{1}{M} \sum_{k=1}^{n} |\mathbb{D}_k| \left[1 + 2 \frac{\partial u_k}{\partial x_1}(\mathbf{a}_k) - \frac{\partial F_k}{\partial x_1}(\mathbf{a}_k) \right]. \qquad (8.2.58)$$

Let all the balls have the same radius $r_k = r_0$. Then, (8.2.58) becomes

$$\sigma_{11}^{(n)} = 1 + \frac{4\pi r_0^3}{3M} \sum_{k=1}^{n} \left[1 + 2 \frac{\partial u_k}{\partial x_1}(\mathbf{a}_k) - \frac{\partial F_k}{\partial x_1}(\mathbf{a}_k) \right]. \qquad (8.2.59)$$

We now proceed to calculate

$$\frac{\partial F_k}{\partial x_1}(\mathbf{a}_k) = \frac{\partial}{\partial x_1} \sum_{m \neq k} \frac{r_0}{|\mathbf{x} - \mathbf{a}_m|} u_m(\mathbf{x}_{(m)}^*) \Big|_{\mathbf{x} = \mathbf{a}_k}. \qquad (8.2.60)$$

First, calculate for $m \neq k$

$$\frac{\partial \mathbf{x}_{(m)}^*}{\partial x_1} \Big|_{\mathbf{x} = \mathbf{a}_k} = \frac{\partial}{\partial x_1} \left[\frac{r_0^2}{|\mathbf{x} - \mathbf{a}_m|^2}(\mathbf{x} - \mathbf{a}_m) + \mathbf{a}_m \right] =$$

$$r_0^2 \left[-2 \frac{a_{k1} - a_{m1}}{|\mathbf{a}_k - \mathbf{a}_m|^4}(\mathbf{a}_k - \mathbf{a}_m) + \frac{(1, 0, 0)}{|\mathbf{a}_k - \mathbf{a}_m|^2} \right] =$$

$$-\frac{r_0^2}{|\mathbf{a}_k - \mathbf{a}_m|^4} \left(a_1^2 - a_2^2 - a_1^3, 2a_1 a_2, 2a_1 a_3 \right),$$

where the following notations are used for brevity

$$a_j := a_{kj} - a_{mj}, \quad j = 1, 2, 3. \qquad (8.2.61)$$

Then,

$$\frac{\partial}{\partial x_1} \left[\frac{r_0}{|\mathbf{x} - \mathbf{a}_m|} u_m(\mathbf{x}_{(m)}^*) \right] \Big|_{\mathbf{x} = \mathbf{a}_k} = -\frac{r_0}{|\mathbf{a}_k - \mathbf{a}_m|^3} a_1 u_m((\mathbf{a}_k)_{(m)}^*) - \qquad (8.2.62)$$

$$\frac{r_0^3}{|\mathbf{a}_k - \mathbf{a}_m|^5} \left[(a_1^2 - a_2^2 - a_1^3) \frac{\partial u_m}{\partial x_1} + 2a_1 a_2 \frac{\partial u_m}{\partial x_2} + 2a_1 a_3 \frac{\partial u_m}{\partial x_3} \right] ((\mathbf{a}_k)_{(m)}^*),$$

where the functions $\frac{\partial u_m}{\partial x_j}$ are calculated at the point $(\mathbf{a}_k)_{(m)}^*$. Ultimately, (8.2.59) takes the form

$$\sigma_{11}^{(n)} = 1 + \frac{4\pi r_0^3}{3M} \sum_{k=1}^{n} \left\{ 1 + 2\frac{\partial u_k}{\partial x_1}(\mathbf{a}_k) + \sum_{m \neq k} \left[\frac{r_0}{|\mathbf{a}_k - \mathbf{a}_m|^3} a_1 u_m((\mathbf{a}_k)_{(m)}^*) + \right. \right. \tag{8.2.63}$$

$$\left. \left. \frac{r_0^3}{|\mathbf{a}_k - \mathbf{a}_m|^5} \left((a_1^2 - a_2^2 - a_3^2)\frac{\partial u_m}{\partial x_1} + 2a_1 a_2 \frac{\partial u_m}{\partial x_2} + 2a_1 a_3 \frac{\partial u_m}{\partial x_3} \right)((\mathbf{a}_k)_{(m)}^*) \right] \right\}.$$

Consider the first-order approximation for equations (8.1.16)

$$u_k(\mathbf{x}) \approx x_1 - c_k - \sum_{l \neq k} \left[\frac{r_0^3(x_1 - a_{l1})}{|\mathbf{a}_k - \mathbf{a}_l|^3} + \frac{r_0(a_{l1} - c_l)}{|\mathbf{a}_k - \mathbf{a}_l|} \right], \quad \mathbf{x} \in \mathbb{D}_k \ (k = 1, 2, \ldots, n). \tag{8.2.64}$$

Let $\{\mathbf{v}\}_1$ denote the first coordinate of the vector $\mathbf{v}$. It follows from (8.1.4) that

$$\{(\mathbf{a}_k)_{(m)}^*\}_1 = \frac{r_0^2}{|\mathbf{a}_k - \mathbf{a}_m|^2}(a_{k1} - a_{m1}) + a_{m1}. \tag{8.2.65}$$

Substitute (8.2.64) into (8.2.63) using (8.2.61) and (8.2.65) to obtain

$$\sigma_{11}^{(n)} = 1 + 3f + \frac{3}{M} \sum_{k=1}^{n} \sum_{m \neq k} \left[r_0 \frac{a_{k1} - a_{m1}}{|\mathbf{a}_k - \mathbf{a}_m|^3} a_{m1} + \right. \tag{8.2.66}$$

$$\left. \frac{r_0^3}{|\mathbf{a}_k - \mathbf{a}_m|^5} (2(a_{k1} - a_{m1})^2 - (a_{k2} - a_{m2})^2 - (a_{k3} - a_{m3})^2) \right] +$$

$$\frac{1}{M} \sum_{k=1}^{n} \sum_{m \neq k} \sum_{l \neq m} r_0^2 \frac{a_{k1} - a_{m1}}{|\mathbf{a}_k - \mathbf{a}_m|^3 |\mathbf{a}_m - \mathbf{a}_l|} a_{l1} -$$

$$\frac{2}{M} \sum_{k=1}^{n} \sum_{l \neq k} c_l r_0 \frac{a_{k1} - a_{l1}}{|\mathbf{a}_k - \mathbf{a}_l|^3} - \frac{1}{M} \sum_{k=1}^{n} \sum_{m \neq k} \sum_{l \neq m} c_l r_0^2 \frac{a_{k1} - a_{m1}}{|\mathbf{a}_k - \mathbf{a}_m|^3 |\mathbf{a}_m - \mathbf{a}_l|} + O(r_0^4),$$

where $f = N\frac{4}{3}\pi r_0^3$ corresponds to the concentration of inclusions in the periodic medium.

Remark 17. The zero approximation in r_0 of (8.2.66) yields the famous Maxwell formula. Higher-order approximations in (8.2.66) require proper definition of the limit sums as n tends to infinity. At present time, it is difficult to say whether the Eisenstein summation corresponds to the physical problem. The general situation is clear for simple periodic structures when Berdichevsky's triply periodic functions, rooted in the Weierstrass approach, can be applied. Moreover, a physical justification for the 3D lattice sums can be applied due to Rayleigh [21], McPhedran et al. [13, 14]. These topics will be outlined below in connection with functional equations.

3. Triply periodic functions

In the present section, we discuss an essentials of triply periodic functions associated with the SC lattice described on page 226, which follows [5]. Let $\mathbf{P} := \omega_j P_j \ (P_j \in \mathbb{Z})$ where the Einstein summation convention over j from 1 to 3 is used. Let x_i denote the i-th coordinate of $\mathbf{x}$. Introduce the function

$$\wp(\mathbf{x}) = \frac{1}{|\mathbf{x}|} + {\sum_{\mathbf{P}}}' \left[\frac{1}{|\mathbf{x}-\mathbf{P}|} - \frac{1}{|\mathbf{P}|} - \frac{P_i x_i}{|\mathbf{x}|^3} - \frac{1}{2} \left(\frac{3 P_i x_i P_j x_j}{|\mathbf{x}|^5} - \frac{|\mathbf{x}|^2}{|\mathbf{P}|^3} \right) \right], \tag{8.3.67}$$

where the term $\mathbf{P} = \mathbf{0}$ is omitted in the summation ${\sum_{\mathbf{P}}}'$. The series (8.3.67) converges uniformly and absolutely in every compact subset of $\mathbb{R}^3 \setminus \cup_P \{P\}$. The functions $\wp(\mathbf{x}) - \frac{2\pi}{3}|\mathbf{x}|^2$ and $\frac{\partial \wp}{\partial x_i}(\mathbf{x}) - \frac{2\pi}{3} x_i$ are triple periodic.

Let $(F)_{|_{i_1, i_2, \dots, i_m}}$ denote the partial derivative of F on $x_{i_1}, x_{i_2}, \dots, x_{i_m}$. Introduce the functions[1]

$$\wp'_{i_1, i_2, \dots, i_m}(\mathbf{x}) = {\sum_{\mathbf{P}}}' \left(\frac{1}{|\mathbf{x}-\mathbf{P}|} \right)_{|_{i_1, i_2, \dots, i_m}}. \tag{8.3.68}$$

and

$$\wp_{i_1, i_2, \dots, i_m}(\mathbf{x}) = \left(\frac{1}{|\mathbf{x}|} \right)_{|_{i_1, i_2, \dots, i_m}} + \wp'_{i_1, i_2, \dots, i_m}(\mathbf{x}). \tag{8.3.69}$$

Consider the functions

$$\wp_i(\mathbf{x}) = \wp(\mathbf{x})_{|_i} = -\frac{x_i}{|\mathbf{x}|^3} - {\sum_{\mathbf{P}}}' \left[(x_i - P_i) \left(\frac{1}{|\mathbf{x}-\mathbf{P}|^3} - \frac{1}{|\mathbf{P}|^3} \right) - 3 P_i \frac{P_j x_j}{|\mathbf{P}|^5} \right] \tag{8.3.70}$$

$$(i = 1, 2, 3)$$

and

$$\wp_{ij}(\mathbf{x}) = \wp(\mathbf{x})_{|_{ij}} = -\frac{\delta_{ij}}{|\mathbf{x}|^3} + 3\frac{x_i x_j}{|\mathbf{x}|^5} - {\sum_{\mathbf{P}}}' \left[\delta_{ij} \left(\frac{1}{|\mathbf{x}-\mathbf{P}|^3} - \frac{1}{|\mathbf{P}|^3} \right) - \right. \tag{8.3.71}$$

$$\left. 3 \left(\frac{(x_i - P_i)(x_j - P_j)}{|\mathbf{x}-\mathbf{P}|^5} - \frac{P_i P_j}{|\mathbf{P}|^5} \right) \right] \quad (i, j = 1, 2, 3).$$

The functions $\wp_{ij}(\mathbf{x})$ are triply periodic. The function $\wp_1(\mathbf{x})$ has the following constant jumps along the unit cell

$$[\wp_1]_1 = \wp_1(\mathbf{x} + \omega_1) - \wp_1(\mathbf{x}) = -\frac{4\pi}{3}, \quad [\wp_1]_j = 0, \ j = 2, 3. \tag{8.3.72}$$

Moreover, $\wp_1(\mathbf{x})$ is an odd function and its modification $\wp'_1(\mathbf{x})$ vanishes at $\mathbf{x} = \mathbf{0}$.

---

[1]Everywhere in this section, prime stands for an omitted term.

4. Functional equations on periodic functions

The system of functional equations (8.1.16) for the periodic media takes the form

$$u_k(\mathbf{x}) = -\sum_{\mathbf{P}}' \sum_{m=1}^{N} \frac{r_m}{|\mathbf{x} - \mathbf{P} - \mathbf{a}_m|} u_m \left(r_m^2 \frac{(\mathbf{x} - \mathbf{P} - \mathbf{a}_m)}{|\mathbf{x} - \mathbf{P} - \mathbf{a}_m|^2} + \mathbf{a}_m \right) + \tag{8.4.73}$$

$$x_1 - c_k, \quad |\mathbf{x} - \mathbf{a}_m| \le r_k \ (k = 1, 2, \dots, N),$$

where $\mathbf{a}_k$ $(k = 1, 2, \dots, N)$ belong to the **0**-cell O. The sum $\sum_{\mathbf{P}}'$ means that the term $\mathbf{P} = \mathbf{0}$ is omitted for $m = k$. Instead of (8.1.17) we have

$$u(\mathbf{x}) = -\sum_{\mathbf{P}} \sum_{m=1}^{N} \frac{r_m}{|\mathbf{x} - \mathbf{P} - \mathbf{a}_m|} u_m \left(r_m^2 \frac{(\mathbf{x} - \mathbf{P} - \mathbf{a}_m)}{|\mathbf{x} - \mathbf{P} - \mathbf{a}_m|^2} + \mathbf{a}_m \right) + x_1, \ \mathbf{x} \in \dot{\mathbb{D}}. \tag{8.4.74}$$

In the case $N = 1$, when one ball of radius r_0 is located at the point **0**, we have $u_1(\mathbf{0}) = 0$ and $c_1 = 0$ because of the symmetry. Equations (8.4.73) and (8.4.74) become

$$u_1(\mathbf{x}) = -\sum_{\mathbf{P}}' \frac{r_0}{|\mathbf{x} - \mathbf{P}|} u_1 \left(r_0^2 \frac{\mathbf{x} - \mathbf{P}}{|\mathbf{x} - \mathbf{P}|^2} \right) + x_1, \quad |\mathbf{x}| \le r_0, \tag{8.4.75}$$

$$u(\mathbf{x}) = -\sum_{\mathbf{P}} \frac{r_0}{|\mathbf{x} - \mathbf{P}|} u_1 \left(r_0^2 \frac{\mathbf{x} - \mathbf{P}}{|\mathbf{x} - \mathbf{P}|^2} \right) + x_1, \ \mathbf{x} \in \dot{\mathbb{D}}, \tag{8.4.76}$$

where $\sum_{\mathbf{P}}'$ does not contain the term with $\mathbf{P} = \mathbf{0}$. Convergence of the triple sum $\sum_{\mathbf{P}}$ has to be properly described because the positive series $\sum_{\mathbf{P}}' |\mathbf{P}|^{-s}$ diverges for $s \le 3$ what implies the absolute divergence of the corresponding functional series [5].

Let us introduce the modified Eisenstein 3D vector function

$$\mathbf{E}(\mathbf{x}) = \sum_{\mathbf{P}} \frac{\mathbf{x} - \mathbf{P}}{|\mathbf{x} - \mathbf{P}|^3}, \quad \mathbf{E}'(\mathbf{x}) = \sum_{\mathbf{P}}' \frac{\mathbf{x} - \mathbf{P}}{|\mathbf{x} - \mathbf{P}|^3}. \tag{8.4.77}$$

We proceed now to give proper expression to the series (8.4.77) directly via Berdichevsky's functions. First, we check that the functions

$$U(\mathbf{x}) = x_1 - r_0^3 \left(-\frac{4\pi}{3} x_1 - \wp_1(\mathbf{x}) \right), \quad U_1(\mathbf{x}) = x_1 - r_0^3 \left(-\frac{4\pi}{3} x_1 - \wp_1'(\mathbf{x}) \right) \tag{8.4.78}$$

approximate the functions $u(\mathbf{x})$ and $u_1(\mathbf{x})$ in such a way that the boundary conditions

$$u(\mathbf{x}) = 0, \quad \frac{1}{r_0} u_1(\mathbf{x}) + 2 \frac{\partial u_1}{\partial r}(\mathbf{x}) = \frac{\partial u}{\partial r}(\mathbf{x}), \quad |\mathbf{x}| = r_0, \tag{8.4.79}$$

are fulfilled within the accuracy $O(r_0^2)$. We have

$$U(\mathbf{x}) = x_1 - r_0^3 \left(-\frac{4\pi}{3} x_1 - \frac{x_1}{r_0^3} - \wp_1'(\mathbf{x}) \right) = O(r_0^3), \quad |\mathbf{x}| = r_0 \tag{8.4.80}$$

and

$$\frac{1}{r_0}U_1(\mathbf{x}) + 2\frac{\partial U_1}{\partial r}(\mathbf{x}) - \frac{\partial U}{\partial r}(\mathbf{x}) = \frac{x_1}{r_0} - r_0^2\left(-\frac{4\pi}{3}x_1 - \frac{\partial \wp_1'}{\partial r}(\mathbf{x})\right) + \tag{8.4.81}$$

$$2\frac{x_1}{r_0} - 2r_0^3\left(-\frac{4\pi}{3}\frac{x_1}{r_0} - \frac{\partial \wp_1'}{\partial r}(\mathbf{x})\right) - \frac{x_1}{r_0} + r_0^3 - \frac{4\pi}{3}\frac{x_1}{r_0} - 2\frac{x_1}{r_0} +$$

$$(2r_0^3 + r_0^2)\wp_1'(\mathbf{x}) = O(r_0^2), \quad |\mathbf{x}| = r_0.$$

In accordance with (8.4.75)–(8.4.77) the same approximation is given by

$$U(\mathbf{x}) = x_1 - r_0^3 E_1(\mathbf{x}), \quad U_1(\mathbf{x}) = x_1 - r_0^3 E_1'(\mathbf{x}). \tag{8.4.82}$$

Therefore, the function $E_1(\mathbf{x})$ must be defined by formula

$$E_1(\mathbf{x}) = -\frac{4\pi}{3}x_1 - \wp_1(\mathbf{x}) \tag{8.4.83}$$

and

$$\mathbf{E}(\mathbf{x}) = \left(-\frac{4\pi}{3}x_1 - \wp_1(\mathbf{x}), -\wp_2(\mathbf{x}), -\wp_3(\mathbf{x})\right). \tag{8.4.84}$$

Remark 18. In the Section 2, the 3D functional equations (8.4.73) and (8.4.75) are constructed for periodic problems. Analogous functional equations for a finite number of inclusions in $\mathbb{R}^3$ have been solved on page 221 exactly in terms of the 3D Poincaré series. However, the direct application of the explicit formulae is impossible until more computationally effective formulae for the 3D Eisenstein series are constructed. This depends on the further simplification of the triply periodic functions.

5. Analytical formulae for the effective conductivity. Discussion and overview of the known results.

In the previous works, the method of series was applied to the 3D periodic problems. In the present section, we outline these results and show their connection to the functional equations as it is done on page 98 concerning 2D periodic problems.

First, we consider the functional equation (8.4.75). Introduce the spherical coordinates (r, θ, ϕ) near the origin and the spherical harmonics

$$Y_{lm}(\theta, \phi) = (-1)^m \left[\frac{2l+1}{4\pi}\frac{(l-m)!}{(l+m)!}\right]^{\frac{1}{2}} P_l^m(\cos\theta)e^{im\phi}. \tag{8.5.85}$$

Here, $P_l^m(\cos\theta)$ denote the associated Legendre functions defined by equations [22, Sec.18.2]

$$P_l^m(x) = \frac{1}{2^l l!}(1-x^2)^{\frac{m}{2}}\frac{d^{l+m}}{dx^{l+m}}(x^2-1)^l. \tag{8.5.86}$$

In the case $m = 4m'$ ($m' \in \mathbb{Z}$) interested for us, the functions $P_l^m(\cos\theta)$ are polynomials. The function $u_1(\mathbf{x})$ can be expressed as the series [13, 14, 20, 21]

$$u_1(\mathbf{x}) = C_0 + \sum_{l=1}^{} \sum_{m=-l}^{m=l} C_{lm} r^l Y_{lm}(\theta, \phi), r < r_0, \tag{8.5.87}$$

where $C_0 = u_1(0) = 0$ and m runs over numbers divided onto 4.

Consider the SC lattice when $\mathbf{P} \in \mathbb{Z}^3$ and the inversions $\mathbf{x}_{(\mathbf{P})}^*$ with respect to the spheres $|\mathbf{x}_{\mathbf{P}}| = r_0$ where $\mathbf{x}_{\mathbf{P}} := \mathbf{x} - \mathbf{P}$. We have $|\mathbf{x}_{(\mathbf{P})}^*| \equiv r_{\mathbf{P}}^* = \frac{r_0^2}{r_{\mathbf{P}}}$. Introduce the local polar coordinate $(r_{\mathbf{P}}, \theta_{\mathbf{P}}, \phi_{\mathbf{P}})$ with the center at $\mathbf{P}$. The expression from the right hand side of (8.4.75) can be written in the form

$$\frac{r_0}{r_{\mathbf{P}}} u_1\left(r_0^2 \frac{\mathbf{x}_{\mathbf{P}}}{r_{\mathbf{P}}}\right) = \sum_{l=1}^{} \sum_{m=-l}^{m=l} C_{lm} \frac{r_0^{2l+1}}{r_{\mathbf{P}}^{l+1}} Y_{lm}(\theta_{\mathbf{P}}, \phi_{\mathbf{P}}). \tag{8.5.88}$$

Introduce the polar coordinates $(R_{\mathbf{P}}, \Theta_{\mathbf{P}}, \Phi_{\mathbf{P}})$ of the point $\mathbf{P}$ with the center at the origin. We use the formula (15) from the paper [14]

$$\frac{P_l^m(\cos\Theta_{\mathbf{P}}) e^{im\Phi_{\mathbf{P}}}}{r_{\mathbf{P}}^{l+1}} = \tag{8.5.89}$$

$$\sum_{l',m'} (-1)^{l+m'} \frac{(l+l'+m-m')!}{(l-m)!(l'+m'')!} \frac{r^{l'}}{R_{\mathbf{P}}^{l+l'+1}} P_{l'}^{m'}(\cos\theta) P_{l+l'}^{m-m'}(\cos\Theta_{\mathbf{P}}) e^{im\phi+(m-m')\Phi_{\mathbf{P}}},$$

where m and m' run from $-l$ to l, and from $-l'$ to l', respectively. Equation (8.5.89) can be written in the form

$$Y_{lm}(\theta_{\mathbf{P}}, \phi_{\mathbf{P}}) = \sum_{l',m'} (-1)^{l+m'} \frac{(l+l'+m-m')!}{(l-m)!(l'+m'')!} \frac{1}{R_{\mathbf{P}}^{l+l'+1}} P_{l+l'}^{m-m'}(\cos\Theta_{\mathbf{P}}) e^{(m-m')\Phi_{\mathbf{P}}} Y_{l'm'}(\theta, \phi).$$

$$\tag{8.5.90}$$

Substitute (8.5.90) into (8.5.88) and the result into the functional equation (8.4.75). Using (8.5.87), we get

$$\sum_{l,m} C_{lm} r^l Y_{lm}(\theta, \phi) = \tag{8.5.91}$$

$$\sum_{l,m} C_{lm} r_0^{2l+1} \sum_{l',m'} (-1)^{l+m'} \frac{(l+l'+m-m')!}{(l-m)!(l'+m'')!} \frac{1}{R_{\mathbf{P}}^{l+l'+1}} P_{l+l'}^{m-m'}(\cos\Theta_{\mathbf{P}}) e^{(m-m')\Phi_{\mathbf{P}}} Y_{l'm'}(\theta, \phi) +$$

$$r_0 \cos\theta \sin\phi = r_0 \sqrt{\frac{2\pi}{3}}(-Y_{11}(\theta, \phi) + Y_{11}(\theta, \phi)),$$

where the formula $x_1 = r_0 \cos\theta \sin\phi = r_0 \sqrt{\frac{2\pi}{3}}(-Y_{11}(\theta, \phi) + Y_{11}(\theta, \phi))$ is used. The coefficients in the same spherical harmonics yield an infinite system of linear algebraic equations. Below, we shall obtain it for general conjugation problems.

It should be noted that the method of functional equations in 3D can be applied only to the Dirichlet-type problem, not to the Neumann and the ideal contact (transmission) problems since the corresponding 3D Poincaré series can be constructed only for the Dirichlet-type problem (for details see [19]).

The series method can be directly applied to the both problems. In particular, to the problem when the ratio of spheres conductivity to the matrix conductivity is finite number σ. One can introduce the constants similar to the contrast parameter

$$\rho_l = \frac{\sigma - 1}{\sigma + \frac{2l}{2l-1}}, \quad l = 1, 2, \ldots. \tag{8.5.92}$$

Following [13, 14], we shall use the lattice sums

$$U_l^m = \sum_{\mathbf{P}}{}' \frac{P_m^l(\cos\Theta_{\mathbf{P}}) \cos\Phi_{\mathbf{P}}}{R_{\mathbf{P}}^{l+1}}. \tag{8.5.93}$$

These sums can be expressed in terms of the sums discussed above in Sections 4 and 3 by using of the relations between the Cartesian coordinates of $\mathbf{P} = (P_1, P_2, P_3)$ and its spherical coordinates $(R_{\mathbf{P}}, \Theta_{\mathbf{P}}, \Theta_{\mathbf{P}})$. The papers [13, 14] contain exact and numerical values of the sums (8.5.93) for various lattices.

For regular locations of spheres, substitution of the series (8.5.87) inside the sphere and the corresponding series outside the sphere yields the system of linear algebraic equations (see for instance (16) from [14])

$$\sum_{l=1}^{\infty} \sum_{m=0}^{2l-1} \left[-\frac{\delta_{ll'}\delta_{mm'}(2l-1+m)!}{\rho_l r_0^{4l-1}(2l-1+m)!\epsilon_m} + \frac{1}{2(2l-1+m)!(2l'-1+m')!} \right. \tag{8.5.94}$$

$$\left. \times ((-1)^{\min(m,m')}\mathcal{U}_{2(l+l'-1)}^{|m'-m|} + \mathcal{U}_{2(l+l'-1)}^{m'+m})B_{2l-1,m} \right] = \delta_{l'0}\delta_{m'0},$$

$$l' = 1, 2, \ldots; \quad m' = 0, 1, \ldots, 2l' - 1.$$

Here, $\epsilon_0 = 1$ and $\epsilon_m = 2$ $(m > 0)$; $\delta_{ll'}$ is the Kronecker symbol; $\mathcal{U}_l^m = (l-m)!U_l^m$. The (8.5.94) is obtained by means of differentiation of the series boundary relations that is equivalent to comparison of the coefficients in the basic functions. However, the differentiation is preferable since it leads to the symmetric system (8.5.94).

After solution to the system (8.5.94) the effective conductivity is calculated by the formula

$$\sigma_e = 1 - 4\pi B_1. \tag{8.5.95}$$

We recapitulate below the final formulae for the effective conductivity obtained

McPhedran et al. [13, 14], as well as by other and researchers [5, 23]. In accordance with our schematics presented on page 1, we discuss approximate analytical solutions referred to the items 1 and 2, not to be confused with numerical solutions referred to the item 3.

All results presented below are applicable to the effective thermal conductivity as shown in [5]. Composite materials with inclusions of high thermal conductivity can be used as thermal conductors, and materials with inclusions of low thermal conductivity may be suggested as thermal insulators.

5.1. Highly conducting inclusions. SC

Keller [9] deduced the asymptotic formula for the simple cubic (SC) lattice of spheres

$$\sigma_e = -\frac{\pi}{2} \log (f_c - f) + O(\delta^0), \quad \text{as } \delta \to 0, \tag{8.5.96}$$

where δ stands for the distance between two neighbor spheres. The threshold $f_c = \frac{\pi}{6}$. This formula was rigorously justified in [6, 10]. Keller's formulae when applied for all concentration works rather well with maximum error of 1.382% at $f = 0.05$, compared with numerical results of [13].

Batchelor and O'Brien [4] extended (8.5.96) by calculation of the constant term B of order δ^0 based on the same Keller's estimation of the flux between two spheres. However, the validity of the constant term was not supported in [6, 10]. We apply asymptotic methods to estimate B below. First, consider Keller's formula in the general form with the higher-order constant correction

$$\sigma_e(f) \sim A \log(f_c - f) + B,$$

with $A = -\frac{\pi}{2}$. We applied for the SC lattice the same Padé technique as described below in the cases of face-centered cubic lattice (FCC) and body-centered cubic lattice (BCC). It turned out that the amplitude B is small, $B \approx -0.01$. Similar conclusion was reached in [13, 14]. In what follows we simply put $B = 0$.

In order to get an low-concentration expression we employ Sangani–Acrivos formula [23],

$$\sigma_{SA}(f) = 1 + \cfrac{3f}{-\cfrac{1.3047\left(0.2305 f^{11/3} + 1\right) f^{10/3}}{1 - 0.4054 f^{7/3}} - 0.07231 f^{14/3} - 0.0105 f^{22/3} - 0.1526 f^6 - f + 1}, \tag{8.5.97}$$

which can be expanded at small volume fractions f,

$$\begin{aligned}
\sigma_{SA}(f) = {}& 1 + 3f + 3f^2 + 3f^3 + 3f^4 + 3.9141 f^{13/3} + 3f^5 + \\
& 7.8282 f^{16/3} + 0.21693 f^{17/3} + 3f^6 + 11.7423 f^{19/3} + \\
& 2.02064 f^{20/3} + 3.4578 f^7 + 15.6564 f^{22/3} + O(f^{23/3}).
\end{aligned} \tag{8.5.98}$$

Already a simple formula which respects (8.5.96) and the two non-trivial terms from

(8.5.98)

$$\sigma_0^*(f) = -\frac{1}{2}\pi \log \left(\frac{\frac{1.19317f}{\frac{\pi}{6}-f} + 0.529078}{\frac{2.27879f^2}{\left(\frac{\pi}{6}-f\right)^2} + \frac{3.2552f}{\frac{\pi}{6}-f} + 1} \right) \qquad (8.5.99)$$

is accurate with maximal error of 0.57%. When corrected with diagonal Padé approximants, first expressed in terms of the variable $y = f^{\frac{1}{3}}$, it becomes very accurate,

$$\sigma_{SC}^*(f) = \sigma_0^*(f)\frac{P(f)}{Q(f)}. \qquad (8.5.100)$$

where

$$P(f) = 1 - 0.635307 \sqrt[3]{f} - 1.72351 f^{2/3} + 3.89054 f - 3.83355 f^{4/3} - 4.82397 f^{5/3}$$
$$-0.445866 f^2 - 4.47204 f^{7/3} + 1.43886 f^{8/3} - 12.8392 f^3 + 1.23565 f^{10/3} + 8.64477 f^{11/3}$$
$$-4.43321 f^4 - 3.1461 f^{13/3} - 0.0914565 f^{14/3} - 6.0514 f^5 + 11.5362 f^{16/3} - 6.28424 f^{17/3}$$
$$-32.7335 f^6,$$

(8.5.101)

and

$$Q(f) = 1 - 0.635307 \sqrt[3]{f} - 1.72351 f^{2/3} + 3.89054 f - 3.83355 f^{4/3} - 4.82397 f^{5/3}$$
$$-0.445866 f^2 - 4.47204 f^{7/3} + 1.43886 f^{8/3} - 12.5157 f^3 + 1.03017 f^{10/3} + 8.08734 f^{11/3}$$
$$-1.14923 f^4 - 9.587 f^{13/3} - 2.65628 f^{14/3} + 4.53203 f^5 - 6.51228 f^{16/3} + 3.43114 f^{17/3}$$
$$-21.9652 f^6.$$

(8.5.102)

The maximum error incurred by (8.5.100) equals 0.072% as compared with numerical results of [13].

Final formula (8.5.100) is compared with (8.5.97) and with Maxwell formula $\sigma_e = \frac{1+2f}{1-f}$, as shown in Fig. 8.2.

5.2. Highly conducting inclusions. FCC

Batchelor and O'Brien [4] extended Keller's results to the high-concentration limit for the FCC arrays of an ideally conducting spherical inclusions

$$\sigma(f) \simeq -\sqrt{2}\pi \log(f_c - f),$$

where the threshold $f_c = \frac{\sqrt{2}\pi}{6} \approx 0.74048$. According to Kepler's conjecture [11] this is the highest threshold (density) that can be achieved by any arrangement of spheres, no matter regular or irregular. The very same highest threshold can also be achieved for hexagonal close-packed (HCP) lattice. The two lattices however differ in how the sheets of spheres are stacked upon one another. Mind that such maximum close packing value is significantly larger than empirical random close packing values about 0.64 making such packing attractive for applications.

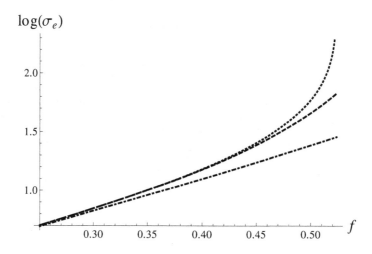

Figure 8.2 Different expressions for σ_e calculated from (8.5.100) (dotted line), (8.5.97) (dashed line) and from Maxwell formula (dot-dashed line).

The contact number per partice of HCP and FCC is 12. Only the triangular (in 2*D*), FCC and HCP lattices are all jammed in every sense defined in [26].

BOX A.8.9 Barlow packings

We remind the elementary distinctions between the FCC, the HCP, and the hybrid close-packed structures. All can be most transparently viewed as stacks of planar triangular arrays of spheres, within which each sphere contacts six neighbors. The triangular layers are stacked on one another, fitting spheres of one layer into "pockets" formed by nearest-neighbor triangles in the layer below. At each such layer addition there are two choices of which set of pockets in the layer below are to be filled. A lower layer with lateral position to be called A, is then surmounted with the next layer in lateral position B or C. A third layer subsequently can revert to lateral position A, or can be C on a second layer B, or B on second layer C. The FCC structure utilizes the repeating pattern ...*ABCABCABC*..., while the HCP case a periodic packing corresponds to ...*ABABABAB*.... Hybrid close-packed structures utilize other A,B,C patterns of lateral positions, never immediately repeating one of these three letters. Since there are two ways to place each layer after the second, there is an uncountable infinity of distinct packing schemes, all with the same density. These are called the Barlow packings and include random stacking variants, i.e., the two ways to place each layer after the second occur with equal probabilities. In the latter case, there is no repeating pattern, as exhibited by the following partial sequence: ...*ABACBACBCA*...

[3, 11, 25]

The higher-order constant correction to Batchelor and O'Brien expression

$$\sigma(f) \sim A \log(f_c - f) + B, \quad (8.5.103)$$

can be considered as well, and it is not small, $B \approx -3.55389$, as deduced from [1].

In order to get a low-concentration expression, we use Berdichevsky formula [5]

$$\sigma_B(f) = \frac{0.145 f^{13/3}}{(1-f)^2} + \frac{0.725 f^{17/3}}{(1-f)^2} + \frac{2f+1}{1-f}. \quad (8.5.104)$$

which can be expanded at small volume fractions f,

$$\sigma_B(f) = 1 + 3f + 3f^2 + 3f^3 + 3f^4 + 0.145 f^{13/3} + 3f^5 + 0.29 f^{16/3} + \\ 0.725 f^{17/3} + 3f^6 + 0.435 f^{19/3} + 1.45 f^{20/3} + 3f^7 + 0.58 f^{22/3} + O(f^{23/3}). \quad (8.5.105)$$

More terms can be deduced since the error of formula (8.5.104) is of $O(f^{17})$. We would like to obtain a formula which respects both (8.5.103) and a number of terms from (8.5.105).

Let us apply the following transformation,

$$z(f) = \frac{\sqrt[3]{\frac{f}{f_c}}}{1 - \sqrt[3]{\frac{f}{f_c}}} \Leftrightarrow f(z) = \frac{z^3 f_c}{(z+1)^3}, \quad (8.5.106)$$

designed to make functions depend only on powers of z. Let us look for the solution in the following form

$$\sigma^*(z) = -\sqrt{2}\pi \, \log[U_F(z)]. \quad (8.5.107)$$

where $U_F(z)$ will be approximated by the Padé approximants $P_{n,n+1}(z)$, which are able to match both asymptotic conditions. Among all approximants, we select the best and find corresponding amplitude B. The following approximant is selected,

$$U_F(z) = P_{10,11}(z) = \frac{P(z)}{Q(z)}, \quad (8.5.108)$$

where

$$P(z) = 6.31124 z^{10} + 17.5453 z^9 + 103.19 z^8 + 298.09 z^7 + 478.833 z^6 + 490.815 z^5 + \\ 340.628 z^4 + 161.38 z^3 + 50.4183 z^2 + 9.41336 z + 0.798453,$$
$$Q(z) = 1.27642 z^{11} + 10.1736 z^{10} + 40.6371 z^9 + 171.858 z^8 + 424.321 z^7 + 636.905 z^6 + \\ 631.595 z^5 + 431.005 z^4 + 202.615 z^3 + 63.145 z^2 + 11.7895 z + 1. \quad (8.5.109)$$

In original variables

$$\sigma^*_{FCC}(f) = -\sqrt{2}\pi \, \log[U_F(z(f))], \quad (8.5.110)$$

and the corresponding amplitude $B = -3.55484$ is very close to the known value from

(8.5.103). Final formula (8.5.110) is accurate with maximal error of 0.314% when compared with numerical results of [14].

5.3. Highly conducting inclusions. BCC

For the BCC lattice same treatment holds with minor differences, i.e., $f_c = \frac{\sqrt{3}\pi}{8}$, $A = -\frac{\sqrt{3}\pi}{2}$, and $B = -0.55$ according to [14]. A slightly larger value, $B = -0.46$ can be deduced from [1]. Both Berdichevksy [5]

$$\sigma_B(f) = \frac{0.387 f^{13/3}}{(1-f)^2} + \frac{0.779 f^{17/3}}{(1-f)^2} + \frac{2f+1}{1-f}, \tag{8.5.111}$$

and Sangani–Acrivos [23]

$$\sigma_{SA}(f) = 1 + \frac{3f}{-\frac{0.129\left(1-0.41286 f^{11/3}\right) f^{10/3}}{1-0.7642 f^{7/3}} - 0.2569 f^{14/3} - 0.00056 f^{22/3} - 0.013 f^6 - f + 1}, \tag{8.5.112}$$

variants, lead to a good convergence and very close results for the effective conductivity. We opt for the latter asymptotic formula to obtain the compact approximation for the conductivity,

$$\sigma^*_{BCC}(f) = -\frac{\sqrt{3}\pi}{2}\log[U_B(z(f))] \tag{8.5.113}$$

with

$$U_B(z) = P_{5,6}(z) = \frac{1.57958 z^5 + 11.7529 z^4 + 16.959 z^3 + 11.751 z^2 + 4.17905 z + 0.692427}{0.63718 z^6 + 5.92969 z^5 + 19.25 z^4 + 25.2422 z^3 + 16.9707 z^2 + 6.03536 z + 1}. \tag{8.5.114}$$

The corresponding amplitude $B = -0.53$ is close to the known values quoted above. The maximum error incurred by (8.5.113) equals 0.6%. This estimate is conservative since it is the last point, deep into the critical regime which brings the largest error and computations here became less reliable [14].

6. Non-conducting inclusions embedded in an conducting matrix. FCC lattice

In 3D, this case should be considered independently from the case of ideally conducting inclusions. Berdichevsky [5] gives the following expression for the conductivity which may be viewed as corrected Maxwell formula,

$$\sigma_B(f) = \frac{2(1-f)}{f+2} - \frac{0.10875 f^{13/3}}{(f+2)^2} - \frac{0.604167 f^{17/3}}{(f+2)^2}. \tag{8.6.115}$$

In the limit case of high concentration of the non-conduction inclusions our best guess is based on the analogy with the critical exponent t for the conductivity and corresponding critical index for the divergence of the effective viscosity of regular arrays [8].

Although a rigorous analogy does not hold for viscosity and conductivity, it seems to hold in practice, including 3D case [7], so that from the result on regular arrays, one concludes that

$$\sigma(f) \sim A(f_c - f), \tag{8.6.116}$$

with unknown amplitude A which has to be estimated in the course of calculations.

Let us construct the formula which combines the two limit-forms, matching (8.6.115) and (8.6.116). After applying the transformation (8.5.106), and using the $DLog$ Padé technique, we can express the effective conductivity

$$\sigma^*_{n,m}(f) = \exp\left(\int_0^{z(f)} P_{n,m}(Z)\,dZ\right).$$

With the following Padé approximants

$$P_{2,3}(z) = -\frac{3.33216z^2}{3.33216z^3 + 6z^2 + 4z + 1},$$

$$P_{4,5}(z) = -\frac{1.4506z^4 - 0.685493z^3 + 3.33216z^2}{1.4506z^5 + 2.49505z^4 + 4.13677z^3 + 5.61245z^2 + 3.79428z + 1}, \tag{8.6.117}$$

one can reconstruct conductivity with the correct limits.

$$\sigma^{*FCC}_{2,3}(f) = \frac{0.0239134\left(\frac{f^{2/3} + 0.67138\sqrt[3]{f} + 1.08624}{\left(0.9047 - \sqrt[3]{f}\right)^2}\right)^{1.14878} e^{-2.20407\tan^{-1}\left(1.25713 + \frac{1}{0.392771\sqrt[3]{f} - 0.35534}\right)}}{\left(\frac{1}{1 - 1.10534\sqrt[3]{f}} - 0.305953\right)^{3.29756}}. \tag{8.6.118}$$

Formula (8.6.118) leads to the value of $A = 1.245$. In the higher order, we get another analytical expression,

$$\sigma^{*FCC}_{4,5}(f) = 0.0076966\exp\left(0.0857123\tan^{-1}\left(0.805702 + \frac{1}{1.54184\sqrt[3]{f} - 1.3949}\right)\right.$$

$$-2.08149\tan^{-1}\left(1.28797 + \frac{1}{0.345837\sqrt[3]{f} - 0.312879}\right)\right)\frac{1}{\left(\frac{1}{1 - 1.10534\sqrt[3]{f}} - 0.226281\right)^{5.49575}}$$

$$\times\left(\frac{f^{2/3} - 1.17567\sqrt[3]{f} + 0.500218}{\left(0.9047 - \sqrt[3]{f}\right)^2}\right)^{0.0643366}\left(\frac{f^{2/3} + 0.991948\sqrt[3]{f} + 1.42866}{\left(0.9047 - \sqrt[3]{f}\right)^2}\right)^{2.18354}. \tag{8.6.119}$$

Formula (8.6.119) brings $A = 1.396$.

Let us check the Hashin–Shtrikman bounds, the upper bound $U(f)$ given by (3.1.1), and lower bound $L(f)$ given by (3.1.2), properly adapted from Chapter 3. In Fig. 8.3 various formulae are shown and all of them agree with the bounds. Although our formulae seems to be more in compliance with both bounds than original Berdichevsky formula. Our derived formulae are very close to each other so we won't invoke here or below (for SC and BCC) the higher-order approximations. It also seems that the

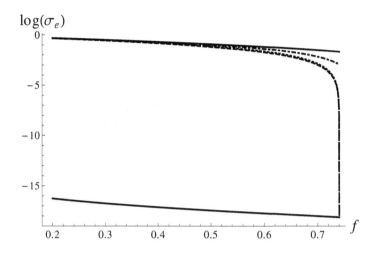

Figure 8.3 The Hashin–Shtrikman lower and upper bounds are shown with solid lines. Different expressions for σ_e calculated from (8.6.119) (dotted line), (8.6.118) (dashed line) and by (8.6.115) (dot-dashed line). Our derived formulae are very close to each other.

DLog technique of predicting the critical index gives reasonable estimates only in the low orders.

7. Non-conducting inclusions embedded in an conducting matrix. SC and BCC lattices

Consider the case of simple cubic lattice (SC). We follow below the same approach as above for the FCC lattice. Berdichevsky [5] gives the following expression for the conductivity of the SC lattice which is corrected Maxwell formula,

$$\sigma_{BSC}(f) = \frac{2(1-f)}{f+2} - \frac{2.93475 f^{13/3}}{(f+2)^2} - \frac{1.22417 f^{17/3}}{(f+2)^2}. \qquad (8.7.120)$$

In the limit case of high concentration of the non-conduction inclusions, we assume the power law for this lattice as well,

$$\sigma_e(f) \sim A(f_c - f),$$

with unknown amplitude A which has to be estimated in the course of calculations.

Let us construct the formula which combines the two limit-form, matching asymptotically the two formulae. Apply transformation (8.5.106), and use the *DLog* Padé technique. Then one can express the effective conductivity in the integral form just as above for FCC lattice. With the following Padé approximants

$$P_{2,3}(z) = -\frac{2.35619z^2}{2.35619z^3+6.z^2+4z+1},$$

$$P_{4,5}(z) = -\frac{0.437526z^4-0.819105z^3+2.35619z^2}{0.437526z^5+0.552808z^4+2.39513z^3+4.79514z^2+3.65236z+1},$$

(8.7.121)

one can reconstruct conductivity with the correct limits.

$$\sigma_{2,3}^{*SC}(f) = \frac{6.73238\left(\frac{f^{2/3}+0.720584\sqrt[3]{f}+0.412311}{\left(0.805996-\sqrt[3]{f}\right)^2}\right)^{0.359356} e^{0.200659\tan^{-1}\left(2.1943+\frac{1}{0.323552\sqrt[3]{f}-0.260782}\right)}}{\left(\frac{1}{0.690945-0.857256\sqrt[3]{f}}+1\right)^{1.71871}}.$$

(8.7.122)

Formula (8.7.122) leads to the following value of amplitude $A = 2.312$. In the higher order, we get another analytical expression,

$$\sigma_{4,5}^{*SC}(f) = \frac{0.00491081\left(\frac{1.x^{2/3}-1.30791\sqrt[3]{x}+0.491325}{\left(0.805996-1.\sqrt[3]{x}\right)^2}\right)^{0.115842}\left(\frac{1.x^{2/3}+1.02508\sqrt[3]{x}+1.46425}{\left(0.805996-1.\sqrt[3]{x}\right)^2}\right)^{2.34887}}{\left(\frac{1}{1-1.2407\sqrt[3]{x}}-0.189417\right)^{5.92942}} \times$$
$$\exp\left[-1.98237\tan^{-1}\left(1.20287+\frac{1}{0.372831\sqrt[3]{x}-0.3005}\right)\right.$$
$$\left.-0.272172\tan^{-1}\left(0.602563+\frac{1}{2.90754\sqrt[3]{x}-2.34347}\right)\right].$$

(8.7.123)

Formula (8.7.123) brings $A = 2.964$. In Fig. 8.4 various formulae are shown and all of them agree with the Hashin–Shtrikman bounds. Although our formulae seems to be more in agreement with both bounds than original Berdichevsky formula.

In complete compliance with the treatment of the SC lattice and the the Hashin–Shtrikman bounds, we present below results for the BCC lattice,

$$\sigma_{2,3}^{*BCC}(f) = \frac{0.627044\left(\frac{f^{2/3}+0.875522\sqrt[3]{f}+0.823235}{\left(0.879441-\sqrt[3]{f}\right)^2}\right)^{1.12756} e^{-0.183753\tan^{-1}\left(1.65742+\frac{1}{0.335809\sqrt[3]{f}-0.295324}\right)}}{\left(\frac{1}{1-1.13709\sqrt[3]{f}}-0.06077\right)^{3.25513}}.$$

(8.7.124)

Formula (8.7.124) leads to $A = 1.447$. In the higher order, we get another analytical expression,

$$\sigma_{4,5}^{*BCC}(f) = 0.00491081\exp\left[-1.98237\tan^{-1}\left(1.20287+\frac{1}{0.341694\sqrt[3]{f}-0.3005}\right)\right.$$
$$\left.-0.272172\tan^{-1}\left(0.602563+\frac{1}{2.66472\sqrt[3]{f}-2.34347}\right)\right]\left(\frac{1}{1-1.13709\sqrt[3]{f}}-0.189417\right)^{-5.92942} \times$$
$$\left(\frac{f^{2/3}-1.4271\sqrt[3]{f}+0.584948}{\left(0.879441-\sqrt[3]{f}\right)^2}\right)^{0.115842}\left(\frac{f^{2/3}+1.11849\sqrt[3]{f}+1.74327}{\left(0.879441-\sqrt[3]{f}\right)^2}\right)^{2.34887}.$$

(8.7.125)

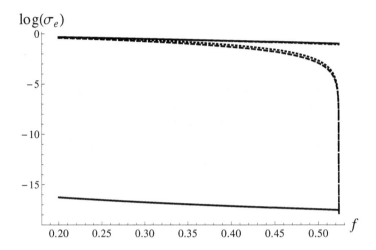

Figure 8.4 The Hashin–Shtrikman lower and upper bounds are shown with solid lines. Different expressions for σ_e calculated from (8.7.123) (dotted line), (8.7.122) (dashed line) and by (8.6.115) (dot-dashed line). Out derived formulae are close to each other. Berdichevsky formula is very close to the upper bound.

Formula (8.7.125) brings $A = 2.282$.

In fact Berdichevsky [5] suggested most general asymptotic formula for the effective conductivity of the three lattices described above. We considered only the two limiting high-contrast cases which are analogous with the critical phenomena. Interesting enough, Berdichevsky approach after resummation works better for FCC than SC.

REFERENCE

1. Andrianov I, Danishevs'kyy V, Tokarzewski S, Quasifractional approximants for effective conductivity of regular array of spheres, Arch.Mech. 2000; 2: 319–327.
2. Axler Sh, Bourdon P, Ramey W, Harmonic Function Theory. 2nd Berlin: Springer Verlag; 2001.
3. Barlow W, Probable Nature of the Internal Symmetry of Crystals, Nature 1883; 29: 186–188
4. Batchelor GK, O'Brien RW, Thermal or Electrical Conduction Through a Granular Material, Proceedings of the Royal Society of London A 1977; 355: 313–333.
5. Berdichevsky VL, Variational Principles of Continuum Mechanics. Moscow: Nauka; 1983 [in Russian].
6. Berlyand L, Kolpakov AG, Novikov A, Introduction to the Network Approximation Method for Materials Modeling. Cambridge: Cambridge University Press; 2012.
7. Bicerano J, Douglas JF, Brune DA, Model for the Viscosity of Particle Dispersions, Polymer Reviews 1999; 39: 561–642.
8. Frankel NA, Acrivos A, On the Viscosity of a Concentrated Suspension of Solid Spheres, Chemical Eng. Science 1967; 22: 847–853.
9. Keller JB, Conductivity of a Medium Containing a Dense Array of Perfectly Conducting Spheres or Cylinders or Nonconducting Cylinders, Journal of Applied Physics 1963; 34: 991–993.
10. Kolpakov AA, Kolpakov AG, Capacity and Transport in Contrast Composite Structures: Asymp-

totic Analysis and Applications. Boca Raton etc: CRC Press Inc.; 2009.

11. Lagarias, JC, The Kepler Conjecture: The Hales-Ferguson Proof. New York etc: Springer; 2011.

12. Lanzani L, Shen Z, On the Robin Boundary Condition for Laplace's Equation in Lipschitz Domains, Communications in Partial Differential Equations, 2004; 29: 91–109.

13. McPhedran RC, McKenzie DR, The conductivity of lattices of spheres I. The simple cubic lattice, Proceedings of the Royal Society of London A 1978; 359: 45–63.

14. McKenzie DR, McPhedran RC, Derrick GH, The conductivity of lattices of spheres II. The body centred and face centred cubic lattices, Proceedings of the Royal Society of London A 1978; 362: 211–232.

15. Mikhlin SG, Integral Equations and Their Applications to Certain Problems in Mechanics, Mathematical Physics and Technology. 2nd New York: Macmillan; 1964.

16. Mityushev V, Generalized method of Schwarz and addition theorems in mechanics of materials containing cavities, Arch. Mech. 1995; 47: 1169–1181.

17. Mityushev VV, Rogosin SV, Constructive methods to linear and non-linear boundary value problems of the analytic function. Theory and applications. Boca Raton etc: Chapman & Hall / CRC; 1999/2000. [Chapter 4].

18. Mityushev V, Transport properties of doubly periodic arrays of circular cylinders and optimal design problems, Appl. Math. Optimization 2001; 44: 17–31.

19. Mityushev V, Rylko N, Boundary value problems, the Poincare series, the method of Schwarz and composite materials, Int. Congres IMACS 97; Berlin. 1; 165–170: 1997.

20. Perrins WT, McKenzie DR, McPhedran RC, Transport properties of regular array of cylinders, Proc. R. Soc.A 1979; 369, 207–225.

21. Rayleigh Lord, On the influence of obstacles arranged in rectangular order upon the properties of medium, Phil. Mag. 1892; 34, 481–502.

22. Riley KF, Hobson MP, Bence SJ, Mathematical Methods for Physics and Engineering. 3d Cambridge: Cambridge University Press; 2006.

23. Sangani AS, Acrivos A, The effective conductivity of a periodic array of spheres, Proceedings of the Royal Society of London A 1983; 386: 263–275.

24. Sretensky LN, Theory of Newtonian Potential. Moscow: Ogiz-Gostekhizdat,; 1946. [in Russian].

25. Torquato S, Stillinger FH, Toward the jamming threshold of sphere packings: Tunneled crystals, Journal Of Applied Physics 2007; 102: 093511.

26. Torquato S, Stillinger FH, Jammed hard-particle packings: From Kepler to Bernal and beyond, Reviews of Modern Physics 2010; 82: 2634–2672.

CHAPTER 9

Random 2D composites

I had suspected for some time now that the Cosmic Command, obviously no longer able to supervise every assignment on an individual basis when there were literally trillions of matters in its charge, had switched over to a random system. The assumption would be that every document, circulating endlessly from desk to desk, must eventually hit upon the right one. A time-consuming procedure, perhaps, but one that would never fail. The Universe itself operated on the same principle. And for an institution as everlasting as the Universe certainly our Building was such an institution the speed at which these meanderings and perturbations took place was of no consequence.

—Stanislaw Lem, Memoirs found in a bathtub

1. Critical properties of an ideally conducting composite materials

This chapter is about finding the effective conductivity of 2D random composites. Two-component composites made from a collection of non-overlapping, identical, ideally conducting circular disks, embedded randomly in an otherwise uniform locally isotropic host (see Fig. 9.1) are considered. The effective conductivity is obtained in the form of series (4.2.26) on page 86 in the concentration f. This series will be the key input to solve the next problems:

1. what quantity should stand for the maximum volume fraction f_c of random composites, and

2. theoretical explanation of the values of critical indices for conductivity and superconductivity denoted by t and s, respectively [77].

The problem of defining the threshold is highly non-trivial, since the random closest packing of hard spheres turned out to be ill-defined, and cannot stand for the maximum volume fraction. It depends on the protocol employed to produce the random packing [78] as well as other system characteristics which can be determined in terms of the optimal graphs [66].

Computational Analysis of Structured Media
http://dx.doi.org/10.1016/B978-0-12-811046-1.50009-5

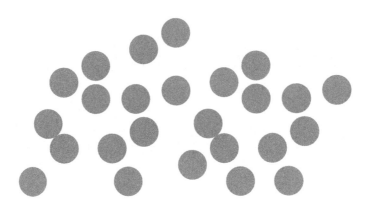

Figure 9.1 Randomly distributed disks.

The problem seems less acute in two dimensions, where various protocols seems to agree on what quantity should stand for the maximum volume fraction of random composites [20, 21, 45, 51, 70]. Namely it is the concentration of $\frac{\pi}{\sqrt{12}} \approx 0.9069$, attained for the regular hexagonal array of disks. The sought value for a long time was thought to be close to 0.82, and considered as random close packing value [15].

It was recognized recently, that it does not correspond to the maximally random jammed state [78]. For volume fractions above 0.82 some local order is present and irregular packing is polycrystalline, forming rather large triangular coordination domains/grains. We apply protocols leading to $f_c = \frac{\pi}{\sqrt{12}}$, although the method can be applied with another protocol with unknown f_c.

All attempts to explain the value of critical indices through geometrical quantities of percolation problem, i.e., universally [19, 77], had failed so far and the indices were considered independent. From the phase interchange theorem [77] it follows that in two dimensions, the superconductivity index is equal to the conductivity index [77].

While it is clear that using expansions in concentration for the conductivity, one should be able to address the two problems, in practice no more than two terms (with necessarily correct values) were available for random systems [19]. No method even such powerful as renormalization, or resummation approaches can draw reliable conclusions systemically, based on such short series [19]. Thus, Choy [19] pointed out that the age-old method of series expansions remained blocked.

This concerns also the whole family of self-consistent methods (SCM) which include Maxwell's approach, effective medium approximations, differential schemes, etc. Bruggeman (see, e.g., [41]), suggested a self-consistent formalism leading to the family of effective medium approximations (EMA) [11, 53, 54, 74, 81]. Popular EMA could be applied to various mixtures [29], even include quantum tunneling in the most intriguing case of nanocomposites [2, 3]. SCMs are valid only for a dilute composites when interactions between inclusions do not matter [62]. The idea to correct a self-consistent method (SCM) result t = s = 1 (in all dimensions) using the series in concentration remained, therefore, practically unattainable (see nevertheless, [82]). An indirect approach to estimating t for resistor networks from resistive susceptibility via scaling relations was suggested in [1]. This approach is based on resummation techniques.

The principal difference between the previous and our approaches lies in the way of introducing randomness only at the last stage of computations (see Chapter 3. For more details see the discussion on page 55). The deterministic problem is reduced to functional equations. In turn from the functional equations the expansions in concentration are obtained in analytical form. Randomness is introduced into the analytical coefficients of such an expansion expressed through the deterministic variables, centers of the inclusions a_m. They are now required to become the random variables. That is, in every formula with a symbolically written a_m, one can safely assume when needed that a_m is a random variable and perform averaging when required.

1.1. Uniform non-overlapping distribution

In order to correctly define the effective conductivity tensor σ_e of random composites, the probabilistic distribution of disks of radius r must be introduced, since already the second-order term of σ_e in concentration depends on the distribution of particles [62]. For macroscopically isotropic composites, already the third-order term begins to depend on the distribution [62]. Consider a uniform non-overlapping distribution when a set of independent and identically distributed points $\mathbf{a}_i$ are located in the plane in such a way that $|\mathbf{a}_i - \mathbf{a}_j| \geq 2r$.

For $r = 0$, it simply becomes a Poisson distribution, and for the maximally possible concentration $f_c = \frac{\pi}{\sqrt{12}}$, the distribution degenerates to the unique location, the hexagonal array. The tensor σ is expressed through the scalar effective conductivity σ as follows $\sigma = \sigma \mathbf{I}$, where $\mathbf{I}$ is the unit tensor.

Mind that we study conductivity of the doubly periodic composite when the host and the inclusions are occupied by materials of conductivity 1 and σ, respectively. Consider sufficiently large number of non-overlapping circular disks of radius r with the centers $\mathbf{a}_k$. The centers of disks are considered as random variables distributed in such a way to generate a set of uniformly distributed non-overlapping disks. Accordingly, the concentration of inclusions $0 \leq f \leq \frac{\pi}{\sqrt{12}} \approx 0.9069$, is the maximal concentration attained for the hexagonal array.

The numerical computations are performed for the hexagonal representative cell $Q_{(0,0)}$. This assumption does not restrict our investigation since the number of inclusions per cell can be taken arbitrary large, hence, the shape of the cell does not have a significant impact on the final result.

The formally defined random variable has to be statistically realized to get numerical results. The main protocol for the data generation is **Random Walks (RW)** based on the Monte Carlo simulations [20, 21]. It can be described in very general terms as follows. At the beginning, the centers $\mathbf{a}_k$ are located at the nodes of the regular hexagonal array $Q_{(0,0)}$ and further randomly moved without overlapping. After sufficiently long random experiment (see for details Section 2 below), the centers form a statistical event satisfying the considered distribution, i.e., a doubly periodic set of uniformly distributed non-overlapping disks is generated. Using these locations of disks the coefficients of σ in f are computed many times and the average is taken.

The second protocol known as **Random Sequential Addition (RSA)** [80] can be used for low concentrations. First random disk is randomly chosen in $Q_{(0,0)}$. More precisely, a point $\mathbf{a}_1$ is randomly chosen in $Q_{(0,0)}$ in accordance with the uniform distribution. This is the center of the first disk $\mathbb{D}_1$. Second point $\mathbf{a}_2$, to be the center of the second disk, is randomly chosen in $Q \backslash \mathbb{D}_1$. Hence, the distribution of the second random point is conditional and depends on the first random point. Continue in this way producing more points $\mathbf{a}_i$, up to some number $i = N$, conditioned that circular regions around all previous points are excluded from Q. This joint random variable for all points correctly determines a probabilistic distribution.

The computer simulations with RSA work only up to concentrations as high as 0.5773, hence is the main RSA limitation in 2D. To overcome the limitation and to penetrate the region of larger concentrations, one has to apply some extrapolation technique.

The effective conductivity tensor of 2D composites with non-overlapping identical disks can be expanded in the concentration

$$\sigma_e(f) = 1 + 2\rho f (1 + A_1 f + A_2 f^2 + \cdots), \tag{9.1.1}$$

where the coefficients A_i up to the sixth order inclusively are given by (4.2.27) from Chapter 3, as the functions of e-sums. The higher-order coefficients A_n can also be written in a closed form. One can find the corresponding polynomials in supplementary to the paper [37].

Consider first the case of a high contrast regular composites, when the conductivity of the inclusions is much larger than the conductivity of the host. That is, the highly conducting inclusions are replaced by the ideally conducting inclusions with infinite conductivity. Below, this expansion is presented in the truncated numerical form by use of the RSA protocol

$$\sigma_e(f) = \sigma^{RSA} = 1 + 2f + 2f^2 + 5.00392 f^3 + 6.3495 f^4 + O(f^9). \tag{9.1.2}$$

The coefficients on f^k ($k = 5, 6, 7, 8$) vanish in (9.1.2) with the precision 10^{-10}. The higher-order terms, or remainder $\Delta(f)$ in (9.1.2), can be specified as well

$$\Delta(f) = 0.0000186711 f^9 + 9.57157 \ 10^{-10} f^{10} + 0.0570669 f^{14} + \\ 27.2148 f^{15} + 7.06377 f^{16} + 1.63666 \ 10^{-6} f^{17}. \tag{9.1.3}$$

It has highly irregular form and does not contribute to the critical properties or general expressions for the conductivity.

In the limiting case of a perfectly conducting inclusions when the conductivity of inclusions tends to infinity, the effective conductivity is also expected to tend to infinity as a power law, as the concentration f tends to the maximal value f_c for the hexagonal array,

$$\sigma_e(f) \simeq A(f_c - f)^{-s}. \tag{9.1.4}$$

The critical superconductivity index (exponent) s is believed to be close to $\frac{4}{3} \approx 1.3$ [26, 44, 68, 75, 77]. This value is known from numerical simulations, while rigorously it can be anywhere between one and two [39]. The critical amplitude A is an unknown non-universal parameter. Various experiments confirm the value of $s = t \approx 1.3$ [10, 18, 49, 75].

In the general case of continuum percolation, the value of the critical index can be much larger than 1.3 [5, 7, 25, 67], in agreement with experiments [54]. It is believed that continuum percolation problems can be mapped into the lattice problems with conducting bonds whose conductivity is drawn from a probability density law [5, 7, 25, 40, 67]. Origin of the continuum percolation goes back to modeling of the communication networks by Gilbert [35]. Most intriguingly, it was suggested in [22] to study a good old conductivity problem for an interdependent networks, treating them as a physical system.

For regular arrays of cylinders (see Chapters 6–7) the index is much smaller, $s = \frac{1}{2}$ [47, 53, 54, 55, 57, 69]. The critical amplitude A is also known with good accuracy for square and hexagonal arrays [38, 47, 55, 57, 69].

Overall effective conductivity of random systems is expected to be higher than in regular systems by order(s) of magnitude as the threshold is approached [14].

Expression (9.1.2) is different from the regular case of superconducting inclusions forming a hexagonal array with the same value of threshold f_c [69, 77]. The power series in the case of hexagonal array [38] is characterized by rather regular behaviour of the coefficients,

$$\sigma_{reg}(f) = 1 + 2f + 2f^2 + 2f^3 + 2f^4 + 2f^5 + 2f^6 + 2.15084 f^7 + 2.30169 f^8 + O(f^9). \tag{9.1.5}$$

Comparing (9.1.2) and (9.1.5), one can see how and where the random effects captured by our procedure contribute most.

Randomness does seem to make the third- and fourth-order coefficients in the series much larger, signifying a stronger coupling between the inclusions already at small concentrations. The starting two (non-trivial) terms are common for the series (9.1.2) and (9.1.5). They are exact and typical also for all effective medium type approximations [4]. One can also anticipate that some different methods will have to be employed for the random composite, compared with regular case considered in previous chapters.

The problem of interest can be formulated mathematically as follows. Given the polynomial approximation (9.1.2) of the function $\sigma_e(f)$, to estimate the convergence radius f_c of the Taylor series of $\sigma_e(f)$, and to determine parameters of the asymptotically equivalent approximation (9.1.4) near $f = f_c$. When such extrapolation problem is solved, one can derive the formula for all concentrations, i.e., solve an interpolation problem. Alternatively in the latter problem, one may assume that the critical behaviour is known in advance and proceed with interpolation from the start.

1.2. Critical point f_c

Start with the technically simple way to estimate the position of a critical point, through the Padé approximants, $P_{n,m}(f)$. Their application does not require any preliminary knowledge of the critical index and identifies the critical point with the position of a simple pole. But in the case of series (9.1.2), the direct application of Padé approximants leads to a poorly convergent results in the diagonal sequence $P_{n,n}$, with the best estimate for the threshold 0.85269, obtained form the approximant $P_{3,3}$. The results do not improve significantly if other sequences are considered. The problem can be attributed to the relative shortage of coefficients in (9.1.2), and to the triviality of the lower-order terms.

In order to compensate for the unchanging values of the coefficients in the starting orders, also a transformed sequences of approximants $P_{n,n}$ could be considered. The most significant improvement is achieved when the original series (9.1.2) are multiplied by the Clausius–Mossotti function $\sigma_{CM}(f) = \frac{1+f}{1-f}$, applying then a diagonal Padé approximants. There is now a better estimate for the critical point, $f_4 = 0.914241$. The percentage error obtained from the approximant $P_{4,4}$, equals 0.8095%. Note, that in order to obtain higher-order approximants, the Padé technique will routinely use trivial, zero values for the missing higher-order coefficients in the expansion (9.1.2). Only a minimal number of such trivial conditions allowing for a reasonable estimates are utilized. The same will apply to calculations with different approximants below.

The estimates does not improve if different types of approximants, other than Padé, are applied and some special efforts are in order to reconstruct the threshold from the series (9.1.2) more accurately. Assuming that f_c is known approximately from the Padé estimates above, let us estimate an improved, or corrected value of threshold, employing general idea of corrected approximants [32].

The technique of corrected approximants allows to treat different groups of coefficients in expansion differently and separately if needed. The solution can be patched together from several different sequences of approximants, not necessarily of the same type. This program will be realized below both for the threshold and critical index. Such approach is potentially more flexible and powerful then "classical" technique employing just one type of approximants. In the this case, one would expect to reach an improvement in the sought quantity by increasing number of terms in the asymptotic expression being involved in the approximant construction (accuracy-through-order).

For the series (9.1.2), it seems natural to treat the two starting terms separately and the rest of the series consider as a correction. Assume also that the initial threshold value is available from previous Padé estimates, and is equal to $f_4 = 0.914241$. Of course, formulation of the initial approximation is a crucial step. Note that using the value of 0.85269 as initial guess will lead to wrong results.

Let us attempt to correct the value of f_4. Also, the value of the critical index $s = \frac{4}{3}$ will be incorporated into the initial approximation. Let us observe that the factorized approximations to the effective conductivity σ can be always represented as a product of two factors: critical part

$$C(f) = \left(1 - \frac{f}{f_c}\right)^{-s}$$

and of the rest, i.e., regular part $R(f)$. As shown by Fuchs, such factorization also holds for non-analytic solutions to the homogeneous linear differential equations [9]. So one can most generally express the threshold

$$f_c = \frac{fC^{\frac{1}{s}}(f)}{C^{\frac{1}{s}}(f) - 1}. \tag{9.1.6}$$

The subsequent steps are described below for the particular case, but without any loss of generality. At first the solution is obtained explicitly as a factor approximant [31, 84]. The simplest factor approximant

$$\sigma_e(f) = \frac{(2.94539f + 1)^{0.183879}}{(1 - 1.0938f)^{\frac{4}{3}}}. \tag{9.1.7}$$

Such approximant satisfy the two non-trivial starting terms from the series (9.1.2), and incorporates the accepted value of $\frac{4}{3}$ for the critical index and the trial value of threshold f_4. Let us look for another solution, supposedly in the same form, but with an exact, yet unknown threshold F_c,

$$\sigma'(f) = (2.94539f + 1)^{0.183879}\left(1 - \frac{f}{F_c}\right)^{-\frac{4}{3}}. \tag{9.1.8}$$

Such solution retains the regular part $R(f) = (2.94539f + 1)^{0.183879}$ from the factor approximant (9.1.7). One can express the new threshold formally just as above,

$$F_c = \frac{f\left(\frac{\sigma'}{R(f)}\right)^{1/s}}{\left(\frac{\sigma'}{R(f)}\right)^{1/s} - 1}, \tag{9.1.9}$$

since $\sigma'(f)$ is also unknown. To make it practical there is only one way, to use for σ' a concrete approximation, namely the series (9.1.2). Instead of a true threshold which is a number, there appears an effective threshold $F_c(f)$, given as an expansion around an approximate threshold $f_0 = f_4$,

$$F_c(f) = 0.914241 - 0.954606f^2 + 0.213995f^3 + \cdots \tag{9.1.10}$$

which should become a number corresponding to the true threshold f_c, as $f \to f_c$. Moreover, let us apply re-summation procedure to the expansion (9.1.10) using again factor approximants $\Phi^*(f)$. Then one can define the sought approximate threshold F_c^* self-consistently,

$$F_c^* = f_0 - 0.954606f^2\Phi^*(F_c^*). \tag{9.1.11}$$

As the threshold is approached, the RHS of (9.1.11) should become the threshold. Since factor approximants are defined as Φ_k^* for arbitrary number of terms k, there appears a sequence of $F_{c,k}^*$. Reasonable solution arises in the six order,

$$\Phi_6^*(f) = (4.75828f + 1)^{0.601565}(1 - (0.548713 - 0.738153i)f)^{-2.31965+3.81508i} \times$$
$$(1 - (0.548713 + 0.738153i)f)^{-2.31965-3.81508i}.$$
$$\tag{9.1.12}$$

Expression (9.1.12) matches (9.1.10) up to the eighth-order terms included. Solving (9.1.11), we obtain $F_{c,6}^* = 0.910181$. In the next even order there is no improvement.

The percentage error of such estimate is 0.3618%. Significant part of the correct threshold value is extracted at the first step, producing reasonable estimate f_0 and then, is corrected a bit using higher orders terms from the series. Mind that the threshold is a purely geometrical quantity, ideally not very sensitive therefore to the value of critical index employed in the calculations.

The technique of corrected threshold can be successfully applied in the case of a regular composite [30], with different way to define the initial approximation to be corrected, see Chapters 6 and 7. It was also tested with good results for the high-temperature series for the 2D Ising model [43]. In those cases, the corresponding series coefficients behave (increase) regularly and standard methods also work well.

1.3. Critical Index s

Conventionally when the threshold is already known, one would first apply the following transformation,

$$z = \frac{f}{f_c - f} \quad \Leftrightarrow \quad f = \frac{z f_c}{z + 1} \tag{9.1.13}$$

to the original series. The transformation maps the segment to a half-line. The transformation (9.1.13) serves to incorporate the information on the threshold to the original series. It also makes it more convenient to calculate with different approximants described in Chapter 5. The most straightforward way to estimate index s is to apply factor approximants [31, 84] (in terms of the variable z), so that possible corrections to the "mean-field" value unity, appear additively, by definition of the factor approximants. Following the standard procedure, the simple factor approximant is written as follows,

$$\sigma_3^* = 1 + b_1 z (b_2 z + 1)^{c_2},$$

where

$$c_2 = -0.01357, \quad b_1 = 1.8138, \quad b_2 = 26.6303,$$

and the critical index is simply $1 + c_2 = 0.996504$.

Already in the next order the value of critical index improves significantly,

$$\sigma_5^* = 1 + \frac{b_1 f \left(\frac{b_3 f}{f_c - f} + 1 \right)^{c_3} \left(\frac{b_2 f}{f_c - f} + 1 \right)^{c_2}}{f_c - f}, \tag{9.1.14}$$

and

$$b_1 = 1.8138, \quad b_2 = 1.33199 + 2.39138i, \quad b_3 = 1.33199 - 2.39138i,$$

$$c_2 = 0.148893 + 0.102399i, \quad c_3 = 0.148893 - 0.102399i.$$

The critical index s is equal to $1 + c_2 + c_3 = 1.297786$. The expression for critical index may be understood as bringing an additive correction to the mean-field value of 1. Such estimate and corresponding expression (9.1.14) for the effective conductivity are already good. To confirm them the different methods are applied below.

Let us see the results of a standard approach to the critical index calculation [6]. To this end, let us again express the original series in terms of z, and to such transformed series $L_1(z)$ apply the *DLog* transformation [6, 33] (differentiate log of $L_1(z)$), and call the transformed series $L(z)$. Applying the Padé approximants $P_{n,n+1}(z)$ to $L(z)$. one can readily obtain the sequence of approximations s_n for the critical index s,

$$s_n = \lim_{z \to \infty} (z P_{n,n+1}(z)). \tag{9.1.15}$$

In the case of (9.1.2), this method turns to be quite accurate. Namely, the best (and only) result is $s_2 = 1.28522$. One can also find corresponding approximant explicitly,

$$P_{2,3}(z) = \frac{15.7323z^2 + 3.89669z + 1.8137993642342}{12.241z^3 + 9.442z^2 + 4.14836z + 1}.$$

(9.1.16)

The effective conductivity can be reconstructed [27], from an effective critical index (also called β-function) approximated in our case by the approximant $zP_{2,3}(z)$,

$$\sigma^*(f) = \exp\left(\int_0^{\frac{f}{f_c - f}} P_{2,3}(z)\, dz\right).$$

(9.1.17)

The integral can be calculated in closed form,

$$\sigma^*(f) = e^{1.25174 - 0.436689\tan^{-1}\left(\frac{2.05578f + 0.389134}{0.9069 - f}\right)} \times$$
$$\left(\frac{0.574015f + 0.386326}{0.9069 - f}\right)^{1.08699} \left(\frac{f^2 - 0.0409192f + 0.186347}{(0.9069 - f)^2}\right)^{0.0991144}.$$

(9.1.18)

Also, the critical amplitude evaluates as 1.57888. Equation (9.1.18) will be compared below with other formula for the effective conductivity valid everywhere. Let us look for the solution initially in the form of a simple pole, also satisfying the two starting non-trivial terms from the series (9.1.2),

$$\sigma_0^*(f) = \frac{(1.35495f + 1)^{0.662269}}{1 - 1.10266f},$$

(9.1.19)

so that our zero approximation with $s^{(0)} = 1$ for the critical index, is typical for various SCMs.

Let us divide then the original series (9.1.2) by $\sigma_0^*(f)$, express the newly found series in term of variable z, then apply *DLog* transformation and call the transformed series $L(z)$. Applying now the Padé approximants $P_{n,n+1}(z)$, one can obtain the following sequence of corrected SCM approximations for the critical index,

$$s_n = s^{(0)} + \lim_{z \to \infty}(z\, P_{n,n+1}(z)).$$

(9.1.20)

The "corrected" values for the critical index can be calculated readily and there is now a unique reasonable estimate, $s_2 = 1.37959$. Effective conductivity can be reconstructed using the complete expression for the effective critical index, employing the approximant $P_{2,3}(z)$,

$$P_{2,3}(z) = \frac{5.98481z^2}{15.7666z^3 + 11.1117z^2 + 4.66455z + 1},$$

(9.1.21)

and

$$\sigma^*(f) = \sigma_0^*(f) \exp\left(\int_0^{\frac{f}{f_c - f}} P_{2,3}(z))\, dz\right).$$

(9.1.22)

The integral can be found analytically, so that $\sigma^*(f) = \sigma_0^*(f)\sigma_e(f)$, with

$$\sigma_e(f) = 1.68244 e^{0.388802\,\tan^{-1}\left(2.19461 + \frac{2.39204}{f - 0.9069}\right)} \times$$
$$\left(\frac{0.9069}{0.9069 - f} + 0.631139\right)^{0.280524} \left(\frac{f(f - 0.00867304) + 0.169156}{(0.9069 - f)^2}\right)^{0.049532}. \qquad (9.1.23)$$

Another way to calculate the index becomes clear from the following considerations. Let one think about a generalization of (9.1.2), i.e., the transition formula from the regular hexagonal array to the random array (9.1.2). We expect to obtain a dependence of the critical index on the degree of randomness. For "zero"-randomness the formula should lead to the regular hexagonal array. For "maximum"-randomness it has to lead to the random composite, the case discussed in the section. All cases with intermediate degrees of randomness are expected to fall in between the two cases. Accordingly, one should be able to describe the regular and random composites within a single formalism.

To this end, let us select the initial approximation to be corrected, as describing a regular hexagonal array of inclusions, namely

$$\sigma_{0,r}^*(f) = \frac{(0.419645 f + 1)^{3.45214}}{\sqrt{1 - 1.10266 f}}. \qquad (9.1.24)$$

This formula incorporates the critical index $1/2$ of the regular hexagonal array, the threshold for the hexagonal array and the two starting, effective medium terms from the series (9.1.5), (9.1.2). Let us divide the original series (9.1.2) by $\sigma_{0,r}^*$, thus extracting the part corresponding to the random effects only. Then express the newly found series in term of variable z, then apply $DLog$ transformation and call the transformed series $L_r(z)$. Let us also process the transformed series

$$L_r(z) = 7.52332 z^2 - 35.008 z^3 + 86.1167 z^4 - 141.937 z^5 + \cdots \qquad (9.1.25)$$

with different approximants, such as iterated roots [32]. One can obtain the following sequence of corrected approximations to the critical index,

$$s_n = \frac{1}{2} + \lim_{z \to \infty}(z\, R_n^*(z)), \qquad (9.1.26)$$

where $R_n^*(z)$ stands for the iterated root of nth order [32], constructed for the series $L_r(z)$ with such a power at infinity that defines constant correction to $s^{(0)}$. Calculations with iterated roots are really easy since at each step only one new coefficient is computed, while keeping all preceding from previous steps. The power at infinity is selected in order to compensate for the factor z and extract the correction to the regular lattice value. Namely,

$$R_1^*(z) = \frac{7.52332 z^2}{(1.55109 z + 1)^3}, \quad R_2^*(z) = \frac{7.52332 z^2}{\left(1.99241 z^2 + (1.55109 z + 1)^2\right)^{3/2}}. \qquad (9.1.27)$$

Correspondingly,

$$\sigma^*(f) = \sigma_{0,r}^*(f) \exp\left(\int_0^{\frac{f}{f_c-f}} R_2^*(z)\, dz \right). \tag{9.1.28}$$

The critical index s is approximated by $s_2 = 1.31561$. Then, (9.1.28) becomes

$$\boxed{\begin{array}{l} \sigma^*(f) = 0.121708\sigma_{0,r}^*(f)\times \\[4pt] \exp\left(\dfrac{(0.64454f-1.38151)f+0.72278}{(f-0.9069)^2\sqrt{\frac{f(f+0.435329)+0.3582}{(f-0.9069)^2}}} - 0.815613\sinh^{-1}\left(\dfrac{2.0171(f+0.494058)}{f-0.9069}\right) \right) \end{array}} \tag{9.1.29}$$

There are indications that physics of a 2D regular and irregular composites is related to the so-called "necks", certain areas between closely spaced disks where the electric current is supposed to occur entirely [14, 47]. Randomness eases the necks formation.

1.4. Final formula for all concentrations

Assume now that the critical index and threshold are both known and derive the formula valid for all concentrations. Such program is less ambitious but still entails calculation of the critical amplitude not known in advance. Let us discuss briefly some formulae for the effective conductivity from [4, 21] valid for all concentrations. The first formula (Eq. (22), [21]) is nothing else but an improved Padé approximant conditioned by appearance of a simple pole at f_c, [21],

$$\sigma_M(f) = \frac{0.014f + 0.001}{f^2 + 0.261f + 0.076} + \frac{3.223}{f - 1.247} - \frac{3.237}{f - 0.9069}. \tag{9.1.30}$$

For sake of comparison, one can also employ Eq. (5) from [4], adjusting it with regard to the threshold and critical index values. It exemplifies a crossover from the diluted regime where SCM is valid, to the percolation regime with typical critical behaviour. as described generally in [34, 83]. The expression below represents properly adapted, quasi-fractional two-point Padé approximant according to [52],

$$\sigma_A = \frac{\dfrac{f^2\left(\frac{f}{0.9069-f}+1\right)^{4/3}}{(0.9069-f)^2} + \dfrac{1.97425f}{0.9069-f} + 1}{\dfrac{0.877834f^2}{(0.9069-f)^2} + \dfrac{0.160454f}{0.9069-f} + 1}. \tag{9.1.31}$$

Our Padé-based suggestion for the conductivity valid for all concentrations in the random case is based on the approach of [8, 30]. To derive an explicit formula, let us first express the original series in terms of z and call the result $L_1(z)$. Apply then another transformation, $L(z) = L_1(z)^{-1/s}$, with $s = \frac{4}{3}$, in order to get rid of the power law behaviour at infinity. Employing Padé approximants asymptotically equivalent to $L(z)$. one can readily obtain the sequence of approximations A_n for the critical amplitude A,

$$A_n = f_c^s \lim_{z \to \infty} (zP_{n,n+1}(z)))^{-s}. \tag{9.1.32}$$

There are only few reasonable estimates for the amplitude,

$$A_0 = 1.32316, \quad A_1 = 1.20082, \quad A_2 = 1.54817.$$

The effective conductivity can be then reconstructed explicitly for $n = 2$,

$$\sigma_p^*(f) = \left(\frac{f(f(0.651542 - 1.16957f) - 0.0407068) + 0.373419}{f(f(f + 1.01058) + 0.519422) + 0.373419} \right)^{-\frac{4}{3}}. \tag{9.1.33}$$

We can also apply the factor approximants to the series $L_1(z)$ directly. There is a convergence now with increasing number of terms. The best factor approximant appears to be given as follows,

$$\Phi_6^*(z) = (B_1 z + 1)^{c_1} (B_2 z + 1)^{c_2} (B_3 z + 1)^{s - c_1 - c_2}, \tag{9.1.34}$$

where

$$B_1 = 1.00734 - 2.05598i, \quad B_2 = 2.2771, \quad B_3 = 1.00734 + 2.05598i,$$

$$c_1 = 0.0705011 - 0.253723i, \quad c_2 = 1.19233.$$

The critical amplitude is simply the limiting value of the approximant $\Phi_6^*(z)$ calculated as $z \to \infty$,

$$A = f_c^s B_1^{c_1} B_2^{c_2} B_3^{s - c_1 - c_2} = 1.49445. \tag{9.1.35}$$

To finalize, one should apply to the formula (9.1.34) an inverse transformation (9.1.13). Expressions (9.1.34) and (9.1.33) appear to be very close.

Significant deviations of the Padé formula (9.1.30) (with typical value of the critical index $s = 1$), compared to our results, start around $f = 0.82$. The two formulae, (9.1.29) and (9.1.33), happen to be very close to each other everywhere, although the former is the result of extrapolation from the low-concentration region, and the latter is an interpolation between the two limiting behaviours. Another formula (9.1.23) gives results slightly higher than (9.1.33), while formula (9.1.18) gives result lower than (9.1.33).

Closed-form expression for the effective conductivity of the regular hexagonal array of disks is given by (7.5.44) from Chapter 7. Since it is supposed to be defined in the same domain of concentrations as in the random case, a comparison can explicitly quantify the role of a randomness (irregularity) of the composite. For random two-dimensional composite the closed-form expression (9.1.29) for the effective conductivity is chosen. In order to estimate an enhancement factor due to randomness, we

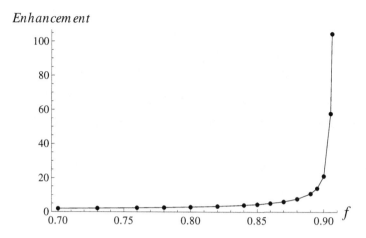

Figure 9.2 The enhancement, or ratio $\frac{\sigma^*(f)}{\sigma_{13}^*(f)}$ of the effective conductivity for the random composite to the effective conductivity of the hexagonal regular array calculated with (9.1.29) and (7.5.44), respectively. The enhancement factor with respect to available numerical data from [69] is shown as well, with dots.

use ratio of (9.1.29) to (7.5.44).

Since the two expressions are defined in the same domain of concentrations, a comparison can explicitly quantify the role of a randomness (irregularity) of the composite. In Figure 9.2, the enhancement factor is shown in the region of high concentrations where the enhancement becomes visible. There is a rapid growth of enhancement as the threshold is approached. In particular, the enhancement factor at $f = 0.906$, is equal to 104.593. Utilization of this effect depends on protocols being able to approach f_c. The Random Walks (RW) protocol described below seems the most promising in this regard.

In summary, a direct approach to the effective conductivity of the random 2D arrangements of an ideally conducting cylinders is developed, based on the series obtained from RSA protocol. Despite its relatively short length, such series turned out to be very informative. Therefore, it is possible to confine our study to closed form expressions for the effective conductivity for all volume fractions, possibly suggesting that the problem of 2D high-contrast random composite is tractable.

As shown in Fig. 9.3, various approximations for the 2D random composite do satisfy comprehensive Hashin–Shtrikman bounds adapted to the 2D case.

We confirm the position of a threshold for the effective conductivity and calculate accurately the value of a superconductivity critical index by different methods. Finally, we obtain a crossover expression (9.1.33), valid for arbitrary concentrations.

The main achievement is a direct (independent on other critical indices) calculation

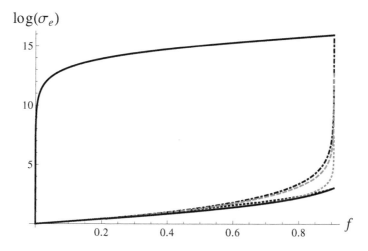

Figure 9.3 The effective conductivity of the random composite given by (9.1.29) is shown with black, dot-dashed line; original series (9.1.2) with the addition of the remainder is shown with black, dotted line; formula (9.1.30) is shown as gray, dashed. Formula (7.5.44) for the regular array (gray, dotted) is brought up for comparison as well. Hashin–Shtrikman bounds are shown with black solid line.

of the critical index for superconductivity $s = \frac{4}{3} \approx 1.3$ from the series in concentration, starting from the equations describing the composite. Randomness does tend to make the series coefficients to become larger compared with regular case. The method of corrected critical index allows thus to correct effectively the value of the critical index given by the large family of self-consistent methods, the most popular among them being ever useful effective medium approximation [19, 62].

Interestingly, the effective viscosity of two-dimensional suspensions also diverges near the threshold f_c with the critical exponent 1.3 [17]. Although a rigorous analogy does not hold for viscosity and conductivity, it seems to hold in practice, including three-dimensional case [16]. Such quantities as a 2D fluid permeability [77], and the effective elastic modulus for the 2D system of antiplane cracks [23], are expected to be characterized by the same value of critical exponent 1.3. It would be both extremely interesting and challenging to explain such universality based on the series in concentration and applying similar techniques. On the other hand, an elastic properties characterized by the effective Young modulus [77], are characterized by more than 3 times larger value of the index analogous to t, while the so-called superelasticity index analogous to s, is quoted with close value of 1.24. Explanation of such differences directly from the series presents another challenge.

2. Random composite: stirred or shaken?

James Bond's (JB) affirmation "Shaken, not stirred" can be explored and verified with numerical experiments for the problem of effective conductivity of composites.

First, the superconductivity critical index s for the conductivity of random non-overlapping disks turns out distinctly different for "shaking" and "stirring" protocols [37].

Second, in the case of stirring modeled by random walks, special formula (9.2.42) is deduced for evolution of the critical index with the normalized time $0 \le \tau \le 1$, proportional to the number of random walks and serving as the disorder measure. Strikingly, the coefficient 0.8 appearing in this formula is very close to the critical index for shaking protocol and 0.5 is the critical index for regular lattices. Shaking and stirring thus are intertwined.

Consider again the most prolific case of an ideally conducting circular disks of radius r, embedded regularly or *randomly* in an otherwise uniform locally isotropic host. The conductivity of the host as usual is normalized to unity. The term *random* can be described from the probabilistic point of view in terms of a parameter τ, quantifying a probabilistic distribution of disks. In particular, regular and, generally, deterministic locations of disks correspond to degenerate distributions with a single probabilistic event. The effective conductivity is presented then in the form of series in the volume fraction f (concentration) of the disks on the plane

$$\sigma^{(\tau)}(f) = 1 + 2f + 2f^2 + a_3^{(\tau)}f^3 + a_4^{(\tau)}f^4 + \cdots, \qquad (9.2.36)$$

with τ-dependent coefficients.

The effective properties of dilute composites are described by famous Clausius–Mossotti formula obtained from the series (9.2.36) by cutting the terms $O(f^3)$. For macroscopically isotropic composites, the second-order term in f does not depend on the location of inclusions while the third-order term does [59, 61, 62, 64]. This implies that any EMA is valid only up to the third-order term. For macroscopically anisotropic composites the EMA can be applied with confidence only within the first-order approximation. That is, it is impossible to write a universal formula independent on τ. We focus our attention on estimation of the coefficients $a_m^{(\tau)}$ for randomly located inclusions and refer to [11, 58, 77] for the theory of corresponding bounds.

Every deterministic or random composite has its series representation $\sigma^{(\tau)}$. The same statement holds for the dependence of critical index on τ. This fact means that any attempt to write a universal formula, i.e., to process the series independent on τ, or to perform any other mathematically equivalent operation without taking into account the microstructure of medium, was not justified.

Let us employ again the technique in the theory of 2D composites suggested in [20, 63] and extensively presented in Chapter 3. It develops the line of thought dedicated to derivation of an arbitrary long series (9.2.36) in f, but also allows to take into account the particles spatial arrangements (e.g., distribution of particles in a matrix). The latter task is accomplished through a direct computer simulations for lattice, or off-lattice continuum percolation models [24], by means of some Monte-Carlo (MC) algorithm (protocol) depending on τ. Mind that the maximum volume fraction depends on the protocol employed to produce the random packing [78, 79].

In particular such rather intuitive "shaking" and "stirring" protocols are available. In the context of 2D problems, shaking and stirring have meaning of generating locations of the infinitely long unidirectional cylinders (fibers) when random perturbations of fibers take place in the section plane perpendicular to fibers. It turns out that the critical index for superconductivity (or conductivity) is protocol dependent as well.

As demonstrated in Chapter 3, the deterministic boundary value problem governed by Laplace's equation, in the case of non-overlapping disks can be solved exactly for arbitrary locations of inclusions. The effective conductivity thus is written as an expression which contains the fundamental translation vectors of the lattice, and physical parameters such as radius of the disk and material constants. The formula is given explicitly in a symbolic, parameter-free form. It is written in the form (9.2.36) presented above, with all coefficients $a_m^{(\tau)}$ being expressed in exact closed form. In practice one should truncate the series, hence an approximate formula arises.

In random case, the local conductivity tensor $\sigma_e(f)$ can be considered as a random function of spatial variables. First, deterministic boundary value problem is solved for arbitrary locations of inclusions, i.e., for all events in the considered probabilistic space by the method of functional equations [61, 62, 64]. When an approximate formula for the deterministic case is deduced, the random case is treated through the ensemble averaging performed through direct MC computations. More precisely, mathematical expectation $\langle a_m^{(\tau)} \rangle$ is calculated in the framework of the fixed probabilistic distribution of disks. Thus, computation of the correlation functions is avoided [77], while $\langle a_m^{(\tau)} \rangle$ is expressed and computed through their weighted moments, the basic sums $e_{m_1 \ldots m_q}$, see formulae (4.2.26) and (4.2.27) on page 86.

The effective properties are expressed through the moments, i.e., the effective conductivity tensor can be written in the form of expansion in the e-sums. The moments depend only on locations of inclusions and yield a direct method for the effective properties computation, which does not involve the correlation functions.

The RVE is defined as the minimal size periodicity cell corresponding to the set of moments calculated for the composite. It follows from simulations for the uniform distribution of a non-overlapping disks, that in order to reach high accuracy in the effective conductivity one needs to solve the corresponding boundary value problems with at least for 81 inclusions per cell, repeated at least 1500 times. The effective properties are obtained after averaging explicit analytical expressions for the deterministic composites over the probabilistic space. The structure of the space reflects on the actual physical means to create randomness, such as shaking and stirring.

We construct a set of truncated series considering τ as a non-negative disorder parameter. $\tau = 0$ corresponds to regular arrays and $\tau = \infty$ to the theoretically disordered location of disks obeying the uniform non-overlapping distribution. This sets of polynomial yields the dependence $s = s(\tau)$ of the critical index on the degree of disorder τ. The effective conductivity is expected to tend to infinity as a power law (9.1.4) with the critical index $s(\tau)$.

The numerical computations for random composites are performed for the hexagonal representative cell bounded by rhombus and serves as the domain Q, where random composite is generated as a probabilistic distribution of non-overlapping disks, by means of some Monte-Carlo algorithm (protocol). The number of inclusions per cell can be taken arbitrary large, still the shape of the cell does somewhat influences the final result. The two algorithms, Random walks and Random shaking are described below.

2.1. Random walks

Random walks (RW) can be considered as a model of mechanical stirring (!) of the inclusion particles with the matrix [63]. Initially, N random points are generated, at first being put onto the nodes of the hexagonal array. Let each point move in a randomly chosen direction with some step. Thus, each center obtains new coordinate. This move is repeated many times, without particles overlap. If particle does overlap with some previously generated, it remains blocked at this step. After a large number of walks (steps) the obtained locations of the centers can be considered as a sought statistical realization, defining random composite. The number of steps T is proportional to real time scale. The coefficients $a_i^{(\tau)}$ for disordered locations do not differ much after 30 steps when good saturation of the results is achieved. Hence, we introduce the time scale $\tau = \frac{1}{30}T$ which is equal to unity when the full disorder is practically achieved in our simulations. All sought properties should be considered as functions of τ. In particular, approximation polynomials for the effective conductivity (truncated power series) acquire the "time" dependence, as well as the critical index $s(\tau)$ and amplitude $A(\tau)$ extrapolated from them. RW protocol can be applied for arbitrary concentrations including those very close to f_c, where $f_c = \frac{\pi}{\sqrt{12}}$ stands also for the maximum volume fraction of 2D composites achieved for the regular hexagonal array of disks.

The resulting series are presented in the truncated numerical form

$$\sigma^{RW}(f) = 1 + 2f + 2f^2 + 5.13057f^3 + 5.9969f^4 + \Delta(f). \qquad (9.2.37)$$

The coefficients on f^k ($k = 5, 6, 7, 8$) are small. The remainder $\Delta(f)$ has highly irregular form and does not contribute to the critical properties or general expressions for the conductivity. The approximation polynomials (9.2.37) for σ^{RW} and (9.1.2) for

σ^{RSA} appear numerically as almost the same. The polynomial σ^{RSA} was constructed by fitting at $f = 0.1, 0.2, \ldots, 0.5$, while σ^{RW} by fitting at $f = 0.3, 0.35, \ldots, 0.85, 0.9$. Always present starting terms, $1 + 2f + 2f^2$, ensure that the region of small concentrations is approximated properly. When the set of points $f = 0.1, 0.2, \ldots, 0.9$ is used to construct σ^{RW}, the results for critical properties appear worse [38], and require more efforts to extract the critical properties. In particular, RSA and RW approximating polynomials could be considered jointly (see [38]).

Let us employ a standard approach to the critical index calculation [6]. To this end, let us apply the transformation (9.1.13) to the original series, to make calculations with different approximants more convenient. To such transformed series, apply the *DLog* transformation. Applying the Padé approximants $P_{n,n+1}(z)$ to the transformed series one can readily obtain the sequence of approximations

$$s_n = \lim_{z \to \infty} (z\, P_{n,n+1}(z)). \tag{9.2.38}$$

The result, $s_2 = 1.24078$ is reasonably good, but can be improved further.

The corrected regular lattice approximation dwells on the idea of corrected approximants [13, 32, 33, 36, 38]. To start one has to select the initial approximation to be corrected, as describing a regular hexagonal array of inclusions, namely

$$\sigma_{0,r}^*(f) = \frac{(0.419645f + 1)^{3.45214}}{\sqrt{1 - 1.10266f}}. \tag{9.2.39}$$

This formula incorporates the critical index $\frac{1}{2}$ of the regular hexagonal array, the threshold for the hexagonal array and the two starting, effective medium terms from the series. Let us divide the original series by $\sigma_{0,r}^*(f)$, extracting the part corresponding to the random effects only. Then express the new series in terms of z, apply *DLog* transformation and call the transformed series $L_r(z)$. One can process the transformed series with different approximants, e.g., iterated roots [32, 33]. The following sequence of corrected approximations to the critical index arises,

$$s_n = s^{(0)} + \lim_{z \to \infty} (z\, R_n^*(z)), \tag{9.2.40}$$

where $R_n^*(z)$ stands for the iterated root of nth order, constructed for the series $L_r(z)$ with such a power at infinity that defines constant correction to the initial approximation $s^{(0)} = \frac{1}{2}$. The second-order iterated root has a simple form

$$R_2^*(z) = \frac{v_0 z^2}{\left(v_2 z^2 + (v_1 z + 1)^2\right)^{\frac{3}{2}}}.$$

The parameters v_i here are computed from the series $L_r(z)$. The power at infinity is selected in order to compensate for the factor z and extract the correction to regular lattice value. The effective conductivity is simply,

$$\sigma^*(f) = \sigma^*_{0,r}(f) \exp\left(\int_0^{\frac{f}{f_c-f}} R_2^*(z)\, dz \right),$$

and can be expressed in a closed albeit long form, see e.g., (9.1.29) on page 260. The results $s_2 = 1.3367$, $A_2 = 1.6651$ are close to RSA beginning with $T = 30$. They tend to depend very weakly on Δ. If we simply set $\Delta = 0$, then they change to $s = 1.3275$ and $A = 1.6837$, supporting the view that only starting four terms are relevant when the critical region is concerned.

2.2. Random shaking

Randomness can be also introduced through the MC simulation by gentle "shaking", through the random locations of the centers of the disks [12, 13]. According to James Bond (JB), stirring should be avoided and shaking preferred. Implicitly is assumed that an observer can distinguish the two ways of randomization applied to preparing a mixture. This corresponds to various technologies of stir casting processes including vibration.

Following [12] consider the unit cell with identical inclusions whose centers are random variables. Each center is uniformly distributed in a disk of the radius d called shaking parameter. Centers of these disks form a hexagonal array on the plane whereas the disks by themselves do not form the periodic array. Hence, we investigate a random shaking of the disks about the periodic hexagonal array. When d approaches zero, we return to the regular composite.

Shaking parameter d does not have to be small and therefore our results are not perturbative. The parameter d is chosen so that the disks cannot touch. Original approach to Random shaking (RS) [12], did not include the threshold for continuum percolation. However, RS also works when disks/cylinders touch each other but do not form a spanning cluster of touching cylinders in the periodic cell, i.e., no infinite chains of touching disks in the plane are allowed.

Heuristically, the shaking geometries provide a reasonable approximation for random mixtures at not very high concentrations. There is not much room for the inclusions to move around, when their density is relatively high. Therefore, the inclusions could naturally form some kind of a random shaking pattern, whereas at small concentrations the random patterns could be very different from the shaking geometries. For the shaking model both linear and quadratic terms of the Maxwell approach do not depend on the random locations, and are determined solely by the total volume fraction and material properties [12]. But already the cubic term depends on the random locations of inclusions. The expansion for $d = \frac{2\pi}{\sqrt{5N}}$ and the number of disks $N = 81$ is

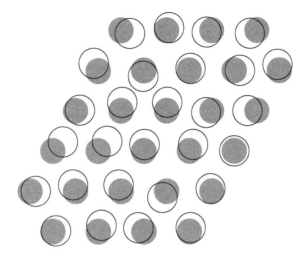

Figure 9.4 Shaken structure. Regular located disks of the radius r (shaded disks). Also shown are the possible location of the centers of inclusions (transparent disks of radius d).

presented in the truncated numerical form

$$\sigma^{RS}(f) = 1 + 2f + 2f^2 + 2.50496f^3 + 1.34794f^4 + 2.28669f^5$$
$$+2.78469f^6 + 2.66817f^7 + 2.31341f^8 + 1.99453f^9 + 1.82074f^{10}$$
$$+1.82006f^{11} + 1.99754f^{12} + 2.34983f^{13} + 2.85936f^{14}$$
$$+3.48676f^{15} + 4.17211f^{16} + 4.84582f^{17} + 5.44337f^{18} + O(f^{19}).$$
$$(9.2.41)$$

up to to the terms of 18th order inclusively. When the radius of disks approaches d, index s tends to 0.5, the critical index of regular composites. Applying the *DLog* Padé technique one finds,

$$s_1 = 0.944643, \quad s_2 = 0.803755, \quad s_3 = 0.79748, \quad s_4 = 0.808613,$$

$$s_5 = 0.440396, \quad s_6 = 0.329953, \quad s_7 = 0.812099, \quad s_8 = 0.816668, \quad s_9 = 0.812114.$$

The critical amplitude for $n = 9$ equals 1.53383. The result s = 0.812114 is weakly sensitive to the value of shaking parameter d used for computations.

Thus, JB prefers the shaken composite ("martini") with the critical index of 0.81 − 0.82 very much different from the stirred result 1.3. His requirement is substantial and can be fulfilled.

2.3. Temporal crossover for RW

We intend to study in detail a transition from the regular hexagonal array to the random array. For RW protocol consider dependence of the critical index and critical amplitude, on the degree of disorder quantified by the time τ. The formula is going to be constructed in such a manner that for "zero" randomness ($\tau = 0$) it is going to behave as the regular hexagonal array. For "maximum" randomness $\tau \to \infty$, we expect to have a random composite. All cases with intermediate degrees of randomness for finite τ, are expected to fall in between the two cases. The critical behaviour of regular composites occurs for f very close to f_c, due to the direct particles contact in the whole area of composite. A relatively simple Laplace equation for the potential, when complemented with a non-trivial boundary conditions in the regular domain of inclusions, behaves critically, with $s = \frac{1}{2}$ without an explicit non-linearity or randomness. On the other hand, randomness dominates in the case of continuum percolation.

The two limiting cases of small τ (quasi-regular composite), and of the large τ (random composite), should be considered separately. Final formula will be obtained by matching the two behaviours, as a "regular-to-random" crossover. In the case of large τ, we apply literally the technique of corrected regular lattice approximation described above, since it is fairly accurate for the random case for the number of steps $T = 30$. For very small τ, we have a regular hexagonal array of disks, and we consider a peculiar critical phenomena using different initial approximation, $\sigma_{0,h}^*(\tau = 0)$, in place of $\sigma_{0,r}^*$. General form of the

$$\sigma_{0,h}^* = \alpha_0 + \alpha_1 \frac{1}{\sqrt{f_c - f}} + \alpha_2 \sqrt{f_c - f} + \alpha_3 (f_c - f)$$

is the same as employed in the preceding Chapters 6-7 and [30, 36, 38], where the coefficients of the series were obtained from the exact formulae. Approximating polynomial for the regular case is obtained following exactly the same procedure as for the random case,

$$\sigma_e(\tau = 0) = 1 + 2f + 2f^2 + 2.01527 f^3 + 1.94715 f^4 + 2.00003 f^5 + 2.08909 f^6 + \cdots,$$

so that instead of the exact values $a_k = 2$ for the six starting coefficients, there is an approximation. The following values for the coefficients in $\sigma_{0,h}^*(\tau = 0)$ follow,

$$\alpha_0 = -6.94364, \quad \alpha_1 = 5.18603, \quad \alpha_2 = 3.33683, \quad \alpha_3 = -0.749575.$$

The calculations of s and A for arbitrary τ, are identical to the random case considered for RW protocol and described above.

Interpolation between the two sets of calculations for the critical index $s(\tau)$ and the interpolation curve are shown and explained in Fig. 9.5. Upper set corresponds to the initial approximation $\sigma_{0,r}^*$, and it overestimates critical index at small τ. Lower set of points was obtained with the initial approximation $\sigma_{0,h}^*(\tau = 0)$ and it underestimates the index at large τ. In the case of stirring modeled by random walks, formula

$$\boxed{s(\tau) = 0.5 + 0.8 \sqrt[3]{\tau}}, \tag{9.2.42}$$

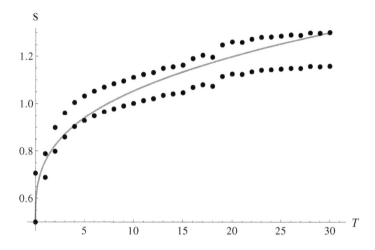

Figure 9.5 Interpolation curve for the critical index $s(\tau) = 0.5 + 0.8\sqrt[3]{\tau}$ (shown as solid line), between the two sets of calculations. Upper set corresponds to the initial approximation $\sigma^*_{0,r}$, while lower set of points to the initial approximation $\sigma^*_{0,h}$. The time scales τ and T are simply related, $\tau = \frac{1}{30}T$.

can be suggested for evolution of the critical index with the normalized time $0 \leq \tau \leq 1$, proportional to the number of random walks and serving as the disorder measure. Strikingly, the coefficient 0.8 is very close to the critical index for shaking protocol and 0.5 is the critical index for regular lattices. Thus, one can think that the regular and random components in 2D percolate independently at the same threshold, not unlike a double percolation but without any added new structural levels [50].

The formula for s is based on the analytical solution to the 2D conductivity problem of randomly distributed disks up to $O(f^{19})$ where f denotes the concentration of inclusions and its extension to special 3D composites.

Good saturation of the results is achieved already at $T = 30$. Thus, for small and moderately large times, the value of index is bounded by its regular and random values, $0.5 \leq t \leq 1.3$. It is believed that physics of a 2D regular and irregular composites is related to the so-called "necks", certain areas between closely spaced disks. Randomness (stirring) adds to the regularly formed necks an additional random component. For very strong disorder this additional contribution is, with a good precision, the part estimated above for the shaking protocol.

Similar approach holds for the critical amplitude $A(\tau)$. The initial approximation $\sigma^*_{0,r}$ underestimates critical amplitude at small τ, while the initial approximation $\sigma^*_{0,h}(\tau = 0)$ overestimates the amplitude at large τ. The amplitude is bounded by its random and regular values, $1.71 \leq A \leq 5.19$. The whole spectrum of composites can

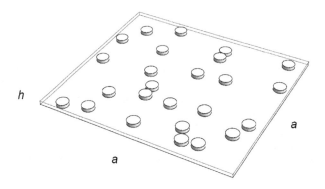

Figure 9.6 Unidirectional cylinders in the 3D cuboid cell of the size $a \times a \times h$.

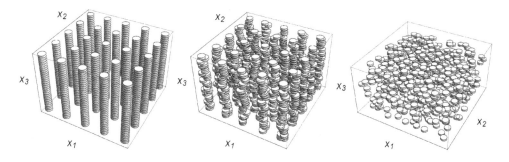

Figure 9.7 3D composites corresponding to regular, shaken and random walks 2D structures.

be realized, not just regular or random cases.

2.4. 3D extensions

The homogenization theory of random media [76] demonstrates that the effective properties of composites can be precisely determined as the mathematical expectation of the effective constants of the statistically representative cells. Consider a 3D cuboid cell $Q_{000} = \{(x_1, x_2, x_3) \in \mathbb{R}^3 : 0 < x_1 < a, 0 < x_2 < a, 0 < x_3 < h\}$ of the size $a \times a \times h$ displayed in Fig. 9.6. It is assumed that unidirectional cylinders have the height h and their axis is parallel to the axis x_3. We have a disks distribution in each section of the cell perpendicular to the axis x_3. It is assumed that this distribution is isotropic in the plane (x_1, x_2). Let the cell Q_{000} with fixed inclusions represent a 3D random composites. Let all the cells $Q_{klm} = \{(x_1, x_2, x_3) \in \mathbb{R}^3 : k < x_1 < a + k, l < x_2 < a + l, m < x_3 < h + m\}$ (k, l, m run over integers) to be obtained by the same random distribution of cylinders but with different statistical realizations. Examples of such composites for regular, shaken and random walk distributions are displayed in Fig. 9.7. According to

[76] all the cells have the same effective conductivity tensor

$$\sigma = \begin{pmatrix} \sigma_e & 0 & 0 \\ 0 & \sigma_e & 0 \\ 0 & 0 & \sigma_z \end{pmatrix}. \tag{9.2.43}$$

The component σ_e coincides with the effective conductivity of the 2D composites discussed above. For long enough cylinders the effective conductivity can be approximated with σ_e as their in-plane locations are modeled by RW or RS protocols.

The critical index is shown to intermediate between the regular and random values. RS protocol is preferable to stirring (RW protocol) because in practice it is easier to control shaking parameter d, then stirring parameter τ. On the other hand, stirring is much better than shaking when one needs to create a random composite. We are confident that modern 3D printing techniques will be able eventually to produce actual composite from any pattern generated by any protocol. Did someone actually try to prepare martini in the printer?

3. 2D Conductivity. Dependence on contrast parameter

Besides of the high-contrast case, one can consider a more general situation, when contrast parameter ρ. Method for derivation of expressions including ρ and f is expounded in Chapter 3. The following series in concentration f and contrast parameter ρ was obtained in [21], up to the six order inclusively,

$$\sigma_e(f,\rho) = 1 + a_1(\rho)f + a_2(\rho)f^2 + a_3(\rho)f^3 + a_4(\rho)f^4 + a_5(\rho)f^5 + a_6(\rho)f^6. \tag{9.3.44}$$

The coefficients depend only on ρ,

$$\begin{aligned}
a_1(\rho) &= 2\rho, \\
a_2(\rho) &= 2\rho^2, \\
a_3(\rho) &= 4.9843\rho^3 - 0.0688\rho^4 - 0.1463\rho^5 - 0.7996\rho^6, \\
a_4(\rho) &= -6.829\rho^3 + 7.3652\rho^4 + 6.3079\rho^5 + 4.517\rho^6, \\
a_5(\rho) &= 4.2139\rho^3 - 12.4218\rho^4 - 10.4599\rho^5 - 7.8602\rho^6, \\
a_6(\rho) &= -0.3462\rho^3 + 7.0868\rho^4 + 6.7108\rho^5 + 5.9897\rho^6.
\end{aligned} \tag{9.3.45}$$

Since there are two parameters f and ρ, we derive the final formula in two resummation steps.

First, let us attempt to guarantee the correct behaviour in concentration and contrast. To this end, the solution for the effective conductivity for $-1 \le \rho \le 1$ and $0 \le f \le f_c = \frac{\pi}{\sqrt{12}}$, is to be sought in the form of approximant satisfying the phase-interchange symmetry [28]. We have simply to apply Keller's phase-interchange relation [48, 77]

$$\frac{1}{\sigma_e(f,\rho)} = \sigma_e(f,-\rho).$$
(9.3.46)

It is valid for the general case of average conductivity of a statistically homogeneous isotropic random distribution of cylinders of one medium in another medium. We develop below an expression which satisfy (9.3.46) by design. The simplest solution in terms of factor approximants is given as follows, which includes both types of critical behaviour, as $\rho \to -1$, and $\rho \to 1$, respectively,

$$\sigma^*(f,\rho) = (B_1(\rho)f + 1)^{\frac{4}{3}}(B_2(\rho)f + 1)^{-\frac{4}{3}},$$
(9.3.47)

with

$$B_1(\rho) = \frac{3\rho}{4}, \quad B_2(\rho) = -\frac{3\rho}{4},$$
(9.3.48)

with the parameters defined from the asymptotic equivalence to the series (9.3.44). The symmetry dictated by (9.3.46) is self-evident, and only the values of thresholds as $\rho \to -1$ and $\rho \to 1$, are wrong.

Thus, second step of resummation is required. It consists in imposing conditions on correct threshold in the both limits. We can accomplish this goal by means of the simplest exponential approximant for the effective thresholds considered as functions of ρ, so that we arrive to the following improved expression

$$\sigma_e^{**}(f,\rho) = (B_1^*(\rho)f + 1)^{\frac{4}{3}}(B_2^*(\rho)f + 1)^{-\frac{4}{3}},$$
(9.3.49)

where

$$B_1^*(\rho) = \frac{3\rho}{4} e^{-0.385406\rho}, \quad B_2^*(\rho) = -\frac{3\rho}{4} e^{0.385406\rho},$$
(9.3.50)

or

$$\sigma_e^{**}(f,\rho) = \left(\frac{4 + 3e^{-0.385406\rho}\rho f}{4 - 3e^{0.385406\rho}\rho f}\right)^{\frac{4}{3}}.$$
(9.3.51)

Formulae based only on Padé approximants [46] do respect the phase-interchange relations, but do not give the correct values for critical indices. Correcting (9.3.51) with the diagonal Padé approximants may be though beneficial. Indeed let us construct the following corrected factor approximant,

$$\sigma_{1c}^{**}(f,\rho) = \sigma_e^{**}(f,\rho)\frac{1+wf}{1-wf} = \frac{(3e^{-0.385406\rho}\rho f+4)^{4/3}(\rho f-\rho f\cosh(0.385406\rho)+1)}{(4-3e^{0.385406\rho}\rho f)^{4/3}(-\rho f+\rho f\cosh(0.385406\rho)+1)},$$
(9.3.52)

where

$$w = w(\rho) = 0.5\left(-e^{-0.385406\rho}\rho - e^{0.385406\rho}\rho + 2\rho\right)$$
(9.3.53)

is obtained from asymptotic equivalence with the series (9.3.44). The form of correc-

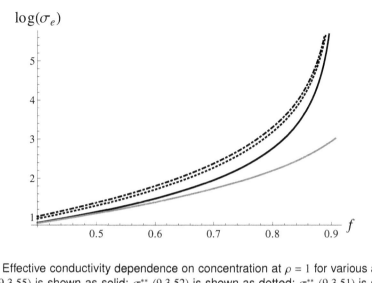

Figure 9.8 Effective conductivity dependence on concentration at $\rho = 1$ for various approxima-tions. σ_{2c}^{**} (9.3.55) is shown as solid; σ_{1c}^{**} (9.3.52) is shown as dotted; σ_e^{**} (9.3.51) is shown with dot-dashed line. Finally the Clausius–Mossotti formula (9.3.54) is shown with gray line.

tion is additionally motivated by the celebrated Clausius–Mossotti (CM) formula valid for small concentrations which respects the phase interchange symmetry.

$$\sigma_{CM}(f,\rho) = \frac{1 + \rho f}{1 - \rho f}. \qquad (9.3.54)$$

Formula (9.3.52) respects this symmetry as well.

Assume even more general form for the correcting approximants,

$$\sigma_{2c}^{**}(f,\rho) = \sigma_e^{**}(f,\rho)\frac{w_1(\rho)\rho f + w_2(\rho)\rho^3 f^2 + 1}{-w_1(\rho)\rho f - w_2(\rho)\rho^3 f^2 + 1}, \qquad (9.3.55)$$

where

$$w_1(\rho) = 1 - \cosh(0.385406\rho), \quad w_2(\rho) = -\frac{0.375 \sinh(0.770811\rho)}{\rho}. \qquad (9.3.56)$$

The symmetry is again preserved.

As usual it is imperative that the formulae agree with celebrated upper and lower bounds in 2D, [1] see [58]. Lower Hashin–Shtrikman bound is given as follows

$$\sigma_{HS}^-(f,\sigma) = 1 + \frac{2f(\sigma - 1)}{(1 - f)(\sigma - 1) + 2}.$$

---

[1] The 3D bounds are given by formulae (3.1.1), (3.1.2) on page 43.

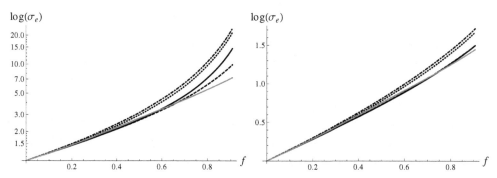

Figure 9.9 Left: Effective conductivity dependence on concentration at $\rho = 0.9$ for various approximations. σ_{2c}^{**} (9.3.55) is shown as solid; σ_{1c}^{**} (9.3.52) as dotted; σ_e^{**} (9.3.51) as dot-dashed. They are compared with Clausius–Mossotti formula (9.3.54) shown with dashed line. Finally the series (9.3.44) is shown with gray line.   Right: Effective conductivity dependence on concentration at $\rho = 0.7$ for various approximations. σ_{2c}^{**} is shown as solid; σ_{1c}^{**} as dotted; σ_e^{**} as dot-dashed. They are compared with formula (9.3.54) shown with dashed line. The series (9.3.44) is shown with gray line.

Upper Hashin–Shtrickman bound is given by the formula

$$\sigma_{HS}^{+}(f,\sigma) = \sigma + \frac{2(1-f)(1-\sigma)\sigma}{f(1-\sigma) + 2\sigma}.$$

In Fig. 9.10, we present the most interesting case when ρ is close to unity, and compare the formula for σ_{2c}^{**} (9.3.55), and the Padé -approximant $P_{2,4}$ [21], with the bounds. In Fig. 9.11, the formulae for σ_{1c}^{**} (9.3.52), and σ_{2c}^{**} (9.3.55) are compared with the bounds and the Padé approximant. The (9.3.55) seems to be the most consistent with bounds.

Let us now proceed slightly different. Assume from the very beginning the simplest plausible dependence of the effective thresholds leading to the correct value of f_c as $\rho \to -1$ and $\rho \to 1$ respectively,

$$B_1^*(\rho) = \frac{\rho}{f_c}, \quad B_2^*(\rho) = -\frac{\rho}{f_c}. \tag{9.3.57}$$

Such dependence follows the CM formula. Let us guarantee the correct thresholds first and asymptotic equivalence next. Using this approach, we obtain a much simpler expression for the effective conductivity

$$\sigma_s^{**}(f,\rho) = \frac{\left(\frac{2\sqrt{3}\rho f}{\pi}+1\right)^{\frac{4}{3}}\left(\frac{\left(\pi-\frac{8}{\sqrt{3}}\right)\rho f}{\pi}+1\right)}{\left(1-\frac{2\sqrt{3}\rho f}{\pi}\right)^{\frac{4}{3}}\left(1-\frac{\left(\pi-\frac{8}{\sqrt{3}}\right)\rho f}{\pi}\right)}, \tag{9.3.58}$$

which gives results very close to the more sophisticated (9.3.55). Moving singularity to the non-physical values of f allows to preserve the form typical for critical regime

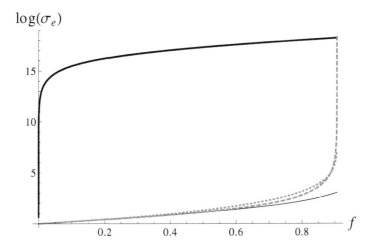

Figure 9.10 Case of very large conductivity of inclusions, $\sigma = 10^8$. $\log(\sigma_e)$ of the effective conductivity is shown as a function of concentration f. The lower and upper bounds are shown with solid lines. The results from (9.3.55) are shown with dashed line. The Padé approximant $P_{2,4}$ [21], is shown with dotted line.

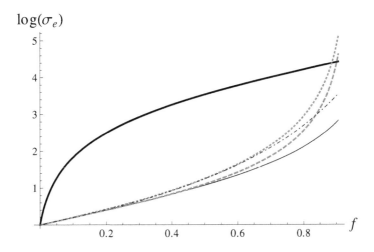

Figure 9.11 Case of large conductivity of inclusions, $\sigma = 10^2$. $\log(\sigma_e)$ of the effective conductivity is shown as a function of concentration f. The lower and upper bounds are shown with solid lines. The results from (9.3.55) are shown with dashed line. The (9.3.52) is shown with dotted line. The Padé approximant $P_{2,4}$ is shown with dot-dashed line.

for all values of ρ, but with the singularity getting blunted.

Remark 19. The approximate formula (9.3.58) is in agreement with Rylko's asymptotic formula (3.5.141) discussed on page 75, since σ_s^{**} as a function on $\varepsilon = 1 - \rho$ is analytic near $\varepsilon = 0$ for any fixed $f < f_c = \frac{\pi}{2\sqrt{3}}$.

Ansatz (9.3.57) is most intuitive and can serve as a guide in generally more complicated situation of the elastic moduli. One can hope to be able to solve corresponding critical problem and then simply modify parameters of the solution to move away to a non-critical situations.

Formula (9.3.58) can be extended to 3D case with proper account for the different critical indices s and t, and amplitudes. Of course, there is no phase-interchange symmetry in 3D to restrict also the coefficients in expansions.

REFERENCE

1. Adler J, Meir Y, Aharony A, Harris AB, Klein L, Low-Concentration Series in General Dimension. J. Stat. Phys. 1990; 58: 511–538.
2. Ambrosetti G, Balberg I, Grimaldi C, Percolation-to-hopping crossover in conductor-insulator composites, Phys. Rev. B 2010; 82: 134201.
3. Ambrosetti G, Grimaldi C, Balberg I, Maeder T, Danani A, Ryser P, Solution of the tunneling-percolation problem in the nanocomposite regime. Phys. Rev B 2010; 81: 155434.
4. Andrianov IV, Danishevskyy VV, Kalamkarov AL, Analysis of the effective conductivity of composite materials in the entire range of volume fractions of inclusions up to the percolation threshold. Composites: Part B Engineering 2010; 41: 503–507.
5. Baker DR, Paul G, Sreenivasan S, Stanley HE, Continuum percolation threshold for interpenetrating squares and cubes, Phys.Rev.E 2002; 66: 046136.
6. Baker GA, Graves-Moris P, Padé Approximants. Cambridge, UK: Cambridge University; 1996.
7. Balberg I, Limits on the continuum-percolation transport exponents, 1998; 57: 13351–13354.
8. Bender CM, Boettcher S, Determination of $f(\infty)$ from the asymptotic series for $f(x)$ about $x = 0$, J. Math.Phys 1994; 35: 1914–1921.
9. Bender CM, Orszag SA, Advanced Mathematical Methods for Scientists and Engineers. Asymptotic Methods and Perturbation Theory. New York: Springer; 1999.
10. Benguigui L, Experimental study of the elastic properties of a percolating system, Phys. Rev. Lett. 1984; 53: 2028–2030.
11. Bergman DJ, The self-consistent effective medium approximation (SEMA): New tricks from an old dog, Physica B Condensed Matter 2007; 394; 344–350.
12. Berlyand L, Mityushev V, Generalized Clauisius-Mossotti Formula for Random Composite with Circular Fibers, J. Stat. Phys. 2001; 102: 115–145.
13. Berlyand L, Mityushev V, Increase and Decrease of the Effective Conductivity of Two Phase Composites due to Polydispersity, J. Stat. Phys. 2005; 118: 481–509.
14. Berlyand L, Novikov A, Error of the network approximation for densely packed composites with irregular geometry, SIAM J. Math. Anal. 2002; 34: 385–408.
15. Berryman JG, Random close packing of hard spheres and disks, Phys.Rev.A 1983; 27: 1053–1061.
16. Bicerano J, Douglas JF, Brune DA, Model for the Viscosity of Particle Dispersions, Polymer Reviews 1999; 39: 561–642.
17. Bouillot JL, Camoin C, Belzon M, Blanc R, Guyon E, Experiments on 2-D suspensions, Adv. Colloid Interface Sci. 1982; 17: 299–305.
18. Chen Y, Schuh S, Effective properties of random composites: Continuum calculations versus map-

ping to a network, Phys.Rev. E 2009; 80: 040103.

19. Choy TC, Effective Medium Theory. Principles and Applications. Oxford. Clarendon Press. 1999.

20. Czapla R, Nawalaniec W, Mityushev V, Simulation of representative volume elements for random 2D composites with circular non-overlapping inclusions, Theoretical and Applied Informatics 2012; 24: 227–242.

21. Czapla R, Nawalaniec W, Mityushev V, Effective conductivity of random two-dimensional composites with circular non-overlapping inclusions, Comput. Mat. Sci. 2012; 63: 118–126.

22. Danziger M, Bashan A, Havlin S, Interdependent resistor networks with process-based dependency, New J. Phys. 2005; 17: 043046.

23. Davis PM, Knopoff L, The elastic modulus, percolation and disaggregation of strongly interacting intersecting antiplane cracks, Proc. Natl. Acad. Sci. USA 2008; 106: 12634–12639.

24. Eischen JW, Torquato S, Determining elastic behavior of composites by the boundary element method, J. Appl. Phys. 1993; 74: 159–170.

25. Feng S, Halperin B, Sen PN, Transport properties of continuum systems near the percolation threshold, Phys. Rev. B 1987; 35: 197–214.

26. Frank DJ, Lobb CJ, Highly efficient algorithm for percolative transport studies in two dimensions, Phys. Rev. B 1988; 37: 302–307.

27. Gluzman S, Karpeev DA, Berlyand LV, Effective viscosity of puller-like microswimmers: a renormalization approach, J. R. Soc. Interface 2013; 10: 20130720.

28. Gluzman S, Karpeev DA, Perturbative Expansions and Critical Phenomena in Random Structured Media, ed. P. Drygas and S. Rogosin, Modern Problems in Applied Analysis. Birkhauser Publishing House, 1–16, 2017.

29. Gluzman S, Kornyshev A, Neimark A, Electrophysical Properties of Metal-Solid Electrolyte Composite, Phys. Rev. B 1995; 52: 927–938.

30. Gluzman S, Mityushev V, Nawalaniec W, Cross-properties of the effective conductivity of the regular array of ideal conductors, Arch. Mech. 2014; 66: 287–301.

31. Gluzman S, Yukalov VI, Sornette D, Self-similar factor approximants, Phys. Rev. E 2003; 67: 026109.

32. Gluzman S, Yukalov VI, Self-similar extrapolation from weak to strong coupling, J.Math.Chem. 2010; 48: 883–913.

33. Gluzman S, Yukalov VI, Extrapolation of perturbation theory expansions by self-similar approximants, Eur. J. Appl. Math. 2014; 25: 595–628.

34. Gluzman S, Yukalov VI, Unified approach to crossover phenomena, Phys. Rev. E 1998; 58: 4197–4209.

35. Gilbert EN, Random plane networks, J. Soc. Indust. Appl. 1961; 9: 533–543.

36. Gluzman S, Mityushev V, Series, index and threshold for random 2D composite, Arch. Mech. 2015; 67: 75–93.

37. Gluzman S, Mityushev V, Nawalaniec W, Sokal G, Random composite: stirred or shaken? Arch. Mech. 2016; 68: 229–241.

38. Gluzman S, Mityushev V, Nawalaniec W, Starushenko G, Effective Conductivity and Critical Properties of a Hexagonal Array of Superconducting Cylinders. ed. P.M. Pardalos (USA) and T.M. Rassias (Greece), Contributions in Mathematics and Engineering. In Honor of Constantin Caratheodory. Springer: 255–297, 2016.

39. Golden K, Convexity and Exponent Inequalities for Conduction near Percolation, Phys.Rev.Lett. 1990; 65: 2923–2926.

40. Golden KM, Critical Behavior of Transport in Lattice and Continuum Percolation Models. Phys.Rev.Lett. 1997; 78: 3935–3938.

41. Goncharenko AV, Generalizations of the Bruggeman equation and a concept of shape-distributed particle composites, Phys. Rev. E 2003; 68: 041108.

42. Grassberger P, Conductivity exponent and backbone dimension in 2d percolation, Physica A 1999; 262, 251–263.

43. He HX, Hamer CJ, Oitmaa J, High-temperature series expansions for the (2+1)-dimensional Isong model, J.Phys.A 1990; 23: 1775–1787.

44. Hong DC, Havlin S, Herrmann HJ, Stanley HE, Breakdown of Alexander-Orbach conjecture for percolation: Exact enumeration of random walks on percolation backbones, Phys. Rev. B 1984; 30: 4083–4086.

45. Jullien R, Sadoc JF, Mosseri R, Packing at Random in Curved Space and Frustration: a Numerical Study, J. Phys. France 1997; 7: 1677–1692.

46. Kalamkarov AL, Andrianov IV, Starushenko GA, Three-phase model for a composite material with cylindrical circular inclusions, Part II: Application of Padé approximants, International Journal of Engineering Science 2014. 78: 178–191.

47. Keller JB, Conductivity of a Medium Containing a Dense Array of Perfectly Conducting Spheres or Cylinders or Nonconducting Cylinders, Journal of Applied Physics 1963; 34: 991–993.

48. Keller JB, A Theorem on the Conductivity of a Composite Medium, J. Math. Phys. 1964; 5: 548–549.

49. Lebovka N, Lisunova M, Mamunya YeP, Vygornitskii N, Scaling in percolation behaviour in conductive-insulating composites with particles of different size, J. Phys. D : Appl. Phys. 2006; 39: 2264–2272.

50. Levon K, Margolina A, Patashinsky AZ, Multiple Percolation in Conducting Polymer Blends, Macromolecules 1993; 26: 4061–4063.

51. Lubachevsky B, Stillinger FH, Geometric properties of random disk packings, J. Stat. Phys. 1990; 60: 561–583.

52. Martin P, Baker GA Jr, Two-point quasifractional approximant in physics, Truncation error, J. Math. Phys. 1991; 32: 1470–1477.

53. McLachlan DS, Sauti G, Chiteme C, Static dielectric function and scaling of the ac conductivity for universal and nonuniversal percolation systems, Phys. Rev. B 2007; 76: 014201.

54. McLachlan DS, Sauti G, The AC and DC Conductivity of Nanocomposites, Journal of Nanomaterials 2007; Volume 2007: 30389.

55. McPhedran RC, Transport Properties of Cylinder Pairs and of the Square Array of Cylinders, Proc.R. Soc. Lond. 1986; A 408: 31–43.

56. O'Neill J, Selsil Ö, McPhedran RC, Movchan AB, Movchan NV, Active cloaking of inclusions for flexural waves in thin elastic plates, Q. J. Mechanics Appl. Math. 2015; 68: 263–288.

57. McPhedran RC, Poladian L, Milton GW, Asymptotic studies of closely spaced, highly conducting cylinders, Proc. R. Soc. A 1988; 415: 185–196.

58. Milton GW, The theory of composites. Cambridge Uni. Press; 2002.

59. Mityushev V, Steady heat conduction of a material with an array of cylindrical holes in the nonlinear case, IMA Journal of Applied Mathematics 1998; 61: 91–102.

60. Mityushev V, Representative cell in mechanics of composites and generalized Eisenstein–Rayleigh sums, Complex Variables 2006; 51, 1033–1045.

61. Mityushev V, Exact solution of the $\mathbb{R}$-linear problem for a disk in a class of doubly periodic functions, J. Appl. Functional Analysis 2007; 2: 115–127

62. Mityushev V, Rylko N, Maxwell's approach to effective conductivity and its limitations, The Quarterly Journal of Mechanics and Applied Mathematics 2013.

63. Mityushev V, Nawalaniec W, Basic sums and their random dynamic changes in description of microstructure of 2D composites, Comput. Mater. Sci. 2015; 97: 64–74.

64. Mityushev VV, Pesetskaya E, Rogosin SV, Analytical Methods for Heat Conduction in Composites and Porous Media, in Cellular and Porous Materials: Thermal Properties Simulation and Prediction. A. Öchsner, G. E. Murch, M. J. S. de Lemos, eds., 121-164; Wiley. 2008.

65. Mityushev V, Rylko N, Boundary value problems, the Poincare series, the method of Schwarz and composite materials, Int. Congres IMACS 97; Berlin. 1; 165–170: 1997.

66. Mityushev V, Pattern formations and optimal packing, Mathematical Biosciences 2016; 274: 12–16.

67. Murat M, Marianer S, Bergman DJ, A transfer matrix study of conductivity and permeability exponents in continuum percolation, J. Phys. A, **19**, 1986; 5: L275–L279.

68. Normand JM, Herrmann HJ, Hajjar M, Precise calculation of the dynamical exponent of two-dimensional percolation, J. Stat. Phys. 1988; 52: 441–446.

69. Perrins WT, McKenzie DR, McPhedran RC, Transport properties of regular array of cylinders, Proc. R. Soc.A 1979; 369: 207–225.

70. Quickenden T, Tan GK, Random packing in two dimensions and the structure of monolayers, Journal of Colloid and Interface Science 1974; 48: 382–393.

71. Rylko N, Representative volume element in 2D for disks and in 3D for balls, Journal of Mechanics of Materials and Structures 2014; 9: 427–439.

72. Rylko N, Transport properties of the regular array of highly conducting cylinders, J. Engrg. Math. 2000; 38: 1–12.

73. Rylko N, Structure of the scalar field around unidirectional circular cylinders, Proc. R. Soc. A 2008; 464: 391–407.

74. Sahimi M, Hughes BD, Scriven LE, Davis HT, Real-Space Renormalization and Effective-Medium Approximation to the Percolation Conduction Problem, Phys. Rev. B 1983; 28: 307–311.

75. Smith LN, Lobb CJ, Percolation in two-dimensional conductor-insulator networks with controllable anisotropy, Phys. Rev. B 1979; 20: 3653–3658.

76. Telega JJ, Stochastic homogenization: convexity and nonconvexity, eds. Castañeda PP, Telega JJ, Gambin B. Nonlinear Homogenization and its Applications to Composites, Polycrystals and Smart Materials, NATO Science Series, 305-346: Dordrecht, Kluwer Academic Publishers, 2004.

77. Torquato S, Random Heterogeneous Materials: Microstructure and Macroscopic Properties. New York. Springer-Verlag: (2002).

78. Torquato S, Stillinger FH, Jammed hard-particle packings: From Kepler to Bernal and beyond, Reviews of Modern Physics 2010; 82: 2634–2672.

79. Torquato S, Truskett TM, Debenedetti PG, Is random close packing of spheres well defined? Phys. Rev. Lett. 2000; 84: 2064–2067.

80. Widom B, Random Sequential Addition of Hard Spheres to a Volume, J. Chem. Phys. 1966; 44: 3888–3894.

81. Wu J, McLachlan DS, Scaling behaviour of the complex conductivity of graphite-boron nitride systems, Phys. Rev. B 1998; 58: 14880–14887.

82. Yukalov VI, Gluzman S, Critical Indices as Limits of Control Functions, Phys.Rev.Lett. 1997; 79: 333–336.

83. Yukalov VI, Gluzman S, Self-similar crossover in statistical physics, Physica A 1999; 273: 401–415.

84. Yukalov VI, Gluzman S, Sornette D, Summation of Power Series by Self-Similar Factor Approximants, Physica A 2003; 328: 409–438.

CHAPTER 10

Elastic problem

*Tradition! And again Tradition! How strong
and how elastic!*

— John Galsworthty, The Forsyte Saga

1. Introduction

Rayleigh [38] in 1892 applied a series expansion method to doubly periodic problems for harmonic functions when one circular inclusion is embedded in a host material. This paper contains the first study of boundary value problems on Riemann surfaces since a doubly periodic problem can be considered as a problem on torus. To the best of our knowledge, the second most important result in this field was obtained by Filshtinsky in 1964 [13, 14, 19] where the biharmonic problem for one circular inclusion was solved by expansions on the elliptic functions. In particular, the effective elastic constants were determined [13].

Muskhelishvili was the first to discover that a stationary 2D elasticity boundary value problem can be reduced to a singular integral equation. The theory of composites based on the MMM principle of Hashin (see page 46) and the theory of homogenization [3] requires as an input the solution to the boundary value problem with many inclusions per the periodicity cell represented by RVE.

A constructive approach based on the integral equations and the series method to computations of the effective properties of the tensor elastic media was presented in [18, 19, 20]. Integral equations first constructed in [18] are efficient for the numerical investigation of a non-dilute composites when the inclusion interactions are taken into account. The method can be effectively applied for the numerical computations when the inclusions are closely spaced.

Solutions to the boundary value problems with finite number of bounded inclusions in the whole plane theoretically can be applied only to a dilute composites through Maxwell's approach. It follows from the fact that the area fraction of a finite set C of bounded inclusions on the plane is equal to zero. The approach was developed for elastic composites by Budiansky [6], Hill [24, 25], Eshelby [12], Christensen [8], Chow and Hermans [7], and many others (see literature in [11, 27, 40]. In order to reach the non-diluted regime one should find the way to solve the problem for infinite number of inclusions.

The dilute elastic composites were studied in many works [6, 7, 8, 24, 25] by solving exactly a simple one-inclusion problem. As expected, such an approach does not agree with the advanced theory of homogenization [3] applied in the another limiting case of closely spaced inclusions. The latter situation was studied in great detail within the framework of structural approximations in the spirit of percolation ideas [1, 4].

In essence, Maxwell defined the effective properties of a composite material by equating the asymptotic far-field solutions of inclusions embedded in an infinite isotropic matrix and of a single equivalent inclusion embedded in the same matrix. Maxwell's approach can be extended to finite clusters of inclusions yielding a formula for the effective tensor of the dilute equal sets C on the plane (see Fig. 3.1 on page 43).

Our approach amounts to a modified Maxwell's approach. It is amended with the MMM principle of Hashin and applied to doubly periodic composites. At this point we consider the problem for infinite number of inclusions. In particular, the problem of the divergent integral [34, 38, 39, 43] is resolved by means of the Eisenstein summation. Thus, one can proceed analytically and derive the power series in concentration for the effective elastic quantities.

In the present chapter, we deduce formulae for two-phase elastic composites with isotropic components with infinite identical unidirectional fibers of circular section. Problems for fibrous composites refer to the plane strain because it serves as an 2D approximation of the 3D fibrous composites.[1] Such an idealized model of arrays of infinitely long, aligned cylinders in a matrix is physically relevant for a fiber-reinforced materials [36].

Let the axis x_3 be parallel to the unidirectional fibres and a section perpendicular to x_3 form on the plane of variables x_1 and x_2 a macroscopically isotropic structure. It is assumed that fibers and matrix are occupied by isotropic elastic materials described by the shear moduli μ_1 and μ, Young's moduli E_1 and E, where the subscript 1 denote the elastic constants for inclusions. We also use the Poisson ratio ν and Muskhelishvili's constant $\kappa = 3 - 4\nu$ for the plane strain [36], and the transverse bulk modulus $k = \frac{\mu}{1-2\nu} = \frac{2\mu}{\kappa-1}$.

Consider the transverse effective moduli μ_e, E_e and ν_e in the plane of isotropy. The longitudinal effective constants are denoted as μ^L, E^L and ν^L. The transverse moduli are related by equation $E_e = 2\mu_e(1 + \nu_e)$. Every transversely isotropic material is described by five independent elastic moduli [29]. However, the considered two-phase fibrous composite at the beginning has only four independent entries μ_1, μ, E_1 and E. Hill [24, 25] proved that transversely isotropic two-phase fibrous composite are described by three independent elastic moduli since two longitudinal effective moduli

---

[1]The alternative plane stress problems deal with elastic thin plates. These two stress-deformation statements are described by slightly different 2D equations [36].

are expressed through others, namely,

$$E^L = E_1 f + E(1 - f) + 4 \left(\frac{\nu_1 - \nu}{\frac{1}{k_1} - \frac{1}{k}} \right)^2 \left(\frac{f}{k_1} + \frac{1 - f}{k} - \frac{1}{k_e} \right), \qquad (10.1.1)$$

$$\nu^L = \nu_1 f + \nu(1 - f) - \frac{\nu_1 - \nu}{\frac{1}{k_1} - \frac{1}{k}} \left(\frac{f}{k_1} + \frac{1 - f}{k} - \frac{1}{k_e} \right), \qquad (10.1.2)$$

where k_e denotes the effective transverse bulk modulus. Relations between 2D and 3D elastic moduli of fibrous composites are discussed in [11]. In the present chapter, we shall determine the transverse effective moduli.

Formulae for dilute composites in the first order approximation in the concentration f were deduced by various self-consistent methods [6, 7, 8, 24, 25, 26, 41]. Though all these formulae are formally different, they are equivalent up to $O(f^2)$ and the effective constants μ_e and k_e can be written in the form [5]

$$\frac{\mu_e}{\mu} = 1 + (\kappa + 1) \frac{\frac{\mu_1}{\mu} - 1}{\kappa \frac{\mu_1}{\mu} + 1} f + O(f^2), \qquad (10.1.3)$$

$$\frac{k_e}{k} = 1 + (\kappa + 1) \frac{\frac{\mu_1}{\mu} - \frac{\kappa_1 - 1}{\kappa - 1}}{\kappa_1 - 1 + 2\frac{\mu_1}{\mu}} f + O(f^2). \qquad (10.1.4)$$

The relations between the 2D moduli and the 3D elastic moduli marked by prime read as follows [11, 41]

$$\nu = \frac{\nu'}{1 - \nu'}, \quad E = \frac{E'}{(1 - \nu')(1 + \nu')}, \quad \mu = \mu', \quad k = k' + \frac{\mu}{3}. \qquad (10.1.5)$$

They are obtained by comparison of the Hook law written in 2D and 3D statements. The two definitions for moduli transpire when the plane elasticity is considered as a pure 2D problem and formally as the full 3D problem, respectively.

Formulae analogous to (10.1.3), (10.1.4) were deduced for other shapes and for multiphase composites (see [27, 40], and works cited therein). A lot of efforts have been exerted to extend formulae (10.1.3), (10.1.4) to high orders through extending the SCM, apparently without any justification. In the present section, we suggest and realize a rigorous method to extend (10.1.3), (10.1.4) to higher-order terms in concentrations following the methodology developed in previous chapters.

We present below the results for effective quantities in the order $O(f^3)$ in order to explicitly demonstrate the principal impossibility to deduce the high-order formulae without using of the information on microstructure. As above such information will be given in terms of the high order basic sums (statistical moments).

Any expression for the effective constants must obey the 2D Hashin-Shtrikman bounds . Let $\mu_1 \geq \mu$ and $k_1 \geq k$. Then, we have [21, 22]

$$\mu^- \leq \mu_e \leq \mu^+, \quad k^- \leq k_e \leq k^+, \tag{10.1.6}$$

where

$$k^- = k + \frac{f}{\frac{1}{k_1-k} + \frac{1-f}{k+\mu}}, \quad k^+ = k_1 + \frac{1-f}{\frac{1}{k-k_1} + \frac{f}{k_1+\mu_1}}, \tag{10.1.7}$$

$$\mu^- = \mu + \frac{f}{\frac{1}{\mu_1-\mu} + \frac{(1-f)(k+2\mu)}{2\mu(k+\mu)}}, \quad \mu^+ = \mu_1 + \frac{1-f}{\frac{1}{\mu-\mu_1} + \frac{f(k_1+2\mu_1)}{2\mu_1(k_1+\mu_1)}}. \tag{10.1.8}$$

The bounds (10.1.7), (10.1.8) can be rewritten through other elastic constants. For instance,

$$\frac{\mu^-}{\mu} = \frac{1 + \frac{\frac{\mu_1}{\mu}-1}{\kappa\frac{\mu_1}{\mu}+1}f}{1 - \frac{\frac{\mu_1}{\mu}-1}{\kappa\frac{\mu_1}{\mu}+1}\kappa f} = 1 + (\kappa+1)\frac{\frac{\mu_1}{\mu}-1}{\kappa\frac{\mu_1}{\mu}+1}f + O(f^2). \tag{10.1.9}$$

One can see that (10.1.3) coincides up to $O(f^2)$ with the low HS bounds μ^-. The same asymptotic equivalence holds for (10.1.4) and k^-.

The effective constants can be estimated by two methods leading to the same result. The first method is based on the doubly periodic problems. It was proposed and constructively applied by using of integral equations [18, 19, 20, 23]. In the present section, we develop the second method following the MMM principle (3.2.4) discussed in Chapter 3.

2. Method of functional equations for local fields

We begin our study with a finite number n of inclusions on the infinite plane. This number n is given in a symbolic form with an implicit purpose to pass to the limit $n \to \infty$ later. Mutually disjoint disks $\mathbb{D}_k := \{z \in \mathbb{C} : |z - a_k| < r\}$ $(k = 1, 2, \ldots, n)$ are considered in the complex plane $\mathbb{C}$. Let $\mathbb{D} := \mathbb{C} \cup \{\infty\} \setminus \left(\cup_{k=1}^n \mathbb{D}_k \cup \partial\mathbb{D}_k\right)$, where $\partial\mathbb{D}_k := \{t \in \mathbb{C} : |t - a_k| = r\}$. We assume that $\partial\mathbb{D}_k$ are oriented clockwise.

The components of the stress tensor can be determined by the Kolosov–Muskhelishvili formulae [36][2]

$$\sigma_{xx} + \sigma_{yy} = \begin{cases} 4\mathrm{Re}\ \varphi_k'(z), & z \in \mathbb{D}_k, \\ 4\mathrm{Re}\ \varphi_0'(z), & z \in \mathbb{D}, \end{cases} \tag{10.2.10}$$

---

[2]New designation $x = x_1$ and $y = x_2$ for the plane coordinates are introduced here in order to distinguish the components σ_{ij} $(i, j = 1, 2)$ of the effective conductivity tensor considered in the previous chapters and the components $\sigma_{xx}, \sigma_{xy} = \sigma_{yx}, \sigma_{yy}$ of the stress tensor discussed in the present chapter. Hence, the used complex variables have the form $z = x + iy$ and $t = x + iy$.

$$\sigma_{xx} - \sigma_{yy} + 2i\sigma_{xy} = \begin{cases} -2\left[\overline{z\varphi_k''(z)} + \overline{\psi_k'(z)}\right], & z \in \mathbb{D}_k, \\ -2\left[\overline{z\varphi_0''(z)} + \overline{\psi_0'(z)}\right], & z \in \mathbb{D}. \end{cases}$$

The strain tensor components ϵ_{xx}, ϵ_{xy}, ϵ_{yy} are determined by the formulae [36]

$$\epsilon_{xx} + \epsilon_{yy} = \begin{cases} \frac{\kappa_1 - 1}{\mu_1}\, \mathrm{Re}\, \varphi_k'(z), & z \in \mathbb{D}_k, \\ \frac{\kappa - 1}{\mu}\, \mathrm{Re}\, \varphi_0'(z), & z \in \mathbb{D}, \end{cases} \tag{10.2.11}$$

$$\epsilon_{xx} - \epsilon_{yy} + 2i\epsilon_{xy} = \begin{cases} -\frac{1}{\mu_1}\left[\overline{z\varphi_k''(z)} + \overline{\psi_k'(z)}\right], & z \in \mathbb{D}_k, \\ -\frac{1}{\mu}\left[\overline{z\varphi_0''(z)} + \overline{\psi_0'(z)}\right], & z \in \mathbb{D}. \end{cases}$$

Let

$$\sigma^\infty = \begin{pmatrix} \sigma_{xx}^\infty & \sigma_{xy}^\infty \\ \sigma_{yx}^\infty & \sigma_{yy}^\infty \end{pmatrix} \tag{10.2.12}$$

be the stress tensor applied at infinity. Following [36] let us introduce the constants

$$B_0 = \frac{\sigma_{xx}^\infty + \sigma_{yy}^\infty}{4}, \quad \Gamma_0 = \frac{\sigma_{yy}^\infty - \sigma_{xx}^\infty + 2i\sigma_{xy}^\infty}{2}. \tag{10.2.13}$$

Then,

$$\varphi_0(z) = B_0 z + \varphi(z), \quad \psi_0(z) = \Gamma_0 z + \psi(z), \tag{10.2.14}$$

where $\varphi(z)$ and $\psi(z)$ are analytical in $\mathbb{D}$ and bounded at infinity. The functions $\varphi_k(z)$ and $\psi_k(z)$ are analytical in $\mathbb{D}_k$ and twice differentiable in the closures of the considered domains. Special attention will be paid to two independent elastic states, namely, the uniform shear stress

$$\sigma_{xx}^\infty = \sigma_{yy}^\infty = 0,\ \sigma_{xy}^\infty = \sigma_{yx}^\infty = 1 \ \Leftrightarrow\ B_0 = 0,\ \Gamma_0 = i \tag{10.2.15}$$

and the uniform simple tension at infinity

$$\sigma_{xx}^\infty = \sigma_{yy}^\infty = 2,\ \sigma_{xy}^\infty = \sigma_{yx}^\infty = 0 \ \Leftrightarrow\ B_0 = 1,\ \Gamma_0 = 0. \tag{10.2.16}$$

The general case of the given stresses at infinity can be obtained by a linear combination of these two external loadings.

The perfect bonding at the matrix-inclusion interface can be expressed by two equations [36]

$$\varphi_k(t) + t\overline{\varphi_k'(t)} + \overline{\psi_k(t)} = \varphi_0(t) + t\overline{\varphi_0'(t)} + \overline{\psi_0(t)}, \tag{10.2.17}$$

$$\kappa_1\varphi_k(t) - t\overline{\varphi_k'(t)} - \overline{\psi_k(t)} = \frac{\mu_1}{\mu}\left(\kappa\varphi_0(t) - t\overline{\varphi_0'(t)} - \overline{\psi_0(t)}\right). \tag{10.2.18}$$

The problem (10.2.17), (10.2.18) is the classic boundary value problem of the plane elasticity. It was discussed by many authors and studied by various methods [23, 36]. Below, we concentrate our attention on its analytical solution.

Introduce the new unknown functions

$$\Phi_k(z) = \left(\frac{r^2}{z - a_k} + \overline{a_k} \right) \varphi'_k(z) + \psi_k(z), \ |z - a_k| \leq r,$$

analytic in $\mathbb{D}_k$ except the point a_k, where its principal part has the form $r^2 (z - a_k)^{-1} \varphi'_k(a_k)$.

Let $z^*_{(k)} = r^2 (\overline{z - a_k})^{-1} + a_k$ denote the inversion with respect to the circle $\partial \mathbb{D}_k$. If a function $f(z)$ is analytic in $|z - a_k| < r$, then $\overline{f(z^*_{(k)})}$ is analytic in $|z - a_k| > r$. The problem (10.2.17), (10.2.18) was reduced in [35] (see equations (5.6.11) and (5.6.16) in Chapter 5), [32] to the system of functional equations

$$\left(\frac{\mu_1}{\mu} + \kappa_1 \right) \varphi_k(z) = \left(\frac{\mu_1}{\mu} - 1 \right) \sum_{m \neq k} \left[\overline{\Phi_m(z^*_{(m)})} - (z - a_m) \overline{\varphi'_m(a_m)} \right] -$$

$$- \left(\frac{\mu_1}{\mu} - 1 \right) \overline{\varphi'_k(a_k)} (z - a_k) + \frac{\mu_1}{\mu} (1 + \kappa) B_0 z + p_0, \ |z - a_k| \leq r, \quad (10.2.19)$$

$$\left(\kappa \frac{\mu_1}{\mu} + 1 \right) \Phi_k(z) = \left(\kappa \frac{\mu_1}{\mu} - \kappa_1 \right) \sum_{m \neq k} \overline{\varphi_m(z^*_{(m)})} +$$

$$\left(\frac{\mu_1}{\mu} - 1 \right) \sum_{m \neq k} \left(\frac{r^2}{z - a_k} + \overline{a_k} - \frac{r^2}{z - a_m} - \overline{a_m} \right) \left[\overline{\left(\Phi_m(z^*_{(m)}) \right)'} - \overline{\varphi'_m(a_m)} \right] +$$

$$\frac{\mu_1}{\mu} (1 + \kappa) B_0 \left(\frac{r^2}{z - a_k} + \overline{a_k} \right) + \frac{\mu_1}{\mu} (1 + \kappa) \Gamma_0 z + \omega(z), \ |z - a_k| \leq r, \ k = 1, 2, \dots, n.$$

$$(10.2.20)$$

where

$$\omega(z) = \sum_{k=1}^{n} \frac{r^2 q_k}{z - a_k} + q_0, \quad (10.2.21)$$

q_0 is a constant and

$$q_k = \varphi'_k(a_k) \left((\kappa - 1) \frac{\mu_1}{\mu} - (\kappa_1 - 1) \right) - \overline{\varphi'_k(a_k)} \left(\frac{\mu_1}{\mu} - 1 \right), \ k = 1, 2, \dots, n. \quad (10.2.22)$$

The unknown functions $\varphi_k(z)$ and $\Phi_k(z)$ $(k = 1, 2, \dots, n)$ are related by $2n$ equations (10.2.19), (10.2.20). One can see that the functional equations do not contain integral operators but contain compositions of $\varphi_k(z)$ and $\Phi_k(z)$ with inversions. These compositions define compact operators in a Banach space [35].

The functions $\varphi(z)$ and $\psi(z)$ are expressed through $\varphi_k(z)$ and $\psi_k(z)$ by formulae

$$\frac{\mu_1}{\mu}(1+\kappa)\varphi(z) = \left(\frac{\mu_1}{\mu}-1\right)\sum_{m=1}^{n}\left[\overline{\Phi_m(z_{(m)}^*)} - (z-a_m)\overline{\varphi_m'(a_m)}\right] + p_0, \ z \in \mathbb{D}, \quad (10.2.23)$$

$$\frac{\mu_1}{\mu}(1+\kappa)\psi(z) = \omega(z) - \left(\frac{\mu_1}{\mu}-1\right)\sum_{m=1}^{n}\left(\frac{r^2}{z-a_m}+\overline{a_m}\right)\left[\left(\overline{\Phi_m(z_{(m)}^*)}\right)' - \overline{\varphi_m'(a_m)}\right] +$$

$$+ \left(\kappa\frac{\mu_1}{\mu}-\kappa_1\right)\sum_{m=1}^{n}\overline{\varphi_m(z_{(m)}^*)}, \ z \in \mathbb{D}. \quad (10.2.24)$$

We are looking for the complex potentials φ_k and ψ_k up to $O(r^6)$ in the form

$$\varphi_k(z) = \varphi_k^{(0)}(z) + r^2\varphi_k^{(1)}(z) + r^4\varphi_k^{(2)}(z) + O(r^6) \quad (10.2.25)$$

and

$$\psi_k(z) = \psi_k^{(0)}(z) + r^2\psi_k^{(1)}(z) + r^4\psi_k^{(2)}(z) + O(r^6). \quad (10.2.26)$$

Remark 20. The analytical dependencies of the complex potentials (10.2.25), (10.2.26) on r^2 near $r = 0$ follow from the compactness of the operators defined by the system of functional equations (10.2.19)-(10.2.20) in a Banach space and by the uniform convergence of successive approximations for small r [32, 35].

Remark 21. On page 292, we introduce a dimensional rectangle of the unit area. This justifies the consideration of the formally small non-dimensional parameter r^2 through the ratio of the disk area πr^2 to the rectangle area.

The functions $\varphi_k^{(s)}$ and $\psi_k^{(s)}$ ($s = 0, 1, 2$) in each inclusion are presented by their Taylor series. It is sufficient to take only first three terms

$$\varphi_k^{(s)}(z) = \alpha_{k,0}^{(s)} + \alpha_{k,1}^{(s)}(z-a_k) + \alpha_{k,2}^{(s)}(z-a_k)^2 + O(|z-a_k|^3), \quad (10.2.27)$$

$$\psi_k^{(s)}(z) = \beta_{k,0}^{(s)} + \beta_{k,1}^{(s)}(z-a_k) + \beta_{k,2}^{(s)}(z-a_k)^2 + O((z-a_k)^3), \ \text{as } z \to a_k. \quad (10.2.28)$$

The precision $O(|z-a_k|^3)$ is taken here following the reasoning to be explained below after equations (10.2.29–(10.2.31).

Introduce an auxiliary constants $\eta_{m,l}^{(s)}$ for brevity

$$\eta_{m,l}^{(s)} = (l+2)r^2\alpha_{m,l+2}^{(s)} + \overline{a_m}(l+1)\alpha_{m,l+1}^{(s)} + \beta_{m,l}^{(s)} \quad s = 0, 1; \ l = 1, 2. \quad (10.2.29)$$

Substitution of (10.2.25) and (10.2.26) in (10.2.19) and (10.2.20) yields

$$\left(\frac{\mu_1}{\mu} + \kappa_1\right) \sum_{p=0,1,2} r^{2p} \varphi_k^{(p)}(z) =$$

$$= \left(\frac{\mu_1}{\mu} - 1\right) \sum_{m \neq k} \sum_{l+s \leq 2} r^{2(s+l)} \overline{\eta}_{m,l}^{(s)}(z - a_m)^{-l} +$$

$$+ \left(\frac{\mu_1}{\mu}(1 + \kappa)B_0 - \left(\frac{\mu_1}{\mu} - 1\right) \sum_{s=0,1,2} r^{2s} \overline{\alpha}_{k,1}^{(s)}\right)(z - a_k) + p_1 + O(r^6) \quad (10.2.30)$$

and

$$\left(\kappa\frac{\mu_1}{\mu} + 1\right) \sum_{p=0,1,2} r^{2p} \left(\overline{a_k}\left(\varphi_k^{(p)}(z)\right)' + \psi_k^{(p)}(z)\right) =$$

$$= -\frac{\mu_1}{\mu} \sum_{s=0,1} r^{2(s+1)} \left(\varphi_k^{(s)}(z)\right)'(z - a_k)^{-1} +$$

$$+ \left(\kappa\frac{\mu_1}{\mu} - \kappa_1\right) \sum_{m \neq k} \sum_{l+s \leq 2} r^{2(l+s)} \overline{\alpha}_{m,l}^{(s)}(z - a_m)^{-l} -$$

$$- \left(1 - \frac{\mu}{\mu_1}\right) \sum_{m \neq k} r^4 \overline{\eta}_{m,1}^{(0)}(z - a_k)^{-1}(z - a_m)^{-2} +$$

$$+ \left(\frac{\mu_1}{\mu} - 1\right) \sum_{m \neq k} \sum_{l+s \leq 1} lr^{2(l+s+1)} \overline{\eta}_{m,l}^{(s)}(z - a_m)^{-l-2} -$$

$$- \left(1 - \frac{\mu}{\mu_1}\right) \sum_{m \neq k} \sum_{l+s \leq 2} lr^{2(l+s)} \overline{\eta}_{m,l}^{(s)}(\overline{a_k - a_m})(z - a_m)^{-l-1} +$$

$$+ \frac{\mu_1}{\mu}(1 + \kappa)\Gamma_0(z - a_k) + \sum_{m \neq k} r^2 q_m(z - a_m)^{-1} + q_0 + O(r^6), \quad (10.2.31)$$

where the sum $\sum_{l+s \leq 2}$ contains three terms with $l = 1, 2$ and $s = 0, 1$ satisfying the inequality $l + s \leq 2$. The sum $\sum_{m \neq k}$ means that m runs over $\{1, 2, ..., n\}$ except k. Other sums are defined analogously. One can check that the higher-order terms $(z - a_k)^l$ (for $l \geq 3$) in (10.2.27), (10.2.28) produce terms of order $O(r^6)$ in (10.2.30), (10.2.31). Such a rule takes place in general case when the terms $(z - a_k)^l$ yield $O(r^{2l})$. To our benefit, the constants p_1, q_0, $\alpha_{m,0}^{(s)}$ and $\beta_{m,0}^{(s)}$ ($m = 1, \ldots, n$, $s = 0, 1, 2$) in (10.2.30)-(10.2.31) could be deleted since they determine parallel translations not contributing to the stress and deformation fields.

The following combinations of elastic constants are introduced for convenience

$$\gamma_1 = \frac{\frac{\mu_1}{\mu} - 1}{\kappa\frac{\mu_1}{\mu} + 1}, \quad \gamma_2 = \frac{\kappa\frac{\mu_1}{\mu} - \kappa_1}{\frac{\mu_1}{\mu} + \kappa_1}, \quad \gamma_3 = \frac{1 - \frac{\mu_1}{\mu}\frac{\kappa-1}{\kappa_1-1}}{1 + 2\frac{\mu_1}{\mu}\frac{1}{\kappa_1-1}}. \tag{10.2.32}$$

Selecting the terms with the same powers r^{2p}, we arrive at the following iterative scheme for equations (10.2.30), (10.2.31). Straightforward computations give the approximate formulae

$$\varphi(z) = r^2\varphi^{(1)}(z) + r^4\varphi^{(2)}(z) + O(r^6), \tag{10.2.33}$$

where

$$\varphi^{(1)}(z) = \overline{\Gamma_0}\gamma_1 \sum_{k=1}^{n} \frac{1}{z - a_k},$$

$$\varphi^{(2)}(z) = 2\sum_{k=1}^{n}\left[B_0\gamma_1\gamma_3 \sum_{m\neq k} \frac{1}{(a_m - a_k)^2} + \overline{\Gamma_0}\gamma_1^2 \sum_{m\neq k} \frac{\overline{a_m - a_k}}{(a_m - a_k)^3}\right]\frac{1}{z - a_k}. \tag{10.2.34}$$

The third order approximation is written below in the case (10.2.15)

$$\varphi^{(3)}(z) = -i\left[3\gamma_1^2 \sum_{l=1}^{n}\sum_{m\neq l} \frac{1}{(a_l - a_m)^4}(z - a_l)^{-2}\right.$$

$$+ \left(4\gamma_1^3 \sum_{l=1}^{n}\sum_{m\neq l}\sum_{m_1\neq m} \frac{a_l - a_m}{(a_l - a_m)^3}\frac{\overline{a_m - a_{m_1}}}{(a_m - a_{m_1})^3}(z - a_l)^{-1}\right. \tag{10.2.35}$$

$$+ 6\gamma_1^2 \sum_{l=1}^{n}\sum_{m\neq l} \frac{1}{(a_l - a_m)^4}(z - a_l)^{-1}$$

$$+ \gamma_1^2\gamma_3 \sum_{l=1}^{n}\sum_{m\neq l}\sum_{m_1\neq m} \frac{1}{(a_l - a_m)^2}\frac{1}{(a_m - a_{m_1})^2}(z - a_l)^{-1}$$

$$\left.\left.- \gamma_1^2\gamma_3 \sum_{l=1}^{n}\sum_{m\neq l}\sum_{m_1\neq m} \frac{1}{(a_l - a_m)^2}\frac{1}{(a_m - a_{m_1})^2}(z - a_l)^{-1}\right)\right].$$

The function $\varphi(z)$ has another form in the case (10.2.16). Its approximations are calculated by formulae

$$\varphi^{(1)}(z) = 0, \quad \varphi^{(2)}(z) = 2\gamma_1\gamma_3 \sum_{l=1}^{n}\sum_{m\neq l} \frac{1}{(a_m - a_l)^2}(z - a_l)^{-1}.$$

$$\varphi^{(3)}(z) = 2\gamma_1\gamma_3 \left(2\gamma_1 \sum_{l=1}^{n} \sum_{m \neq l} \sum_{m_1 \neq m} \frac{a_l - a_m}{(a_l - a_m)^3} \frac{1}{(a_m - a_{m_1})^2} (z - a_l)^{-1} \right. \tag{10.2.36}$$

$$\left. - \sum_{l=1}^{n} \sum_{m \neq l} \frac{1}{(a_m - a_l)^3} (z - a_l)^{-2} \right).$$

3. Averaged fields in finite composites

3.1. Averaged shear modulus

In the present section, we calculate the averaged shear modulus $\mu_e^{(n)}$ of the considered finite composite with n inclusions on the plane. In order to calculate $\mu_e^{(n)}$ it is sufficient to consider the uniform shear stress (10.2.15). Introduce the average over a sufficiently large rectangle Q_n containing all the inclusions $\mathbb{D}_k$

$$\langle w \rangle_n = \frac{1}{|Q_n|} \int_{Q_n} w \, d\mathbf{x}, \tag{10.3.37}$$

where $d\mathbf{x} = dx_1 dx_2$. The macroscopic shear modulus can be computed by means of (10.3.37)

$$\mu_e^{(n)} = \frac{\langle \sigma_{12} \rangle_n}{2\langle \epsilon_{12} \rangle_n}. \tag{10.3.38}$$

Further, Q_n will be extended to infinity ($n \to \infty$) and we shall arrive at the macroscopic shear moduli $\mu_e = \lim_{n \to \infty} \mu_e^{(n)}$. The stress tensor components are calculated by (10.2.10), and the deformation tensor components by (10.2.11).

Instead of direct computations according to formula (10.3.38) in terms of the complex potentials, it is simpler to compute another quantity

$$P_n := \left\langle \frac{1}{2}(\sigma_{11} - \sigma_{22}) + i\sigma_{12} \right\rangle_n. \tag{10.3.39}$$

and then to take its imaginary part to compute $\langle \sigma_{12} \rangle_n$. Using the definition of the average (10.3.37) and formulae (10.2.10), we obtain

$$P_n = -\frac{1}{|Q_n|} \left\{ \int_{\mathbb{D}} \left[z\overline{\varphi_0''(z)} + \overline{\psi_0'(z)} \right] d\mathbf{x} + \sum_{k=1}^{n} \int_{\mathbb{D}_k} \left[z\overline{\varphi_k''(z)} + \overline{\psi_k'(z)} \right] d\mathbf{x} \right\}. \tag{10.3.40}$$

Green's formula in complex form will be used below

$$\int_{\mathbb{D}} \frac{\partial w(z)}{\partial \overline{z}} \, d\mathbf{x} = \frac{1}{2i} \int_{\partial \mathbb{D}} w(t) \, dt. \tag{10.3.41}$$

The boundary of $\mathbb{D}$ can be decomposed as follows $\partial \mathbb{D} = \partial Q_n - \sum_{k=1}^{n} \partial \mathbb{D}_k$ where ∂Q_n

and $\partial \mathbb{D}_k$ are positively oriented. Application of (10.3.41) to (10.3.40) yields $P_n = P'_n + P''_n$ where

$$P'_n = -\frac{1}{2i|Q_n|} \sum_{k=1}^{n} \int_{\partial \mathbb{D}_k} \left[t\overline{\varphi'_k(t)} + \overline{\psi_k(t)} - t\overline{\varphi'_0(t)} - \overline{\psi_0(t)} \right] dt, \qquad (10.3.42)$$

$$P''_n = -\frac{1}{2i|Q_n|} \int_{\partial Q_n} \left[t\overline{\varphi'_0(t)} + \overline{\psi_0(t)} \right] dt.$$

Green's formula (10.3.41) for $w = \bar{z}$ yields the area formula

$$\frac{1}{2i} \int_{\partial Q_n} \bar{t} \, dt = |Q_n|. \qquad (10.3.43)$$

Lemma 2. *Let Q_n be a rectangle of size $a \times b$ and a function $g(z)$ be analytic in Q'_n the complement of $Q_n \cup \Gamma_n$ to the extended complex plane $\widehat{\mathbb{C}} = \mathbb{C} \cup \{\infty\}$. Then,*

$$\lim_{\substack{a \to \infty \\ b \to \infty}} \frac{1}{ab} \int_{\partial Q_n} g(t) \, d\bar{t} = 0. \qquad (10.3.44)$$

Proof. It follows from the estimation

$$\left| \frac{1}{ab} \int_{\partial Q_n} g(t) \, d\bar{t} \right| \leq \max_{t \in \partial Q_n} |g(t)| \frac{2(a+b)}{ab} \qquad (10.3.45)$$

and boundedness of $|g(z)|$ in Q'_n. $\qquad \qquad \square$

It follows from (10.2.14), (10.3.43) and Lemma 1 that $\lim_{n \to \infty} P''_n = -\overline{\Gamma_0}$. Using the boundary condition (10.2.17), we obtain

$$P'_n = -\frac{1}{2i|Q_n|} \sum_{k=1}^{n} \int_{\partial \mathbb{D}_k} [\varphi_0(t) - \varphi_k(t)] \, dt = -\frac{1}{2i|Q_n|} \sum_{k=1}^{n} \int_{\partial \mathbb{D}_k} \varphi(t) dt. \qquad (10.3.46)$$

Here, we used equations $\int_{\partial \mathbb{D}_k} \varphi_k(t) dt = 0$ and $\int_{\partial \mathbb{D}_k} B_0 t \, dt = 0$ following from the Cauchy integral theorem. Therefore,

$$\lim_{n \to \infty} P_n = -\overline{\Gamma_0} + iA \quad \Rightarrow \quad \frac{1}{2} \langle \sigma_{12} \rangle = \text{Im} \, \Gamma_0 + \text{Re} \, A, \qquad (10.3.47)$$

where

$$A = \lim_{n \to \infty} \frac{1}{2|Q_n|} \sum_{k=1}^{n} \int_{\partial \mathbb{D}_k} \varphi(t) dt. \qquad (10.3.48)$$

Similar manipulations can be performed for

$$R_n = R'_n + R''_n = \left\langle \frac{1}{2}(\epsilon_{11} - \epsilon_{22}) + i\epsilon_{12} \right\rangle_n, \tag{10.3.49}$$

where

$$R'_n = -\frac{1}{|Q_n|} \frac{1}{G} \int_{\mathbb{D}} \left[z\overline{\varphi''_0(z)} + \overline{\psi'_0(z)} \right] d\mathbf{x} \tag{10.3.50}$$

$$= -\frac{1}{2i|Q_n|} \sum_{k=1}^{n} \int_{\partial \mathbb{D}_k} \left[\frac{1}{\mu_1} \left(\kappa_1 t \overline{\varphi'_k(t)} + \overline{\psi_k(t)} \right) - \frac{1}{\mu} \left(\kappa t \overline{\varphi'_0(t)} + \overline{\psi_0(t)} \right) \right] dt$$

and

$$R''_n = -\frac{1}{2i|Q_n|\mu} \int_{\partial Q_n} \left[t\overline{\varphi'_0(t)} + \overline{\psi_0(t)} \right] dt. \tag{10.3.51}$$

Using the boundary conditions (10.2.18), we get

$$\lim_{n\to\infty} R_n = \frac{1}{\mu}(-\overline{\Gamma_0} - i\,\kappa A) \;\Rightarrow\; \langle \epsilon_{12} \rangle = \frac{1}{\mu}(\operatorname{Im}\Gamma_0 - \kappa \operatorname{Re} A). \tag{10.3.52}$$

Substituting (10.3.47), (10.3.52) into (10.3.38) and taking the limit as n tends to infinity we obtain

$$\frac{\mu_e}{\mu} = \frac{\operatorname{Im}\Gamma_0 + \operatorname{Re} A}{\operatorname{Im}\Gamma_0 - \kappa \operatorname{Re} A}. \tag{10.3.53}$$

3.2. Averaged bulk modulus

It is sufficient to take the particular external stresses (10.2.16) to calculate the effective bulk modulus k_e. Following the previous Section 3.1, we calculate the averaged 2D bulk modulus by formula [9]

$$k_e^{(n)} = \frac{2\operatorname{Re} V_n}{\operatorname{Re} W_n}. \tag{10.3.54}$$

where $V_n = \frac{1}{4}\langle \sigma_{11} + \sigma_{22} \rangle_n$ and $W_n = \langle \epsilon_{11} + \epsilon_{22} \rangle_n$ are expressed in terms of the complex potentials by formulae (10.2.10) and (10.2.11). Therefore,

$$V_n = \frac{1}{|Q_n|} \left[\int_{\mathbb{D}} \varphi'_0(z)\, d\mathbf{x} + \sum_{l=1}^{n} \int_{\mathbb{D}_l} \varphi'_l(z)\, d\mathbf{x} \right] \tag{10.3.55}$$

and

$$W_n = \frac{1}{|Q_n|} \left[\frac{\kappa - 1}{\mu} \int_D \varphi_0'(z) \, d\mathbf{x} + \frac{\kappa_1 - 1}{\mu_1} \sum_{l=1}^{n} \int_{D_l} \varphi_l'(z) \, d\mathbf{x} \right]. \tag{10.3.56}$$

Green's formula will be used below in a complex form

$$\int_D \frac{\partial w(z)}{\partial z} \, d\mathbf{x} = -\frac{1}{2i} \int_{\partial D} w(t) \, d\bar{t}. \tag{10.3.57}$$

Using (10.2.14) and (10.2.16), we transform the integral

$$\frac{1}{|Q_n|} \int_D \varphi_0'(z) \, d\mathbf{x} = \frac{1}{|Q_n|} \int_D \, d\mathbf{x} + \frac{1}{|Q_n|} \int_D \varphi'(z) \, d\mathbf{x}.$$

The first integral tends to $(1 - f)$ as $n \to \infty$. The second integral becomes

$$\frac{1}{|Q_n|} \int_D \varphi'(z) \, d\mathbf{x} = -\frac{1}{2i|Q_n|} \int_{\partial Q_n} \varphi(t) d\bar{t} + \sum_{l=1}^{n} \frac{1}{2i|Q_n|} \int_{\partial D_l} \varphi(t) d\bar{t}.$$

It follows from Lemma 2 that $\lim_{n \to \infty} \frac{1}{2i|Q_n|} \int_{\partial Q_n} \varphi(t) d\bar{t} = 0$.

Using (10.3.57), we transform the next integral

$$\frac{1}{|Q_n|} \int_D \varphi_l'(z) \, d\mathbf{x} = -\frac{1}{2i|Q_n|} \int_{D_l} \varphi_l(z) \, d\bar{t}.$$

Consider the two limits

$$B = \lim_{n \to \infty} \frac{1}{2i|Q_n|} \sum_{l=1}^{n} \int_{\partial D_l} \varphi(t) d\bar{t}, \quad C = -\lim_{n \to \infty} \frac{1}{2i|Q_n|} \sum_{l=1}^{n} \int_{\partial D_l} \varphi_l(t) d\bar{t}. \tag{10.3.58}$$

Then, the limit of (10.3.54) as n tends to infinity can be written in the form

$$\frac{k_e}{k} = \frac{1 - f + B + C}{1 - f + B + \frac{k}{k_1}C}. \tag{10.3.59}$$

4. Roadmap to composites represented by RVE

4.1. Finite sums

The analytical formulae for the effective constants are written in the previous section as limits of the averaged stresses and deformations for a finite number of inclusions n in the plane. These formulae hold for any number n given in a symbolic form. Therefore, one can pass to the limit $n \to \infty$, but this passage must be properly justified and calculated.

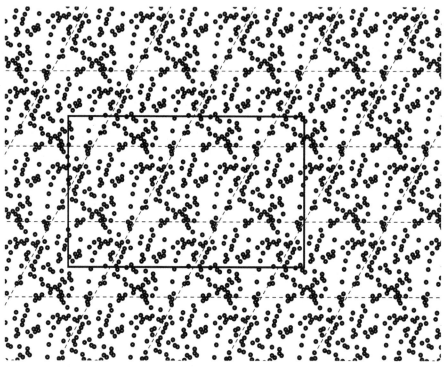

Figure 10.1 Schematic presentation of the infinite number of disks on the plane with the rectangle Q_n containing a finite number of disks. Q_n extends and embraces the corresponding points and the boundary of Q_n tends to the infinity, as $n \to \infty$.

Consider infinite number of the mutually disjoint disks $\mathbb{D}_k = \{z \in \mathbb{C} : |z - a_k| < r\}$ ($k = 1, 2, \ldots$) on the complex plane. Let the centers a_k be ordered in such a way that

$$|a_1| \leq |a_2| \leq |a_3| \leq \cdots . \tag{10.4.60}$$

Let $\mathbb{D}$ be the complement of the closed disks $|z - a_k| \leq r$ ($k = 1, 2, \ldots$) to the complex plane. As above let Q_n denote a rectangle containing first n disks $\mathbb{D}_1, \mathbb{D}_2, \ldots, \mathbb{D}_n$; $F_n = Q_n \setminus \cup_{k=1}^{n} (\mathbb{D}_k \cup \partial \mathbb{D}_k)$ (see Fig. 10.1). Let the finitely connected domains F_n tend to $\mathbb{D}$ as $n \to \infty$, i.e., ∂Q_n tends to the infinite point. The concentration of inclusions is introduced as the limit

$$f := \lim_{n \to \infty} \frac{n\pi r^2}{|Q_n|} = N\pi r^2 \quad \Rightarrow \quad |Q_n| \sim \frac{n}{N}, \text{ as } n \to \infty, \tag{10.4.61}$$

where N is the average number of inclusions per unit area, i.e., the area $|Q_n| \sim \frac{n}{N}$ as $n \to \infty$. This is equivalent to introduction of a dimensionless length scale.

First, consider the pure geometrical sums arisen in the formulae above

$$e_2(n) = \frac{1}{n} \sum_{k=1}^{n} \sum_{m \neq k} \frac{1}{(a_k - a_m)^2}, \qquad (10.4.62)$$

$$e_3^{(1)}(n) = \frac{1}{n} \sum_{k=1}^{n} \sum_{m \neq k} \frac{\overline{a_k - a_m}}{(a_k - a_m)^3}, \qquad (10.4.63)$$

$$e_4(n) = \frac{1}{n} \sum_{k=1}^{n} \sum_{m \neq k} \frac{1}{(a_k - a_m)^4}, \qquad (10.4.64)$$

$$e_{22}(n) = \frac{1}{n} \sum_{k=1}^{n} \sum_{m \neq k} \sum_{l \neq m} \frac{1}{(a_k - a_m)^2} \frac{1}{\overline{(a_m - a_l)}^2}, \qquad (10.4.65)$$

$$\widetilde{e}_{22}(n) = \frac{1}{n} \sum_{k=1}^{n} \sum_{m \neq k} \sum_{l \neq m} \frac{1}{(a_k - a_m)^2} \frac{1}{(a_m - a_l)^2}, \qquad (10.4.66)$$

$$e_{33}^{(1)}(n) = \frac{1}{n} \sum_{k=1}^{n} \sum_{m \neq k} \sum_{l \neq m} \frac{\overline{a_k - a_m}}{(a_k - a_m)^3} \frac{a_m - a_l}{(a_m - a_l)^3}. \qquad (10.4.67)$$

The conditional convergence of (10.4.62) and (10.4.63), as $n \to \infty$, is the principal obstacle preventing a constructive application of the Hashin's MMM principle and derivation of the analytical formulae for the effective constants. The limit of (10.4.62) is discussed in Example 3 on page 87 where it is justified that

$$e_2 := \lim_{n \to \infty} e_2(n) = \lim_{n \to \infty} \frac{1}{n} \sum_{k=1}^{n} \sum_{m \neq k} \frac{1}{(a_k - a_m)^2} = \pi \qquad (10.4.68)$$

for macroscopically isotropic composites.

4.2. The limit of sums

We now proceed to investigate the limit given by formula (10.4.63) for macroscopically isotropic composites represented by a doubly periodic structures.

First, let the centers a_k form a regular lattice, for instance, the square array $Q = \left\{ \frac{p}{\sqrt{N}} + \frac{iq}{\sqrt{N}} : p, q \in \mathbb{Z} \right\}$. Introduce the designation $P = p\omega_1 + q\omega_2$ for shortness. Then, the average number of disks per cell is $N = 1$ and (10.4.63) becomes the conditionally convergent lattice sum

$$S_3^{(1)} = \sum_{P \neq 0}^{e} \frac{\overline{P}}{P^3}, \tag{10.4.69}$$

where $\sum_{P \neq 0}^{e}$ stands for the Eisenstein summation over the integer numbers p and q except the term with $p = q = 0$.

Consider the function introduced by Natanzon [37] and investigated by Filshtinsky [18, 19]

$$\wp_1'(z) = -2 \sum_{P \neq 0} \left(\frac{\overline{P}}{(z - P)^3} + \frac{\overline{P}}{P^3} \right). \tag{10.4.70}$$

The series (10.4.70) is absolutely convergent. Using Eisenstein summation, we introduce the function

$$E_3^{(1)}(z) = \sum_{P}^{e} \frac{\overline{z - P}}{(z - P)^3}. \tag{10.4.71}$$

The functions (10.4.70) and (10.4.71) are related by equation

$$E_3^{(1)}(z) = -\frac{1}{2} \bar{z} \wp'(z) + \frac{1}{2} \wp_1'(z) + S_3^{(1)}, \tag{10.4.72}$$

where $\wp(z)$ denotes the Weierstrass elliptic function. Filshtinsky [19] expressed Natanzon's function $\wp_1'(z)$ in terms of the elliptic functions

$$\pi \wp_1'(z) = \frac{1}{3} \wp''(z) + [\zeta(z) - (S_2 - \pi)z] \wp'(z) - 2(S_2 - \pi)\wp(z) - 10 S_4. \tag{10.4.73}$$

Formula (10.4.73) is obtained from Filshtinsky's formulae (15) and (25) from Appendix 2 of [19] by using of Legendre's identity and formula (3.4.124). Substitution of (10.4.73) into (10.4.72) yields

$$\begin{aligned} E_3^{(1)}(z) = &-\frac{1}{2} \bar{z} \wp'(z) + \frac{1}{6\pi} \wp''(z) + \frac{1}{2} \left[\frac{\zeta(z)}{\pi} - \left(\frac{S_2}{\pi} - 1 \right) z \right] \wp'(z) \\ &- \left(\frac{S_2}{\pi} - 1 \right) \wp(z) - \frac{5}{\pi} S_4 + S_3^{(1)}. \end{aligned} \tag{10.4.74}$$

The latter formula is simplified for the hexagonal cell when $S_2 = \pi$ and $S_4 = 0$

$$E_3^{(1)}(z) = -\frac{1}{2} \bar{z} \wp'(z) + \frac{1}{6\pi} \wp''(z) + \frac{1}{2\pi} \zeta(z) \wp'(z) + S_3^{(1)}. \tag{10.4.75}$$

The passage to the limit in (10.4.63) for statistically homogeneous composites can be accomplished for doubly periodic composites. Below, we use designations from page 48 for the doubly periodic lattice Q. The set of centers of mutually disjoint disks

$$\mathcal{A} = \{a_k + P, k = 1, 2, \ldots, N; P = p\omega_1 + q\omega_2, \ p, q \in \mathbb{Z}\} \qquad (10.4.76)$$

generate a double periodic structure. This set $\mathcal{A}$ can be reordered in accordance with (10.4.60).

Introduce the designation

$$\langle F(z) \rangle = \frac{1}{N^2} \sum_{k=1}^{N} \sum_{m=1}^{N} F(a_k - a_m). \qquad (10.4.77)$$

The limit for doubly periodic structures can be calculated by formula

$$e_3^{(1)} = \lim_{n \to \infty} e_3^{(1)}(n) = \langle E_3^{(1)}(z) \rangle, \qquad (10.4.78)$$

where $E_3^{(1)}(0) := S_3^{(1)}$ similar to (3.4.137). It was proved in [43] that for the hexagonal lattice

$$\boxed{S_3^{(1)} = \frac{\pi}{2}}. \qquad (10.4.79)$$

Numerical simulations with high accuracy justify for macroscopically isotropic composites the conjecture

$$\boxed{e_3^{(1)} = \frac{\pi}{2}}. \qquad (10.4.80)$$

Let l be a natural number and $j = 0, 1$. Introduce the following generalization of $E_3^{(1)}(z)$

$$E_l^{(j)}(z) = \sum_{P}^{e} \frac{\left(\overline{z - P} \right)^j}{(z - P)^l}. \qquad (10.4.81)$$

The superscript j for $j = 0$ will be omitted below, i.e., we write $E_l^{(0)}(z) = E_l(z)$ for brevity. The limit of (10.4.64) for the doubly periodic structures is also met in the conductivity theory

$$e_4 = \lim_{n \to \infty} e_4(n) = \frac{1}{N^3} \sum_{k=1}^{N} \sum_{m=1}^{N} E_4(a_k - a_m), \qquad (10.4.82)$$

$$e_{22} = \lim_{n \to \infty} e_{22}(n) = \frac{1}{N^3} \sum_{k=1}^{N} \sum_{m=1}^{N} \sum_{l=1}^{N} E_2(a_k - a_m)\overline{E_2(a_m - a_l)}. \qquad (10.4.83)$$

Analogously [10]

$$e_{33}^{(1)} = \lim_{n \to \infty} e_{33}^{(1)}(n) = \frac{1}{N^3} \sum_{k=1}^{N} \sum_{m=1}^{N} \sum_{l=1}^{N} E_3^{(1)}(a_k - a_m) \overline{E_3^{(1)}(a_m - a_l)}, \qquad (10.4.84)$$

$$\widetilde{e}_{22} = \lim_{n \to \infty} \widetilde{e}_{22}(n) = \frac{1}{N^3} \sum_{k=1}^{N} \sum_{m=1}^{N} \sum_{l=1}^{N} E_2(a_k - a_m) E_2(a_m - a_l). \qquad (10.4.85)$$

5. Effective constants

The integral (10.3.48) can be calculated explicitly up to $O(f^3)$ by using approximations of the function $\varphi(z)$ given by (10.2.33), (10.2.34). First, we consider the terms up to $O(f^2)$. Applying the residue theorem for any fixed k we obtain up to $O(r^6)$

$$\frac{1}{2} \int_{\partial \mathbb{D}_k} \varphi(t)\, dt = \pi i\, r^2 \gamma_1 \overline{\Gamma_0} + 2\pi i\, r^4 \left[B_0 \gamma_1 \gamma_3 \sum_{m \neq k} \frac{1}{(a_m - a_k)^2} + \Gamma_0 \gamma_1^2 \sum_{m \neq k} \frac{a_m - a_k}{(a_m - a_k)^3} \right]. \qquad (10.5.86)$$

Substitution of (10.5.86) into (10.3.48) yields up to $O(r^6)$

$$\mathrm{Re}\, A = \lim_{n \to \infty} \frac{1}{|Q_n|} \left\{ n\pi r^2 \gamma_1 \mathrm{Im}\, \Gamma_0 - 2n\pi r^4 \left[B_0 \gamma_1 \gamma_3 \, \mathrm{Im}[e_2(n)] + \gamma_1^2 \, \mathrm{Im}[\Gamma_0 e_3^{(1)}(n)] \right] \right\}. \qquad (10.5.87)$$

Using of (10.4.68) and (10.4.80) give within the considered accuracy

$$\mathrm{Re}\, A = \mathrm{Im}\, [\Gamma_0] \left(N\pi r^2 \gamma_1 - N^2 \pi^2 r^4 \gamma_1^2 \right). \qquad (10.5.88)$$

Application of (10.4.61) yields

$$\mathrm{Re}\, A = \mathrm{Im}\, [\Gamma_0] \left(f\gamma_1 - f^2 \gamma_1^2 \right) + O(f^3). \qquad (10.5.89)$$

Then, (10.3.53) implies that

$$\frac{\mu_e}{\mu} = \frac{1 + f\gamma_1}{1 - \kappa f\gamma_1} + O(f^2), \qquad \frac{\mu_e}{\mu} = \frac{1 + f\gamma_1 - f^2 \gamma_1^2}{1 - \kappa \left(f\gamma_1 - f^2 \gamma_1^2 \right)} + O(f^3). \qquad (10.5.90)$$

Remarkably the first order approximaton (10.5.90) coincides with the Hashin–Shtrikman lower bound (10.1.9) for $\mu_1 \geq \mu$ in the form $\frac{\mu^-}{\mu} = \frac{1 + f\gamma_1}{1 - \kappa f\gamma_1}$. It is possible to check that high-order iterations applied to the functional equations (10.2.19), (10.2.20) generate the power terms $(f\gamma_1)^j$ which result in the following formula

$$\frac{\mu_e}{\mu} = \frac{1 + 2\gamma_1 f}{1 + (\kappa - 1)\gamma_1 f} + O(f^3), \qquad (10.5.91)$$

within the same precision! Noting that $\kappa \geq 1$ consider $\frac{\mu^-}{\mu}$ as an increasing function of the argument $\gamma_1 f$. Then, the higher-order formulae, i.e., the second formula (10.5.90) and formula (10.5.91), both give lower values than less precise exact lower bound $\frac{\mu^-}{\mu}$. Therefore, the higher-order formulae give less precise results than the Hashin–Shtrikman first-order expression. This fact demonstrates that the concentration expansion for elasticity is not so well understood as for the conductivity problem. Perhaps, a sufficiently large number of terms in f may clarify this question.

We found that the effective constant μ_e does not depend on the external stresses, i.e., on B_0 and Γ_0. The same holds for k_e. This fact was justified rigorously in [9] by another method. Therefore, it is sufficient to calculate the third-order approximation for $\varphi(z)$ under assumption (10.2.15)

$$\frac{1}{2} \int_{\partial \mathbb{D}_k} \varphi^{(3)}(t) \, dt = \pi \left[4\gamma_1^3 \sum_{m \neq l} \sum_{m_1 \neq m} \frac{a_k - a_m}{(a_k - a_m)^3} \frac{\overline{a_m - a_{m_1}}}{(a_m - a_{m_1})^3} \right.$$

$$\left. + 6\gamma_1^2 \sum_{m \neq k} \frac{1}{(a_l - a_m)^4} + \gamma_1^2 \gamma_3 \sum_{m \neq l} \sum_{m_1 \neq m} \frac{1}{(a_k - a_m)^2} \left(\frac{1}{(a_m - a_{m_1})^2} - \frac{1}{(a_m - a_{m_1})^2} \right) \right].$$

This integral gives the next approximation to (10.5.89) in the case $\operatorname{Im} \Gamma_0 = 1$

$$\operatorname{Re} A = f\gamma_1 - f^2\gamma_1^2 + f^3\gamma_1^2 \left(4\gamma_1 \operatorname{Re} e_{33}^{(1)} + 6\operatorname{Re} e_4 + \gamma_3 \operatorname{Re} (\widetilde{e_{22}} - e_{22}) \right) + O(f^4),$$

$$(10.5.92)$$

where the sums (10.4.84), (10.4.85) are used. In the case of the regular hexagonal array $e_{33}^{(1)} = \left(S_3^{(1)} \right)^2 = \frac{\pi^2}{4}$ by (10.4.79), $\widetilde{e_{22}} = e_{22} = S_2^2 = \pi^2$ and $e_4 = S_4 = 0$ as shown in Chapter 3, (3.4.126). Similarly in the case of square array one can use the same results but with different $e_4 = S_4 \approx 3.151211$ (see Chapter 3, (3.4.127)).

Equation (10.3.53) implies the following formula for composites with arbitrary locations of inclusions in the periodicity cell (RVE)

$$\boxed{\frac{\mu_e}{\mu} = \frac{1 + f\gamma_1 - f^2\gamma_1^2 + f^3\gamma_1^2 \left(4\gamma_1 \operatorname{Re} e_{33}^{(1)} + 6\operatorname{Re} e_4 + \gamma_3 \operatorname{Re} (\widetilde{e_{22}} - e_{22}) \right)}{1 - \kappa \left[f\gamma_1 - f^2\gamma_1^2 + f^3\gamma_1^2 \left(4\gamma_1 \operatorname{Re} e_{33}^{(1)} + 6\operatorname{Re} e_4 + \gamma_3 \operatorname{Re} (\widetilde{e_{22}} - e_{22}) \right) \right]} + O(f^4)}$$

$$(10.5.93)$$

The centers a_k, hence the sums $e_{33}^{(1)}$ and e_4 can be considered as random variables. Therefore, $e_{33}^{(1)}$ and e_4 can be computed for every prescribed probabilistic distribution of a_k in the RVE as it is done for the conductivity problem. Formula (10.5.93) demonstrates that it is impossible to determine the third-order approximation for μ_e of random composites without the statistical (probabilistic) data described by the e-sums $e_{33}^{(1)}$ and e_4. Of course, different distributions yield different values of the e-sums, hence, different μ_e.

Similar manipulations can be performed for the effective bulk modulus k_e. The low-order approximations for (10.3.58) has the form

$$B = 2\gamma_1\gamma_3 f^3 + O(f^4), \quad C = \gamma_2 f - 2\gamma_1\gamma_2\gamma_3 f^3 + O(f^4).$$ (10.5.94)

Using (10.5.94) and (10.3.54), we obtain

$$\boxed{\frac{k_e}{k} = \frac{1 + (\gamma_2 - 1)f - 2\gamma_1\gamma_3(\gamma_2 - 1)f^3}{1 + \left(\frac{k}{k_1}\gamma_2 - 1\right)f - 2\gamma_1\gamma_3\left(\frac{k}{k_1}\gamma_2 - 1\right)f^3} + O(f^4)}.$$ (10.5.95)

Again, the first-order approximation

$$\boxed{\frac{k_e}{k} = \frac{1 + (\gamma_2 - 1)f}{1 + \left(\frac{k}{k_1}\gamma_2 - 1\right)f} + O(f^2)}$$ (10.5.96)

coincides with the lower Hashin-Shtrikman bound $\frac{k^-}{k}$ from (10.1.7).

REFERENCE

1. Andrianov IV, Manevitch LI, Asymptotology: Ideas, Methods, and Applications. Dordrecht, Boston, London: Kluwer Academic Publishers; 2002.
2. Baker GA, Graves-Moris P, Padé Approximants. Cambridge, UK: Cambridge University; 1996.
3. Bakhvalov NS, Panasenko GP, Homogenization: Averaging processes in periodic media. Nauka, Moscow (1984) (in Russian); English transl. Dordrecht /Boston /London Kluwer; 1989.
4. Berlyand L, Kolpakov AG, Novikov A, Introduction to the Network Approximation Method for Materials Modeling. Cambridge: Cambridge University Press; 2012.
5. Bertoldi K. Bigoni D, Drugan WJ, Structural interface in linear elasticity. Part II: Effective properties and neutrality Journal of the Mechanics and Physics of Solids, 2007; 55: 35–63.
6. Budiansky B, On the elastic moduli of some heterogeneous material, Journal of the Mechanics and Physics of Solids, 1965; 13: 223–227.
7. Chow TS, Hermans JJ, The Elastic Constants of Fiber Reinforced Materials, Journal of Composite Materials, 1969; 3: 382396.
8. Christensen RM, Mechanics of Composite Materials. New York: J. Wiley; 1979.
9. Drygas P, Mityushev V, Effective elastic properties of random two-dimensional composites, International Journal of Solids and Structures 2016; 97-98: 543–553.
10. Drygas P, Generalized Eisenstein functions, Journal of Mathematical Analysis and Applications 2016; 444: 1321–1331.
11. Eischen JW, Torquato S, Determining elastic behavior of composites by the boundary element method, J. Appl. Phys. 1993; 74: 159–170.
12. Eshelby JD, The determination of Elastic Field of Ellipsoidal Inclusion and Related Problems, Proc. R. Soc. A London 1957; 241: 376–396.
13. Filishtinskii LA, Stresses and displacements in an elastic sheet weakened by a doubly periodic set of equal circular holes, Journal of Applied Mathematics and Mechanics 1964; 28: 530–543.
14. Filihtinskii LA, Toward a solution of two–dimensional doubly periodic problems of the theory of elasticity. Candidate's thesis. In Russian Novosibirsk. 1964.
15. Gluzman S, Mityushev V, Series, index and threshold for random 2D composite, Arch. Mech. 2015; 67: 75–93.
16. Gluzman S, Mityushev V, Nawalaniec W, Starushenko G, Effective Conductivity and Critical Properties of a Hexagonal Array of Superconducting Cylinders. ed. P.M. Pardalos (USA) and T.M. Rassias (Greece), Contributions in Mathematics and Engineering. In Honor of Constantin Caratheodory. Springer: 255–297, 2016.

17. Gluzman S, Mityushev V, Nawalaniec W, Cross-properties of the effective conductivity of the regular array of ideal conductors, Arch. Mech. 2014; 66: 287–301.
18. Grigolyuk EI, Filshtinsky LA, Perforated Plates and Shells. Moscow, Nauka; 1970 [in Russian].
19. Grigolyuk EI, Filshtinsky LA, Periodical Piece–Homogeneous Elastic Structures. Moscow: Nauka; 1991 [in Russian].
20. Grigolyuk EI, Filshtinsky LA, Regular Piece-Homogeneous Structures with defects. Moscow: Fiziko-Matematicheskaja Literatura; 1994 [in Russian].
21. Hashin Z, Analysis of composite materials a survey, Journal of Applied Mechanics 1983; 50: 481–505.
22. Hashin Z, Shtrikman S, A variational approach to the theory of the effective magnetic permeability of multiphase materials, J. Appl. Phys. 1962; 33: 3125–3131.
23. Helsing J, An integral equation method for elastostatics of periodic composites, J. Mech. Phys. Solids 1995; 6: 815–828.
24. Hill R, Theory of mechanical properties of fibre-strengthened materials: I. Elastic behaviour, J Mech Phys Solids 1964; 12: 199–212.
25. Hill R, Theory of mechanical properties of fibre-strengthened materials: III. Self-consistent model, J Mech Phys Solids 1965; 13: 189–198.
26. Jun S, Jasiuk I, Elastic moduli of two-dimensional composites with sliding inclusions – a comparison of effective medium theories, International Journal of solids and Structures 1993; 30: 2501–2523.
27. Kanaun SK, Levin VM, Self–Consistent Methods for Composites. Springer Science + Business Media B.V. 2008.
28. Landauer R, Electrical conductivity in inhomogeneous media. In Garland, J.C., Tanner, D.B. (eds.) Electrical, transport and optical properties of inhomogeneous media. 2–43: New York, American Institute of Physics. 1978.
29. Lekhnitskii SG, Theory of Elasticity of an Anisotropic Elastic Body. Moscow: Mir; 1981. [Chapter 1].
30. Milton GW, The Theory of Composites. Cambridge University Press; 2002.
31. Mityushev V, Transport properties of two-dimensional composite materials with circular inclusions, Proc. R. Soc. London 1999; A455: 2513–2528.
32. Mityushev V, Thermoelastic plane problem for material with circular inclusions, Arch. Mech. 2010; 52, 6: 915–932.
33. Mityushev V, Transport properties of doubly periodic arrays of circular cylinders and optimal design problems, Appl. Math.Optim. 1999; 44: 17–31.
34. Mityushev V, Transport properties of finite and infinite composite materials and Rayleigh's sum, Arch. Mech. 1997; 49: 345–358.
35. Mityushev VV, Rogosin SV, Constructive methods to linear and non-linear boundary value problems of the analytic function. Theory and applications. Boca Raton etc: Chapman & Hall / CRC; 1999/2000. [Chapter 4].
36. Muskhelishvili NI, Some Basic Problems of the Mathematical Theory of Elasticity. Reprint of the 2rd English edition Dordrecht: Springer-Science + Business Media; 1977.
37. Natanzon VY, On the stresses in a stretched plate weakened by identical holes located in chessboard arrangement, Mat.Sb. 1935; 42: 616–636.
38. Rayleigh Lord, On the influence of obstacles arranged in rectangular order upon the properties of medium, Phil. Mag. 1892; 34, 481–502.
39. Rylko N, Transport properties of the regular array of highly conducting cylinders, J. Engrg. Math. 2000; 38: 1–12.
40. Torquato S, Random Heterogeneous Materials: Microstructure and Macroscopic Properties. New York. Springer-Verlag: (2002).
41. Thorpe MF, Jasiuk I, New results in the theory of elasticity for two-dimensional composites, Proceedings of the Royal Society of London A 1992; 438: 531–544.
42. Wang J, Crouch SL, Mogilevskaya SG, A complex boundary integral method for multiple circular holes in an infinite plane, Engineering Analysis with Boundary Elements 2003; 27: 789–802.
43. Yakubovich S, Drygas P, Mityushev V, Closed-form evaluation of 2D static lattice sums, Proc Roy. Soc. London 2016; A472: 20160662. DOI: 10.1098/rspa.2016.0510.

Table 11.1:   Table of the main analytical formulae. A cross-section is displayed for fibrous composites

1		Effective conductivity for the general case of regular array (see formula (4.2.29) on page 87).
2		Contrast expansion up to $O(\rho^4)$ for the effective conductivity of macroscopically isotropic composites (see formula (4.2.38) on page 90). Ideally conducting inclusions (see formula (9.1.29) on page 260 and (9.1.33) on page 261). Arbitrary contrast parameter of inclusions (see formula (9.3.58) on page 276).
3		Effective conductivity of square array in the high-contrast case (see formulae (6.6.61) on page 179 and (6.7.81) on page 184). Arbitrary contrast parameter of inclusions (see formula (6.8.90) on page 187).
4		Effective conductivity of hexagonal array in the high-contrast case (see formula (7.5.44) on page 207). Arbitrary contrast parameter of inclusions (see formula (7.9.77) on page 216).
5		Effective conductivity of simple cubic (SC) lattice in the high-contrast case of high-conducting inclusions (see formula (8.5.100) on page 239) and in the high-contrast case of ideally insulating inclusions (see formula (8.7.123) on page 245).
6		Effective conductivity of face-centered cubic (FCC) lattice in the high-contrast case of high-conducting inclusions (see formula (8.5.110) on page 241) and in the high-contrast case of ideally insulating inclusions (see formula (8.7.125) on page 245).

Table 11.1:Table of the main analytical formulae (cont.)

7	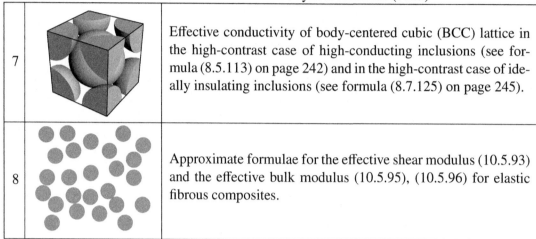	Effective conductivity of body-centered cubic (BCC) lattice in the high-contrast case of high-conducting inclusions (see formula (8.5.113) on page 242) and in the high-contrast case of ideally insulating inclusions (see formula (8.7.125) on page 245).
8		Approximate formulae for the effective shear modulus (10.5.93) and the effective bulk modulus (10.5.95), (10.5.96) for elastic fibrous composites.

INDEX

Printed in the United States
By Bookmasters